FRAGMENTATION PROCESSES

Revolutionary advances in experimental techniques and spectacular increases in computer power over recent years have enabled researchers to develop a much more profound understanding of the atomic few-body problem. One area of intense interest has been the study of fragmentation processes.

Covering the latest research in the field, this edited text is the first to provide a focused and systematic treatment of fragmentation processes, bringing together contributions from a range of leading experts. As well as tackling the more established coincident study of electron-impact ionization, the (e,2e) process, this book also guides the reader through topics such as molecular fragmentation, ion–atom collisions and multiphoton processes.

Combining a broad range of topics with an equal mix of theoretical and experimental discussion, this is an invaluable text for graduate students and researchers in atomic collisions, laser physics and chemistry.

COLM T. WHELAN is a Professor of Physics and an Eminent Scholar at Old Dominion University, USA. He received a Ph.D. in Theoretical Atomic Physics from the University of Cambridge in 1985 and an Sc.D. in 2001. He is a Fellow of both the American Physical Society and the Institute of Physics (UK). He has edited five previous books and has written over 150 research papers, mostly on atomic collision physics.

FRAGMENTATION PROCESSES

Topics in Atomic and Molecular Physics

Edited by

COLM T. WHELAN
Old Dominion University

CAMBRIDGE
UNIVERSITY PRESS

CAMBRIDGE UNIVERSITY PRESS
Cambridge, New York, Melbourne, Madrid, Cape Town,
Singapore, São Paulo, Delhi, Mexico City

Cambridge University Press
The Edinburgh Building, Cambridge CB2 8RU, UK

Published in the United States of America by Cambridge University Press, New York

www.cambridge.org
Information on this title: www.cambridge.org/9781107007444

First published 2013

Printed and bound in the United Kingdom by the MPG Books Group

A catalogue record for this publication is available from the British Library

Library of Congress Cataloguing in Publication data
Fragmentation processes : topics in atomic and molecular physics / edited by Colm T. Whelan,
Old Dominion University.
pages   cm
Includes bibliographical references and index.
ISBN 978-1-107-00744-4
1. Few-body problem.   2. Ion-atom collisions.   3. Nuclear fragmentation.
I. Whelan, Colm T., editor of compilation.
QC174.17.P7F72   2012
530.14 – dc23   2012035049

ISBN 978-1-107-00744-4 Hardback

Contents

Contributors

Lorenzo Avaldi *CNR-Istituto di Metodologie Inorganiche e dei Plasmi, Area della Ricerca di Roma 1, CP10, I-00015 Monterotondo Scalo, Italy*

Philip L. Bartlett *ARC Centre for Antimatter-Matter Studies, Murdoch University, Perth 6150, Australia*

Itzik Ben-Itzhak *James R. Macdonald Laboratory, Cardwell Hall, Kansas State University, Manhattan, KS 66506-2604, USA*

S. J. Brawley *Department of Physics and Astronomy, University College London, Gower Street, London WC1E 6BT, UK*

D. A. Cooke *Department of Physics and Astronomy, University College London, Gower Street, London WC1E 6BT, UK*

I. A. Ivanov *Research School of the Physical Sciences and Engineering, The Australian National University, Canberra ACT 0200, Australia*

Á. Kövér *Institute of Nuclear Research of the Hungarian Academy of Science, (ATOMKI), Debrecen, Hungary*

A. S. Kheifets *Research School of the Physical Sciences and Engineering, The Australian National University, Canberra ACT 0200, Australia*

G. Laricchia *Department of Physics and Astronomy, University College London, Gower Street, London WC1E 6BT, UK*

Julian Lower *Institut für Kernphysik, Max-von-Laue-Str. 1, 60438 Frankfurt am Main, Germany*

Andrew James Murray *Photon Science Institute, School of Physics and Astronomy, University of Manchester, Manchester M13 9PL, UK*

M. McGovern *Department of Applied Mathematics and Theoretical Physics, The Queen's University, Belfast, BT7 1NN, UK*

Giovanni Stefani *Dipartimento di Fisica and CNISM Universitá Roma Tre, Via della Vasca Navale 84, I-00146 Rome, Italy*

Andris T. Stelbovics *ARC, Center for Antimatter-Matter Studies, Curtin University, Perth, WA 6102, Australia*

Masahiko Takahashi   *Institute of Multidisciplinary Research for Advanced Materials, Tohoku University, 2-1-1 Katahira, Sendai 980-8577, Japan*

H. R. J. Walters   *Department of Applied Mathematics and Theoretical Physics, The Queen's University, Belfast, BT7 1NN, UK*

Colm T. Whelan   *Physics Department, Old Dominion University, Norfolk, Virginia, USA*

Masakazu Yamazaki   *Institute of Multidisciplinary Research for Advanced Materials, Tohoku University, 2-1-1 Katahira, Sendai 980-8577, Japan*

Preface

In the past few years, revolutionary advances in experimental techniques and spectacular increases in computer power have offered unique opportunities to develop a much more profound understanding of the atomic few-body problem. One area of intense effort is the study of fragmentation processes – break-up processes – which are studied experimentally by detecting in coincidence the collisional fragments with their angles and energies resolved. These experiments offer a unique insight into the delicacies of atomic and molecular interactions, being at the limit of what is quantum mechanically knowable; the fine detail that is revealed would be swamped in a less differential measurement. The challenge for the theorist is to develop mathematical and computational techniques which are of sufficient ingenuity and sophistication that they can elucidate the Physics observed in existing measurements and give direction to the next generation of experiments. Fragmentation processes are studied by those interested in electron and photon impact ionization, heavy particle collisions, collisions involving antimatter, as well as molecular collisions.

Colm T. Whelan
31 January 2012

1

Direct and resonant double photoionization: from atoms to solids

LORENZO AVALDI AND GIOVANNI STEFANI

1.1 Introduction

Electron–electron correlation plays a crucial role in determining physical and chemical properties in a wide class of materials that exhibit fascinating properties including, for example, high-temperature superconductivity, colossal magnetoresistance, metal insulator or ferromagnetic anti-ferromagnetic phase transitions, self assembly and quantum size effects. Furthermore, electron–electron correlation governs the dynamics of charged bodies via long-range Coulomb interaction, whose proper description constitutes one of the more severe tests of quantum mechanics.

Nevertheless, the effects due to correlation remain rather elusive for almost all of the experimental methods currently used to investigate matter in its various states of aggregation. Indeed, being related to processes with two active electrons, like satellite structures in photoemission (i.e., ionization processes with one ejected and one excited electron), or double ionization events, they influence marginally the spectral responses of the target, that are primarily determined by single and independent particle behaviours. Hence the experimental effort devoted in the last 30 years to develop a new class of experiments, whose spectral response is determined mainly by the correlated behaviour of electron pairs.

The common denominator of this class of experiments is the study of reactions whose final state has two holes in the valence orbitals and two unbound electrons in the continuum. It is exactly through interaction of these holes and electron pairs that correlation shapes the cross section of the double ionization processes.

The archetypal processes that create hole pairs are the direct valence double photoionization (DPI) and the core-hole Auger decay. In both cases, the Coulomb interaction between valence electrons is responsible for the promotion either of

Fragmentation Processes: Topics in Atomic and Molecular Physics, ed. Colm T. Whelan. Published by Cambridge University Press. © Cambridge University Press 2013.

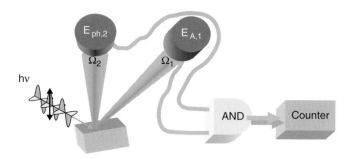

Figure 1.1 Scheme of an electron–electron coincidence experiment.

valence electron pairs to the continuum, in the DPI case, or for the core-hole decay with emission of an Auger electron that is paired to the photoelectron from core ionization. In both cases, experiments designed to study these events detect the unbound electron pairs after energy and momentum selection.

The prototypal scheme for such experiments is **1-photon IN** and **2-electrons OUT**, i.e.,

$$h\nu + A \rightarrow A^{++}(v^{-1}v^{-1}) + e^{-} + e^{-}$$

in the case of DPI, and

$$h\nu + A \rightarrow A^{+}(c^{-1}) + e^{-} \rightarrow A^{++}(v^{-1}v^{-1}) + e^{-} + e^{-}$$

in the case of Auger decay.

It is evident that both processes share a similar, sometimes identical, final doubly charged ion state. These two schemes also clarify the characteristic observables of this class of experiments, i.e., the energy and momentum (emission angle) distributions of the final electron pair.

These distributions are determined by measuring the kinetic energy and momentum of the individual electrons and by correlating them in time in order to discriminate the pairs generated in the same event from the ones generated in independent events. This is done by detecting the final electron pairs with an apparatus whose scheme of principle is given in Fig. 1.1.

Since their first introduction by Haak *et al.* [1], 1-photon IN and 2-electrons OUT experiments have been applied extensively to atoms, molecules and, more recently, solids. The experimental challenge is due to the small cross section of these processes. This handicap has been overcome on the one hand by the advent of the high intensity third-generation synchrotron radiation sources [2], and on the other hand by the development of highly sophisticated set-ups [3] which allow the simultaneous detection of electrons with several energies over a large solid angle. Together with the complementary approach where one or two electrons are detected in coincidence with the recoil ion [4], these experiments have yielded detailed information on electron correlations in both bound and unbound states. A few relevant study cases on either aggregation state are presented and discussed in the following.

1.2 Direct double photoionization

Two electron systems, like the He atom and H_2 molecule, represent the most suited targets to investigate the dynamics of the direct emission into the continuum of two electrons following the absorption of a VUV or soft X-ray photon. Therefore Sections 1.2.1 and 1.2.2 will be devoted to the presentation of selected experimental results in He and H_2 and to the present understanding of direct DPI. For atoms heavier than He, the complexity of their initial states is reflected in a large variety of patterns for the coincidence angular distributions produced in a DPI event. However, the basic understanding of the DPI process in these systems can be traced back to the one in He, thus we have decided to limit this chapter to two electron systems. A review of the DPI results in the heavier rare gases can be found in [5].

Section 1.2.3 is devoted to the description of the results of direct DPI on solids and shows how DPI discloses the possibility to investigate directly the electron correlation in the bands, as well as providing evidence of the exchange-correlation hole.

1.2.1 The He atom

The archetypal system to study direct DPI is the He atom. In this two-electron system the absorption of a photon with energy larger than 79.004 eV [6], the He double-ionization potential I^{2+}, may lead to a bare nucleus and two free electrons:

$$h\nu + \text{He} \rightarrow \text{He}^{2+} + e^- + e^-. \tag{1.1}$$

In the simplest case where the incoming photon is linearly polarized along a well-defined direction, $\boldsymbol{\varepsilon}$, then the latter represents the natural quantization axis. The final state is fully characterized by the momenta $\mathbf{k}_1$ and $\mathbf{k}_2$ or, alternatively, by their kinetic energies E_1 and E_2 and the spherical angles $\Omega_1(\theta_1, \phi_1)$, $\Omega_2(\theta_2, \phi_2)$ of the two electrons, where $\theta_i (i = 1, 2)$ is the angle with respect to $\boldsymbol{\varepsilon}$ and ϕ_i the angle with respect to the plane which contains the axes of polarization and of the photon beam. The excess energy $E = h\nu - I^{2+}$ is shared between the two electrons according to $E = E_1 + E_2$. From dipole selection rules in (1.1) the symmetry of the electron pair is well defined, $^1P^0$, and the quantum numbers of the electron pair are $L = 1$, $M = 0$, $S = 0$ and parity $\pi = $ odd.

The quantities which describe the process (1.1) are: its total cross section σ, the singly differential cross section $\frac{d\sigma}{dE_1}(0 \leq E_1 \leq E)$, which gives the energy distribution of the ejected photoelectrons, the doubly differential cross section $\frac{d^2\sigma}{dE_1 d\Omega_1}$, which describes the angular distribution of each one of the two photoelectrons, and finally the triply differential cross section $\frac{d^3\sigma}{dE_1 d\Omega_1 d\Omega_2}$, which describes the correlated motion of the electron pair for a fixed energy sharing $R = \frac{E_2}{E_1}$. The questions to be answered concern the shape and absolute value of the total DPI cross section, the way the two ejected electrons share the excess energy E and the shape of the correlated angular distribution of the two electrons.

The shape of the total cross section for DPI of He and its ratio to the one of single photoionization have been well characterized both experimentally and theoretically [7]. The cross section increases from threshold ($\sigma = 1.021 \pm 0.005$ kb at $E = 1$ eV [8]) up to a broad maximum at about $E = 20$ eV, where it has a value of about 10 kb [9] and then it decreases continuously. The increase from threshold to about 2 eV above it follows a power law E^n, with an exponent $n = 1.05 \pm 0.02$ [8], in good agreement with $n = 1.056$, as predicted by Wannier [10] from a classical analysis of the dynamics of double escape. The small value of the cross section and the behaviour near threshold are a clear indication of the role of electron correlation in the process, and indicate that close to threshold the double escape of two electrons is very unlikely, as each of them tends to be attracted back to the nucleus.

While a plethora of measurements of the $\frac{d^3\sigma}{dE_1 d\Omega_1 d\Omega_2}$ at different E and R values have been reported [5], less attention has been paid to the measurements of $\frac{d\sigma}{dE_1}$ and $\frac{d^2\sigma}{dE_1 d\Omega_1}$, which have been measured mainly via time-of-flight techniques [11–13]. The energy distribution $\frac{d\sigma}{dE_1}$ is flat in the threshold region and assumes a U shape at higher E. The flatness in the threshold region is a direct consequence of the ergodic character of the dynamics in this energy region, while the U shape near the maximum of the σ and at higher E means that the unequal energy sharing becomes more likely than the equal sharing. The doubly differential cross section,

$$\frac{d^2\sigma}{dE_1 d\Omega_1} = \frac{d\sigma}{dE_1} \frac{[1 + \beta(E_1) P_2(\cos\theta_1)]}{4\pi}, \qquad (1.2)$$

is the product of the singly differential cross section by an angular factor depending upon the asymmetry parameter β and the second-order Legendre polynomial P_2. The measurement of the evolution of β with E, which depends on the radial matrix element for the dipole-allowed transitions, is a sensitive probe of the photoionization dynamics. Wehlitz *et al.* [11] in a series of measurements at $10 \leq E \leq 41$ eV showed that β varies with the ratio $\frac{E_2}{E_1}$. This is consistent with the analysis performed by Maulbetsch and Briggs [14], who also proved that only in the case where this ratio is kept fixed as $E \to 0$ then β approaches the limiting value of -1, expected from a classical analysis of the dynamics of double escape [15, 16].

Finally, for a radiation linearly polarized along the $\boldsymbol{\varepsilon} = \varepsilon \mathbf{x}$ axis, taking into account the invariance with respect to the rotation about a preferential symmetry axis [15], the triply differential cross section for emission of the two electrons is given exactly by:

$$\frac{d^3\sigma}{dE_1 d\Omega_1 d\Omega_2} = |a_g(E_1, E_2, \theta_{12})(\cos\theta_1 + \cos\theta_2)$$
$$+ a_u(E_1, E_2, \theta_{12})(\cos\theta_1 - \cos\theta_2)|^2, \tag{1.3}$$

where $a_g(E_1, E_2, \theta_{12})$ and $a_u(E_1, E_2, \theta_{12})$ are two complex amplitudes (respectively symmetric and antisymmetric in the exchange $E_1 \leftrightarrow E_2$), which resume the dynamics of the pair. Accordingly, these amplitudes only depend on the energies and mutual angle θ_{12} of the two electrons. The other factors $(\cos\theta_1 \pm \cos\theta_2)$ result from the description of the photon–atom interaction within the dipole approximation. They are known as angular factors as they derive from the $L = 1$, $M = 0$ character of the final state. Whenever another type of polarization is used in the experiment, Eq. (1.3) is to be changed in its angular factors, but the amplitudes remain the same. The above expression allows the classification of the experiments into equal ($E_1 = E_2$) and unequal ($E_1 \neq E_2$) sharing subsets. For equal sharing, $\frac{d^3\sigma}{dE_1 d\Omega_1 d\Omega_2}$ reduces to:

$$\frac{d^3\sigma}{dE_1 d\Omega_1 d\Omega_2} = |a_g(E/2, E/2, \theta_{12})|^2 (\cos\theta_1 + \cos\theta_2)^2. \tag{1.4}$$

Equation (1.4) clearly shows that back-to-back emission in equal sharing is forbidden, as can be immediately verified, because $\mathbf{k}_2 = -\mathbf{k}_1$ implies $\theta_2 = \pi - \theta_1$, and therefore $\frac{d^3\sigma}{dE_1 d\Omega_1 d\Omega_2} = 0$. Another point can now be outlined: the amplitudes $a_{g,u}$ themselves should rather be maxima for back-to-back emission ($\theta_{12} = \pi$), because the electron–electron repulsion should drive the electrons to opposite directions with respect to the ion. From these simple ideas it was predicted [15], before any experiment on He became feasible, that from the competition of the two factors the differential cross section in Eq. (1.4) should have a 'butterfly' shape in polar coordinates (Fig. 1.2). If the first factor in Eq. (1.4) is replaced by a Gaussian function, which has the property to be maximum at $\theta_{12} = \pi$ as expected for physical reasons,

$$|a_g(E/2, E/2, \theta_{12})|^2 = a(E)\exp[-4\ln2(\theta_{12} - \pi)^2/\theta_{1/2}^2], \tag{1.5}$$

where $a(E)$ is a scaling factor and $\theta_{1/2}$ the width of the Gaussian at half maximum, then all experiments can be fitted as a function of a single parameter: $\theta_{1/2}$. The

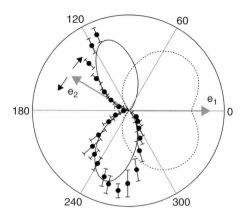

Figure 1.2 Triply differential cross sections at $E = 20$ eV, and equal sharing $E_1 = E_2 = 10$ eV, from Turri *et al.* [17]. The full curve is a fit using Eqs. (1.4) and (1.5) and $\theta_{1/2} = 90 \pm 3°$. The dotted curve uses the same expressions but the value of $\theta_{1/2}$ deduced from the Wannier law.

Gaussian function in Eq. (1.5) is supported by the theory of Wannier, but only in the threshold region [18]. However, Kheifets and Bray [19] numerically proved that the Gaussian ansatz appears to be a suitable representation of the a_g amplitude up to $E = 80$ eV and experimentally a significant breakdown of this representation has not been reported so far. The width parameter $\theta_{1/2}$ gives the strength of angular correlations between the two electrons at equal sharing: these correlations are the highest when $\theta_{1/2}$ is small, the lowest when $\theta_{1/2}$ is large. The fits to the measured $\frac{d^3\sigma}{dE_1 d\Omega_1 d\Omega_2}$ (Fig. 1.3) leads to values of $\theta_{1/2}$, which increase smoothly from $57 \pm 4°$ at $E = 0.1$ eV to $120 \pm 4°$ at $E = 80$ eV. At the highest energy $E = 450$ eV, the equal sharing DPI is observed to be very weak [23], and has not been measured. The predicted value of the width from Wannier theory is $\theta_{1/2} = 91\ E^{0.25}$ (units in degrees for $\theta_{1/2}$ and eV for E), which, as expected, is consistent with experiments only at the lowest energies $E = 0.1$ and 0.2 eV, but overestimates the experimental values already at $E = 1$ eV and gives much too large values at higher energy, as illustrated in Fig. 1.3. The results of the equal energy sharing experiments can be summarized by saying that the angular correlations repelling the two electrons into opposite directions are very intense at threshold, as expected from the theory of Wannier, and then progressively relax as E increases, until equal sharing DPI becomes practically negligible at very high E. These high-energy experiments, where the three-body effects in the final state become negligible, are the ones proposed in the seventies by Neudatchin *et al.* [24] to investigate directly electron correlations in the initial state and therefore to discriminate among different bound state wave functions, which adopt different representation of electron correlations. Unfortunately, this unique property of the $\frac{d^3\sigma}{dE_1 d\Omega_1 d\Omega_2}$ could not be

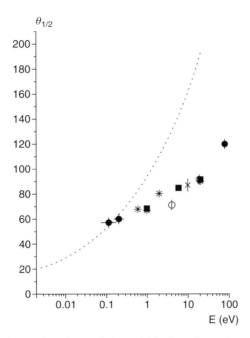

Figure 1.3 Experimental values of the width $\theta_{1/2}$ from fits of Eqs. (1.4) and (1.5) to the measured equal sharing triply differential cross sections. The dotted curve shows the $\theta_{1/2} = 91\ E^{0.25}$ Wannier law. Experimental data by Huetz and Mazeau [20] (dots with uncertainty on the energy scale), Dawber *et al.* [21] (stars), Huetz *et al.* [22] (open circles), Schwarzkopf *et al.* [2] (full squares) and Turri *et al.* [17] (dots).

explored experimentally due to the vanishing values of the cross section, too small even for the present multi-angle and multi-energy set-ups [23].

For unequal sharing, both amplitudes are expected to contribute, leading to more complicated shapes, as shown in Figs. 1.4 and 1.5. However, Eq. (1.3) still provides a guide to disentangle the various effects, and methods have been developed to extract the amplitudes from experiments. This is especially instructive as the amplitudes are the natural quantities to be calculated from *ab initio* theories. Before addressing this topic, here is a qualitative description of the evolution of the shape of the $\frac{d^3\sigma}{dE_1 d\Omega_1 d\Omega_2}$ as E increases. In the region near threshold, Lablanquie *et al.* [3] and Dawber *et al.* [21] reported no significant changes or new features for R up to 12.3. The measured $\frac{d^3\sigma}{dE_1 d\Omega_1 d\Omega_2}$ look almost the same at all the different energy sharings: a node continues to exist at $\theta_{12} = 180°$ even for $E_1 \neq E_2$ and indeed the data look very similar to the equal energy case. These observations demonstrate that a_u remains small compared to a_g and a_g is almost independent of R. Qualitatively, these findings depend on the fact that, close to threshold, double escape of the two electrons is very unlikely, as each of them tends to be attracted

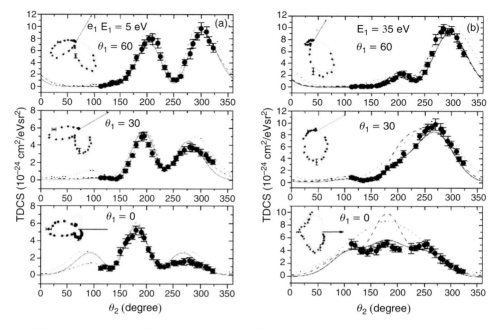

Figure 1.4 Triply differential cross sections from Bolognesi *et al.* [25] at $E = 40$ eV, and $\theta_1 = 0°, 30°, 60°$. (a) $E_1 = 5$ eV ($R = 7$); (b) complementary kinematics: $E_1 = 35$ eV ($R = 0.14$). The CCC and 3C calculations are plotted as dashed and dotted lines, respectively. The full line is a representation of the TDCS according the parameterization by Cvejanovic and Reddish [26]. Both cartesian coordinates and polar coordinates (small insets) are used.

back to the nucleus and the system has to maintain itself along the Wannier ridge ($r_1 \approx r_2$). At larger E, both amplitudes are now significant and they lead to the complicated shapes displayed in Figs. 1.4 and 1.5. A large number of experiments have been performed in this energy domain. The two examples selected and shown in Figs. 1.4 and 1.5 illustrate quite well the advantages and disadvantages of the different experimental methods. The momentum imaging COLTRIMS method [28] gives the absolute $\frac{d^3\sigma}{dE_1 d\Omega_1 d\Omega_2}$ for all angles (only $\theta_1 = 30°$ is shown in Fig. 1.5) and all R values at once [27]. The multi-analyzer technique [25, 29] (Fig. 1.4) usually produces less data on a relative scale only, but each of them can be monitored individually to reach the desired statistics, and indeed the figure shows that excellent statistics have been achieved.

The main effect of the antisymmetric amplitude a_u is that the back-to-back emission is no longer forbidden, as shown in Fig. 1.4. Moreover, considering the two complementary kinematics obtained by exchanging $E_1 \leftrightarrow E_2$ and the properties of the amplitudes, one observes that the difference between the two

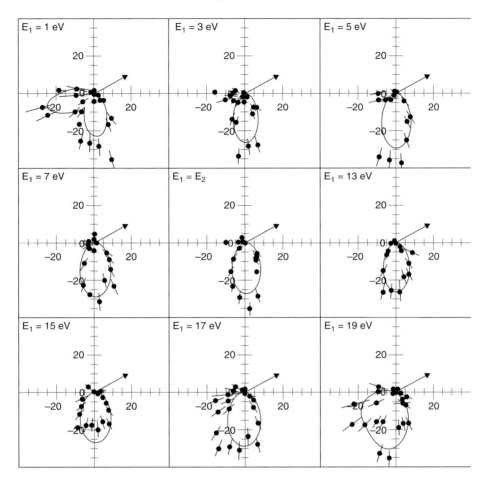

Figure 1.5 Absolute triply differential cross sections (barn $eV^{-1}sr^{-2}$) measured by Bräuning *et al.* [27] at $E = 20$ eV, and $\theta_1 = 30°$ with the COLTRIMS momentum imaging technique. Nine values of E_1 are reported, and CCC calculations are compared with the experiments on absolute scale (full line).

triply differential cross sections,

$$\Delta = \frac{d^3\sigma(E_2, E_1)}{dE_1 d\Omega_1 d\Omega_2} - \frac{d^3\sigma(E_1, E_2)}{dE_1 d\Omega_1 d\Omega_2} = |a_\mathrm{g}| \, |a_\mathrm{u}| \cos \delta \left(\cos\theta_2^2 - \cos\theta_1^2\right), \quad (1.6)$$

is completely determined by the phase difference δ between the two amplitudes. This is responsible for the different shapes that are observed when fixing in space either the highest, or the lowest energy electron (Fig. 1.4). At higher energy ($h\nu = 530$ eV) Knapp *et al.* [23, 30, 31] showed that the dynamics of the electron pair becomes much simpler. This is not surprising as, in general, high kinetic energies of the particles mean that their interaction becomes less important, and therefore three-body effects in the final state tend to be washed out. Two-step mechanisms have been envisaged for the double photoionization process in this region. The first one is the shake-off, where almost all energy is transferred to a quickly removed electron. This leaves a state which is not an eigenstate of the ion, and therefore has a component within the double continuum. Consequently, the second electron can be removed, but with low correlation to the first one. The experimental results show that this shake-off process dominates over the other proposed mechanism, named TS1 [32], in which the first high-energy electron kicks out a second one, leading to an angular correlation around $\theta_{12} = 90°$, as in hard spheres elastic collisions.

Berakdar and Klar [33] and Berakdar *et al.* [34] predicted that He $\frac{d^3\sigma}{dE_1 d\Omega_1 d\Omega_2}$ measured with circularly polarized radiation should display a helicity dependence, i.e., a non-vanishing circular dichroism, CD. This is because, due to parity conservation, the helicity of the incident photon is transferred to the three-body system, the He atom, and the continuum spectrum of this excited system depends on the helicity of the absorbed photon. Using the same representation as for Eq. (1.3), but taking the quantization axis along the photon beam, the $\frac{d^3\sigma}{dE_1 d\Omega_1 d\Omega_2}$ for right, R, and left, L, circular polarizations is given by:

$$\frac{d^3\sigma_L}{dE_1 d\Omega_1 d\Omega_2} = |(a_g(E_1, E_2, \theta_{12}) + a_u(E_1, E_2, \theta_{12})) \sin\theta_1 e^{-i\varphi_1}.$$

$$+ (a_g(E_1, E_2, \theta_{12}) - a_u(E_1, E_2, \theta_{12})) \sin\theta_2 e^{-i\varphi_2}|^2. \quad (1.7)$$

$$\frac{d^3\sigma_R}{dE_1 d\Omega_1 d\Omega_2} = |(a_g(E_1, E_2, \theta_{12}) + a_u(E_1, E_2, \theta_{12})) \sin\theta_1 e^{+i\varphi_1}$$

$$+ (a_g(E_1, E_2, \theta_{12}) - a_u(E_1, E_2, \theta_{12})) \sin\theta_2 e^{+i\varphi_2}|^2. \quad (1.8)$$

The circular dichroism (CD) is given by the difference:

$$CD = \frac{d^3\sigma_L(E_2, E_1)}{dE_1 d\Omega_1 d\Omega_2} - \frac{d^3\sigma_R(E_1, E_2)}{dE_1 d\Omega_1 d\Omega_2}$$

$$= -4 |a_g(E_1, E_2, \theta_{12})| |a_u(E_1, E_2, \theta_{12})| \sin\delta \sin\theta_1 \sin\theta_2 \sin(\varphi_2 - \varphi_1), \quad (1.9)$$

where δ is the phase difference between the two amplitudes, as in Eq. (1.6). At equal energy sharing, $|a_u| = 0$ and the CD cancels. This is not surprising as the momenta of the two electrons are then indistinguishable and cannot constitute, with the photon beam, an oriented system. Equation (1.9) shows that the CD is directly related to the phase difference between the two amplitudes. The same holds for the difference Δ of the triply differential cross sections under linear polarization (Eq. (1.6)). It is worth noting that these quantities are both necessary for a complete determination of δ. They also constitute a severe test of *ab initio* theories, as they directly probe the phase difference δ. A few measurements of the CD are available [13, 35–39]. An interesting observation can be done by looking at Eq. (1.9). CD should display 'dynamic nodes', i.e., CD $= 0$, whenever the two amplitudes have a phase difference $\delta = 0$ or π. The existence of 'dynamic nodes' firstly have been predicted for certain values of E, R and θ_{12} by calculations based on the 3C theory [34] and then experimentally proven by Bolognesi *et al.* [39].

As shown above, the dynamics of DPI is completely described by the complex amplitudes $a_g(E_1, E_2, \theta_{12})$ and $a_u(E_1, E_2, \theta_{12})$. These are the basic quantities that are calculated by the theories and then used to reconstruct any particular triply differential cross section. Thus the extraction of the moduli and relative phase of the symmetric and antisymmetric complex amplitudes from the experimental data at a fixed incident energy and energy sharing between the electrons is of considerable interest, because it allows a direct comparison among different sets of data, between theory and experiment as well as between different theories. Most attempts have relied upon parameterization of a_g and a_u with various degrees of complexity and approximation [36, 40, 41]. Two approaches, which do not rely on any approximation, have been proposed. Krässig [42] introduced a formalism that makes use of the full set of coincidence events obtained in a COLTRIMS absolute measurement. This method has been successfully applied to the set of measurements by Bräuning *et al.* [27]. However, the four specific experimental geometries needed are not easily achieved by most of the experimental apparatuses used in DPI experiments. By contrast, Bolognesi *et al.* [43] proposed a procedure which needs only three determinations of the TDCS at the same relative angle θ_{12} that can be applied with no restrictions to any set of experimental data taken with linear polarization radiation. By solving a set of three non-linear equations, the $|a_g|^2$, $|a_u|^2$ and $\cos \delta$ are obtained. However, this procedure enables one to determine only the cosine of the relative phase between the two complex amplitudes. The sign of the phase remains undetermined, unless the circular dichroism is measured in additional experiments with circularly polarized light. This has been done by Bolognesi *et al.* [39] in a set of experiments in unequal energy ($E_1 = 3.5$ eV, $E_2 = 44.5$ eV) using both linear and circular polarized radiation (Fig. 1.6). In this way a 'complete quantum mechanical experiment' has been achieved, where the basic quantities (amplitudes and phase) determining the process are obtained. The a_g, a_u and δ extracted in this way have then been used to predict all the other observables of the same process [39].

Double photoionization is the archetypal example of a three-body Coulomb problem, and as such it represents a non-separable problem, which cannot be given exact analytical solutions. Methods in quantum mechanics have been developed to give nearly exact solutions for the He three-body ground and excited discrete states, however, the high-lying doubly excited states and the double continuum

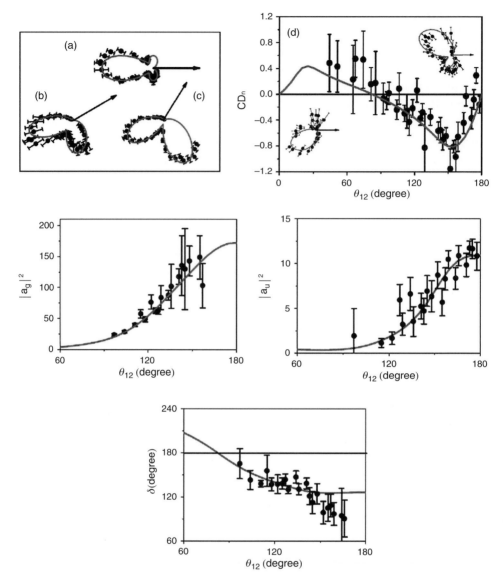

Figure 1.6 He triply differential cross sections for $E_1 = 3.5$ eV, $E_2 = 44.5$ eV and $\theta_1 = 0°$ (a), $30°$ (b) and $60°$(c) from Bolognesi *et al.* [39]. The normalized CD_n [39] versus θ_{12} is shown in panel (d). The triply differential cross sections obtained with the two helicities of the radiation are reported in the top-right and bottom-left corners of the same panel. In the bottom panels, the $|a_g|^2$, $|a_u|^2$ and δ as obtained from the experiments are reported. The experimental triply differential cross sections, CD_n, amplitudes and relative phase are compared with the predictions of the CCC model (solid line).

states had to wait for the development of powerful computers to be accessible by numerical calculations. Briggs and Schmidt [44] have reviewed the first methods, labeled 3C (three-Coulomb), 2SC (two-screen-Coulomb) and CCC (convergent close coupling), which have been applied to the DPI of helium, and Malegat [45] has discussed the more recent TDCC (time dependent close coupling [46]), ECS (exterior complex scaling [47]) and HRM-SOW (hyphersherical R-matrix with semiclassical outgoing waves [48]) theories. Results from CCC calculations are displayed in Fig. 1.5 as full lines, and appear to be in good agreement with the shapes as well as the absolute values of the TDCSs measured at $E = 20$ eV, $\theta_1 = 30°$, and nine different values of R. In Fig. 1.4, CCC calculations are reported as dashed lines and also show an overall agreement in shape with the high statistics data at $E = 40$ eV, for $R = 0.14, 7$ and $\theta_1 = 0°, 30°, 60°$. The only noticeable exception is $R = 0.14$, $\theta_1 = 0°$ where the calculation predicts a shape with a maximum at $\theta_2 = 180°$, which is not confirmed by the experiment. In the specific case of $\theta_1 = 0°$, unequal sharing kinematics is the most stringent test for theories, because, as it can be understood on the basis of Eq. (1.3), it disentangles the contributions of the a_g and a_u amplitudes more than the other kinematics, and the calculated shape of the $\frac{d^3\sigma}{dE_1 d\Omega_1 d\Omega_2}$ can only be correct if both amplitudes are correct, with no possible cancellation effects as shown by Selles *et al.* [49].

All the most recent theories (ECS, TDCC and HRM-SOW) have been applied successfully to the DPI of helium. One of main achievements in all cases is the agreement with the measured energy distributions, which are flat in the threshold region and become U-shaped at higher energy, as already mentioned.

To summarize, it can be said that the most recent theories have achieved a quite remarkable degree of agreement with all types of experiments ($\frac{d\sigma}{dE_1}$, $\frac{d^2\sigma}{dE_1 d\Omega_1}$, $\frac{d^3\sigma}{dE_1 d\Omega_1 d\Omega_2}$) in the $20 < E < 40$ eV range. In addition, CCC calculations at much higher energy, where the two-step processes dominate, are also fully consistent with experimental results. The low-energy region is more demanding and time consuming for these numerical methods, as the vicinity of threshold imposes to extend the calculations to very large distances between the electrons and the ion. Only very recently [50], a calculation has been performed at $E = 0.1$ eV, in good agreement with the experiments of Huetz and Mazeau [20]. Finally, almost all this theoretical work has been developed under the dipole approximation, with only one exception [51]. In principle, the quadrupole and higher order effects can only occur when the wavelength λ of the incoming photon is comparable to the atomic size, i.e., in the X-ray region, but exceptions to the rule have been found for single ionization at much lower photon energies [52]. Istomin *et al.* [51] have predicted that measurable quadrupole effects should exist in the DPI of helium down to $E = 40$ eV (photon energy of 119 eV) and this new result is awaiting experimental confirmation, which will require precise $\frac{d^3\sigma}{dE_1 d\Omega_1 d\Omega_2}$ measurements in a plane containing the photon beam, where these effects introduce a backward–forward asymmetry.

1.2.2 The H_2 molecule and the four-body problem

The hydrogen molecule is the other two-electron system where direct double photoionization occurs. The two-centre nuclear potential in the initial state and the fact that the doubly charged ion is formed in a dissociative state make DPI in H_2 significantly more complex than in He and introduce new physical effects.

In the first measurements only the two electrons were detected and therefore the TDCS integrated over all the molecular orientations was measured. The equal energy-sharing measurements [53–55] showed a butterfly shape similar to that of helium, except for a slightly higher degree of angular correlation and the observation that the node for back-to-back emission is partly filled. A 'helium-like' generalization of Eq. (1.3) has been derived by Feagin [56] and Reddish and Feagin [57]. In this model, the amplitudes are split into Σ and Π components, corresponding to the two components of the polarization vector ε in the molecular frame, parallel and perpendicular to the internuclear axis, respectively. Then the TDCS in equal energy sharing is found to be the sum of two terms, the first with a $(\cos\theta_1 + \cos\theta_2)^2$ angular dependence as for He (Eq. 1.4) and the second one with a $(\cos\theta_{12}/2)^2$ factor. Therefore the back-to-back emission ($\theta_{12} = \pi$) in equal sharing is still forbidden, but the TDCS does not vanish on a cone ($\theta_2 = \pi - \theta_1$, $0 < \phi_2 < 2\pi$), as it does for helium. This explains the partial filling of the node, due also to the finite acceptance angles of experiments, as confirmed in more detail by recent momentum imaging measurements [58]. These findings are fully consistent with the more general selection rules derived by Maulbetsch and Briggs [59] for He and extended to H_2 by Walter and Briggs [60].

Subsequent experiments for unequal energy sharing have also shown similarities as well as visible differences with He TDCS [61, 62]. TDCSs for the two targets at the same excess energy $E = 25$ eV above threshold (IP^{2+} is 51.08 eV at the H_2 equilibrium distance of 1.4 a.u.), for various values of $R = \frac{E_2}{E_1}$ and for $\theta_1 = 0°$ are displayed and compared in Fig. 1.7. The shapes look almost the same at the highest value $R = 24$, then they begin to differ as R decreases, and the triple-lobe structure seen in helium at $R = 5$ and 2.57 is completely washed out in the molecule (bottom panel of Fig. 1.7). These findings were attributed to the average over molecular orientation. This prompted the investigation of the TDCSs for fixed-in-space molecules by means of momentum imaging techniques, where both ions and electrons are detected and energy- and angle-analyzed. Since the photofragmentation process is rapid compared to molecular rotation, the molecular orientation is obtained from the measured ion momenta, while the kinetic energy

FRAGMENTATION PROCESSES: TOPICS IN ATOMIC AND
MOLECULAR PHYSICS; ED. BY COLM T. WHELAN.
                                        Cloth   268 P.
CAMBRIDGE: CAMBRIDGE UNIV PRESS, 2013

ED: OLD DOMINION UNIVERSITY. NEW COLLECTION.

LCCN 2012-035049
  ISBN 1107007445   **Library PO#** FIRM ORDERS

                                   List   140.00 USD
   8395 NATIONAL UNIVERSITY LIBRAR   **Disc**   14.0%
   App. Date 12/10/14   COLS-SCI   8214-08 **Net**   120.40 USD

SUBJ: 1. FEW-BODY PROBLEM. 2. ION-ATOM COLLISIONS.
  3. NUCLEAR FRAGMENTATION.

CLASS QC174.17   DEWEY# 530.14   LEVEL ADV-AC

YBP Library Services

FRAGMENTATION PROCESSES: TOPICS IN ATOMIC AND
MOLECULAR PHYSICS; ED. BY COLM T. WHELAN.
                                        Cloth   268 P.
CAMBRIDGE: CAMBRIDGE UNIV PRESS, 2013

ED: OLD DOMINION UNIVERSITY. NEW COLLECTION.

  LCCN 2012-035049
  ISBN 1107007445   **Library PO#** FIRM ORDERS

                                   List   140.00 USD
   8395 NATIONAL UNIVERSITY LIBRAR   **Disc**   14.0%
   App. Date 12/10/14   COLS-SCI   8214-08 **Net**   120.40 USD

SUBJ: 1. FEW-BODY PROBLEM. 2. ION-ATOM COLLISIONS.
  3. NUCLEAR FRAGMENTATION.

CLASS QC174.17   DEWEY# 530.14   LEVEL ADV-AC

He   **D$_2$**

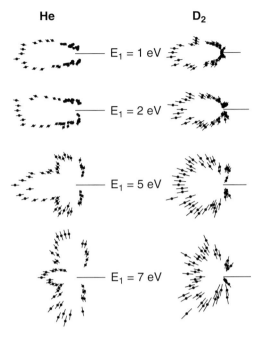

Figure 1.7 Comparison of TDCSs for He and D$_2$ from Seccombe *et al.* [62], at $E = 25$ eV, $\theta_1 = 0°$ and (from top to bottom) $E_1 = 1$ eV, $R = 24$; $E_1 = 2$ eV, $R = 11.5$; $E_1 = 5$ eV, $R = 4$; $E_1 = 7$ eV, $R = 2.57$.

release provides direct information on the internuclear distance at the moment of the double ionization. A few experiments where both ions and electrons are collected in coincidence have been reported [58, 63–65]. These experiments clearly showed that the emission pattern depends on the molecular orientation. As an example, in Figs. 1.8 and 1.9 the absolute fully differential cross section (FDCS) in the coplanar geometry and equal energy sharing at $E = 25$ eV measured by Gisselbrecht *et al.* [64] are shown. In Fig. 1.8, one electron is always detected at 90° with respect to ε, while the orientation of the molecular axis is varied from $\theta_N = 90°$ (a) to 0° (d). In Figs. 1.8(a) and (d), which correspond to the Π and Σ orientation of the molecule with respect to the polarization axis of the incident radiation, the FDCSs have a characteristic two-lobe structure due to the Coulomb repulsion between the two electrons. The two lobes are slightly closer together in Fig. 1.8(a) than in 1.8(d), which indicate a greater degree of electron correlation in the case of Π orientation. The absolute value decreases by about a factor of four from Π to Σ orientation. The other two figures clearly show that the electron–electron angular distribution does depend on the molecular orientation. The two-lobe patterns become strongly asymmetric and there is a reduction of the yield orthogonal to the bond direction. The effect of the molecular orientation is further proven in Fig. 1.9, where results in the orthogonal geometry are reported, i.e., for a geometry where one electron is

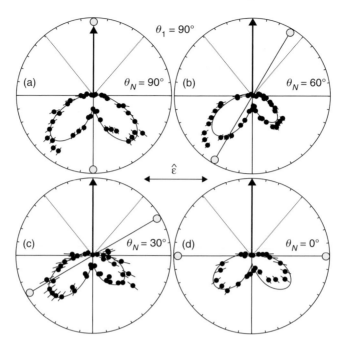

Figure 1.8 Absolute H_2 FDCSs [64] in the 'coplanar' geometry for four 'in plane' molecular orientations as indicated, all with the first electron at $\theta_1 = 90°$ for $E_1 = E_2 = 12.5 \pm 2.5$ eV. The radii of the circles give the absolute scales in millibarns/eV^2/sr^3: 120 (a); 120 (b); 60 (c); 30 (d). The angular step in θ_2 is 5° (a,b,c) and 10° (d). The angular bandwidths are: $\Delta\theta_1 = \pm 15°$; $\Delta\phi_{12} = \pm 20°$; $\Delta\theta_N = \pm 20°$ (a,b,c), $\pm 30°$ (d); $\Delta\phi_{1N} = \pm 45°$. The full lines give the fit from the helium-like model [57] folded by the experimental bandwidths.

perpendicular to the plane containing the other electron and $\boldsymbol{\varepsilon}$. In such a geometry, the mutual angle θ_{12} is a constant 90°, thus the electron correlations are 'frozen'. In Figs. 1.9(b) to 1.9(d), three different molecular orientations are shown, and the corresponding results for He are shown in Fig. 1.9(a). In He, the shape of the angular distribution is given exactly by a $\cos^2(\theta_2)$ distribution, while in H_2 the observed shapes are all different from each other and from the $\cos^2(\theta_2)$ shape expected from the 'helium-like' model. This is because in the molecular case only the projection of the angular momentum onto the nuclear axis is conserved, and the outgoing pair may have several angular momenta and may not have axial symmetry around the ε_π component, as assumed in the 'helium-like' model.

The purely Coulombic character of the potential curve of the H_2 dication (Fig. 1.10) results in energetic protons, whose kinetic energy depends on the internuclear distance at the moment of the double ionization. Thus, selecting the measured FDCS as a function of the ion kinetic energy allows one to probe

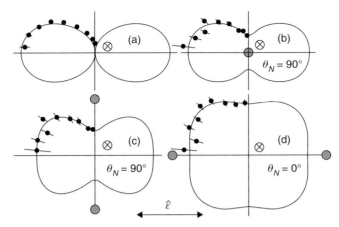

Figure 1.9 Absolute FDCSs [64] for H_2 (b,c,d) – compared to that of He (a) – for $E_1 = E_2 = 12.5 \pm 2.5$ eV (a,b,c) and ± 4 eV (d), in the 'orthogonal' geometry with first electron at $\theta_1 = 90°$, perpendicular to the plane of the figure, which contains the second electron. The molecule is oriented: (b) perpendicular to this plane, (c,d) in this plane. The four quadrants are related by symmetries and equivalent data have been added. The full lines are: (a) $(\cos\theta_2)^2$; (b,c,d) fits with partial waves up to $l = 2$, normalized to the data and the absolute scale is given by their maxima at 5.8 barns/eV/sr^2 (a) and 22 (b), 13.9 (c), 6.9 (d) millibarns/eV^2/sr^3. The angular bandwidths are: $\Delta\theta_1 = \pm 20°$ (a,b,c), $\pm 30°$ (d); $\Delta\phi_{12} = \pm 20°$; $\Delta\theta_N = \pm 25°$ (b,c); $\pm 30°$ (d); $\Delta\phi_{1N} = \pm 45°$.

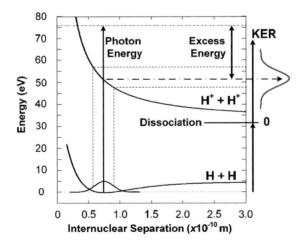

Figure 1.10 Schematic potential-energy diagram of the initial and final states for PDI of H_2. During double ionization the ground vibrational state is projected onto the upper repulsive curve, whose dissociation limit is at 31.67 eV, resulting in a broad range of kinetic energies for the final proton pair.

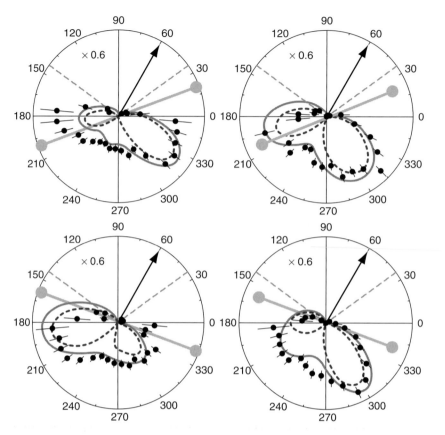

Figure 1.11 H_2 FDCSs [65] in the 'coplanar' geometry for three 'in plane' molecular orientations, $\theta_N = 20°$ and $160°$, all with the first electron at $\theta_1 = 90°$ and for $E_1 = E_2 = 12.5 \pm 10$ eV. Two KER values, 16.5 eV (left) and 23.5 eV (right) corresponding to approximately $R = 1.6$, 1.2 a_0, respectively, are shown for each θ_N angle. The angular step in θ_2 is $10°$. The dotted lines indicate a dead sector, symmetric with respect to the first electron, for the detection of the second electron. The bandwidths are: $\Delta E_{KER} = \pm 2$ eV; $\Delta \theta_N = \pm 10°$; $\Delta \theta_1 = \pm 20°$; $\Delta \phi_{12} = \pm 45°$; $\Delta \phi_{1N} = \pm 60°$. The experimental data are arbitrarily normalized to the TDCC results convoluted over the experimental bandwidths (solid); unaveraged TDCC results for the stated (θ_1, θ_N) values (dashed) have the scaling factors indicated.

the interplay between the electronic and nuclear motion in the DPI of H_2. The first experimental evidence of these 'kinetic energy release, KER, effects' has been provided by Weber *et al.* [58,63] in the perpendicular geometry. Reddish *et al.* [65] analyzed the effect in the coplanar geometry where the four-body interaction is completely probed. An example of their results is given in Fig. 1.11. Despite the large angle bandwidths of the experiment, a clear effect is observed, in particular for molecular orientation $\theta_N = 160°$. The effect is even more spectacular looking at the theoretical prediction (dashed curves in Fig. 1.11) without the convolution with the experimental bandwidths. A detailed analysis of the KER effects [65] has led to the following physical interpretation. The main trend is due to the electron–ion rather than electron–electron interactions. The relative strength of the higher partial waves in the Σ amplitude accounts for the observation that the most significant effects occur for large internuclear distances and the Σ electronic states, sensitive to changes in the internuclear distance, lies near the internuclear axis, while the Π states, mostly insensitive to changes in the internuclear distance, are concentrated in the plane between the nuclei.

The four-body problem raised by the DPI of the H_2 is also an exciting challenge on the theoretical side. Advanced methods developed for He, like for example the CCC model [66], once extended to H_2 using a single-centre expansion for the molecular ground state, were able to describe the shape of the angular distribution for randomly oriented molecules, and to give some hints of the effects due to the axis orientation and the internuclear distance, but did not achieve a satisfactory representation of the experimental FDCS. On the contrary, this has been obtained by computational models, which takes into account the full four-body problem, like the exterior complex scaling (ECS) [67–70] and time dependent close coupling, TDCC, models [71, 72]. The results of the two calculations are in close agreement with each other [72] and therefore can provide a benchmark for the shape and amplitudes of the FDCS. Moreover, once averaged over the experimental uncertainties, they are also in good agreement with the experimental results so far obtained with momentum imaging techniques [58, 63–65]. It will be interesting, in the future, to investigate the theoretically predicted isotope effects [60] between the homonuclear H_2, D_2 and the heteronuclear HD molecule, for which the two repelling ions have different velocities.

1.2.3 Direct DPI in solids and surfaces

Most of the attention in the investigation of DPI in gas phase has been focused on the effects of final-state correlations. However, it would be of great relevance for the application of DPI to solids to establish this technique as an initial-state correlation spectroscopy. Indeed, DPI experiments were originally proposed in the late 1970s [24, 73] within the framework of the independent particle scheme and were conceived to highlight initial-state correlation. Within this approximation the DPI cross section factorizes in such a way that [74, 75] the spectral density of the two-particle Green's function can be extracted from the energy separation spectrum and the Fourier amplitude of the pair correlator from the angular distribution of the two unbound final electrons. It has been speculated [76] that this latter quantity should be very sensitive to different types of correlated initial wave functions.

More recently, Berakdar [77,78] and Fominykh *et al.* [79] have attempted a general approach to DPI from solids in terms of two-electron Green's functions and two-electron bound states. They derived an approximate expression for the correlated electron-pair yield measured in a DPI experiment by incorporating the screened Coulomb interaction between the two outgoing electrons in a dynamically screened effective one-electron potential. Furthermore, it was pointed out that the spectrum of photoexcited electron pairs is one of the few experimental methods expected to give direct information on exchange-correlation interactions, electronic band structures and pair photoelectron diffraction in metals [80]. This theory was applied to predict the results of DPI experiments on quantum dots [81] and Cu(111) surface [82]. All of these works point to establish that DPI is a suitable tool for the study of electron correlations in solids.

Pioneering DPI experiments in condensed matter were performed at low photon energy (20–56 eV) on Van der Waals rare gas solids [83]. They allowed the detection of electron pairs in coincidence without energy and angle discrimination. The measured pair yield was surprisingly high when compared with theoretical DPI cross sections. Thus a single photoionization event followed by cascade electron impact ionization, rather than direct DPI valence processes was supposed to be responsible for such a copious pair production. More than a decade later the first experimental evidence for direct DPI on the valence band of a solid [84] was performed. The set-up used for this latter experiment featured a time-of-flight electron spectrometer that enabled the measurement of the energy distribution of correlated pairs of electrons emitted from clean Cu(100) and Ni(100) surfaces upon excitation with linearly polarized 45 eV photons. The selected electron energies implied that the initial state of the DPI process was located in the vicinity of the Fermi level, thus excluding the possibility of cascade energy-loss processes in favour of nearly pure direct two-electron photoemission with negligible perturbation from multiple scattering processes. Even though interpretation of DPI in solids is more cumbersome than in gases, good agreement between theory and the aforementioned experiments was achieved. Both in theory and experiments the energy-sharing distributions display a minimum for symmetric conditions for the ejected pair. This behaviour directly results from the dependence of the dipole transition matrix element from the scalar product of the light polarization vector with the sum of the momentum vectors of the two outgoing electrons. In the experiment, performed with light at normal incidence, the electric vector was parallel to the surface, and the electrons were detected in directions symmetric with respect to the surface normal, hence the cross section vanishes for equal energy sharing. Similar results have been obtained in the case of Ni(001) [85] where a direct double-ionization mechanism has been considered as opposed to a two-step process involving a direct single photoemission followed by a collision of the photoelectron with a band electron. A subsequent DPI experiment conducted on C_{60} deposited on a Cu surface [86] allowed the determination of an electron correlation energy of 1.6 eV, in good agreement with the one found by Auger spectroscopy [87]. In DPI experiments, the correlation energy is found to influence the energy distribution of the electron pairs through the electrostatic interaction of the two holes generated on the same site at the same time. The tendency of the two charged particles to avoid each other leads to the well-established theoretical concept of an exchange-correlation hole. In spite of the wide use made in solid-state physics of this concept, no direct experimental proof for its validity was available until DPI experiments on NaCl(100) [88] and Cu(111) [89, 90] were performed. These experiments demonstrate that the exchange-correlation hole manifests itself with a depletion zone in the DPI cross section that is centred around the direction of one electron of the emitted pair, as clearly shown in Fig. 1.12. These kind of experiments are 'signal hungry' and, similarly to those performed in the gas phase, have been shown to benefit from the use of imaging techniques in efficiently detecting electron pairs emitted from the surface [91].

1.3 Indirect double photoionization

In targets other than He and H_2, double ionization may also proceed via an indirect process. In such a case, the absorption of the radiation results in the formation of an intermediate excited state of the singly charged ion embedded in the double ionization continuum. The intermediate state eventually decays with a non-radiative process into the double continuum. This process is well known in inner-shell photoionization, where the rearrangement of the target may lead to a doubly charged ion via the emission of an Auger electron. The same process occurs also in the valence region where, for example, the $A^+(ns^2np^4n'l$ or $nsnp^5n'l)$ satellite states of a rare gas may decay to the lower lying $A^{2+}(np^4)$ states via a valence Auger decay [92,93]. Whenever the intermediate state can be treated as a well-defined isolated resonance in the double ionization continuum and the competing direct double ionization is neglected, then the indirect process can be regarded as a sequential one, consisting of two incoherent steps. In such a case, Auger electron–photoelectron coincidence spectroscopy (APECS) can be used to provide a deep insight into the ionization and decay processes [94]. Indeed, these kind of experiments have been proposed as 'complete experiments' [95], where the transition amplitudes and relative phases of either the photoionization or the Auger decay can be obtained experimentally. This has been proved, for example, in the case of Ar $2p$ photoionization by Bolognesi *et al.* [96].

Moreover, APECS produces a better and unambiguous spectroscopic characterization of the decay electron spectra [1,97]. This has found noticeable applications

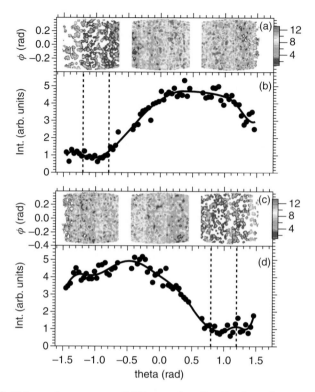

Figure 1.12 False-colour maps of 2D angular distributions (upper panels) and resulting electron yield profiles (lower panels) for pairs of electrons with energy $E_{fast} = 23$ eV, $E_{slow} = 12$ eV excited with 50 eV linearly polarized photons. The direction of the fixed in angle 'fast' electron is centred either at $\Theta = -1$ rad, for panels (a) and (b), or $\Theta = 1$ rad, for panels (c) and (d). The line profiles of the intensity maps shown in (a) and (c) are plotted in panels (b) and (d), respectively. The dashed vertical lines mark the boundaries of the angles accepted by the 'fast' electron detector. The coincidence yield depletion detected always in the Θ_{fast} direction is interpreted as direct evidence for the correlation exchange hole [90].

in the case of Auger spectra of both molecules and solids. Some examples of these applications of APECS are reported in Sections 1.3.1 and 1.3.2.

The tunability, high resolution and intense flux of the new generation of synchrotron radiation sources have opened up the possibility to perform experiments where the conditions for a two-step do not hold. Thus a full set of quantum phenomena that were predicted [98–100] has been investigated experimentally in Ne [101–104], Xe [105–109], Mg [110] and also the Cu(111) surface [111, 112]. The results of a series of experiments on the indirect double photoionization of Ne performed by Schaphorst *et al.* [101] and [102–104] and on the Cu(111) surface by van Riessen *et al.* [111] and Stefani *et al.* [112] will be taken as examples of these phenomena and described in Section 1.3.3.

1.3.1 Auger photoelectron coincidence spectroscopy (APECS) applied to molecules

The interpretation of Auger spectra of molecules presents a quite complicated problem because they involve a large number of dication states and contain information not only on the spectroscopic properties of the molecule, but also on the dynamics of molecular ionization and de-excitation [113–115]. The process involves three molecular states, namely the ground state of the neutral species, the state of the singly charged ion with its inner hole and the final state of the doubly charged ion. These states differ in their potential energy curves. Therefore, molecular Auger spectra are often characterized by band-like structures resulting from the interplay of vibrational excitation in the states of the singly and doubly charged ions. Thus the non-coincidence experimental Auger spectra, which result from the sum over all levels of vibrational excitation in the photoionized state, cannot be directly compared with the detailed and differentiated analysis achieved by theoretical models. Moreover, in polyatomic molecules the presence of two or more atoms of the same element increases the complexity of the spectra due to the superposition of the decay processes originating from atoms of the same species, but in different molecular sites. These spectra are shifted in energy and their relative intensities are determined by the different sensitivity of each atom to electronic relaxation and correlation effects depending on the occupied site. Via the selection of a photoelectron of well-defined kinetic energy, APECS makes it possible to disentangle the contributions to the Auger spectrum from different vibrational levels [116–118], spin-orbit components [119] and satellite states of a core level [120] and from different sites in the molecule [121]. Therefore, APECS adds 'state and site' selectivity to the conventional Auger measurements. In the following paragraphs, an example of site selectivity in N_2O and another one of vibrational state selectivity in CO are described.

   N_2O is one of the simplest molecules, with three inequivalent atoms and two inequivalent bonds. Therefore, it represents a prototypal example to investigate the effects of atom-site selectivity. In the linear N_2O molecule the 1s ionization energies of the central, N_c, and terminal, N_t, nitrogen atoms differ by 3.9 eV, with the smaller ionization potential, N_t 1s, located at 408.7 eV [122]. Thus, the two photoelectron peaks can also be easily resolved with the moderate energy resolution typical of coincidence experiments. The N 1s Auger spectrum extends from 320 to 375 eV in the kinetic energy of the Auger electron with several bands. In Fig. 1.13, the Auger electron–photoelectron coincidence spectra selecting the N_c

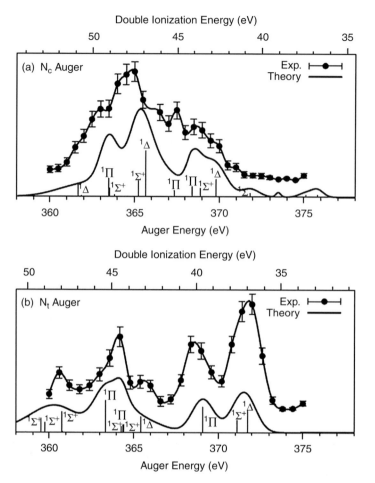

Figure 1.13 The partial N_c (a) and N_t (b) Auger spectra of N_2O obtained via Auger electron–photoelectron coincidence spectroscopy compared with the theoretical calculations. The line connecting the experimental points is obtained by cubic spline interpolation. The most important dicationic states contributing to the Auger spectrum are shown as vertical bars of height proportional to their estimated intensity.

(a) and N_t (b) sites, measured at 423.3 eV, which is close to the maximum of both the N_t and N_c 1s partial ionization cross sections [123], are shown together with the results of the theoretical calculations performed in the frame of the second-order algebraic diagrammatic construction method, ADC(2) [124, 125]. The theoretical spectra were obtained by Gaussian convolution of the calculated discrete manifold, with intensities estimated by an atomic two-hole population analysis. The width of the individual Gaussians and their energy shift with respect to the vertical double ionization energies were determined by nuclear dynamics analysis [121, 125]. The

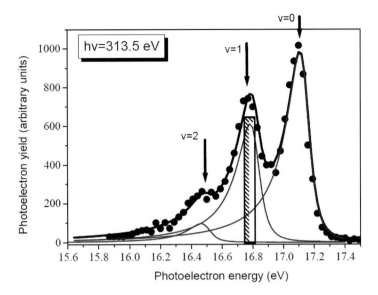

Figure 1.14 The PES of $C(1s^{-1})$ in CO measured at $h\nu = 313.5$ eV and $\theta = 60°$. The PCI lineshape (thick solid line) has been fitted to the experimental data (dots). The contributions of the three vibrational states (thin solid lines) are shown. The dashed area represents the energy resolution of the photoelectron analyzer in the coincidence measurements, while the three arrows indicate the energies of the photoelectrons selected for the APECS experiments.

N_t and N_c Auger spectra are evidently very different. The latter shows a broad region of intensity below 370 eV, while the former displays more numerous and separated bands, with a broad, very intense peak at higher kinetic energy, namely 371.8 eV. This peak is due essentially to the two lowest dicationic singlet states, $^1\Delta$ and $^1\Sigma^+$, which give extremely weak contributions to the N_c spectrum above 375 eV. The theoretical interpretation of the spectra [121] makes clear that the same doubly ionized states that contribute appreciably to both spectra, do so in quite different regions, not just because of the shift in the photoelectron lines but also as a consequence of the very different impact of nuclear dynamics effects. A reliable reproduction of the spectra ignoring such effects is impossible.

The second example consists of the investigation of the partial Auger yield for vibrationally selected CO^+ ($C1s^{-1}$) states [116]. The $C(1s^{-1})$ photoemission spectrum, PES, of CO (Fig. 1.14) displays a vibrational progression with a few members. Their spacing is about 300 meV, thus CO is a good candidate for the coincidence experiments, whose low count rates hamper high-resolution measurements. In the figure, the separate contributions of the three vibrational states, obtained by fitting a post-collision interaction, PCI, lineshape to the experimental data [126] and the kinetic energy of the photoelectrons where the APECS spectrum has been

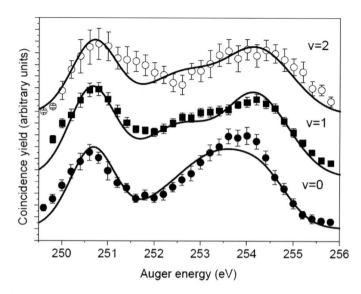

Figure 1.15 Auger electron spectrum measured in coincidence with photoelectrons generated in the photoionization of the $C(1s^{-1})$ $v = 0$, 1 and 2 states. The solid lines are the calculated coincidence spectra (see Fig. 1.16 and text) convoluted with the apparatus response function. The convoluted spectra have been rescaled to the experiments at 250.6 eV.

measured, are shown. The measured coincidence Auger spectra are shown in Fig. 1.15. These spectra cannot be considered the pure partial Auger spectra of each vibrational state of the intermediate cation. Indeed, due to the asymmetric PCI lineshape, which at these energies produces a prominent tail towards lower kinetic energies (see Fig. 1.14), only in the case of the highest kinetic energy the photoelectron yield from just the $C(1s^{-1})$ $v = 0$ state is measured. In the other two cases, always a mixed yield from more than one vibrational state is measured. For example, in the case of the lowest kinetic energy only 40% of the measured yield comes from the $C(1s^{-1})$ $v = 2$ state. Two different behaviours are observed in the coincidence Auger spectra of Fig. 1.15. There is a feature that always appears at about 250.6 eV. Then the other features in the region $251.5 - 256$ eV of the spectrum change shape and display a dispersion with the vibrational excitation of the intermediate state. To interpret these observations the expected coincidence spectra have been calculated using the potential energy curves of the $a^1\Sigma^+$, $b^1\Pi$, $A^3\Sigma^+$ and $d^1\Sigma^+$ states of CO^{2+} derived in Püttner *et al.* [118]. The results of the calculation are reported in Fig. 1.16. In the three panels of this figure the calculated partial Auger spectra for the $C(1s^{-1})$ $v = 0$, 1 and 2 states are shown. While the bands due to the $a^1\Sigma^+$, $b^1\Pi$ and $A^3\Sigma^+$ states display a noticeable change in shape as v changes from 0 to 2, the intensity of the band due to the $d^1\Sigma^+$ state always peaks at almost

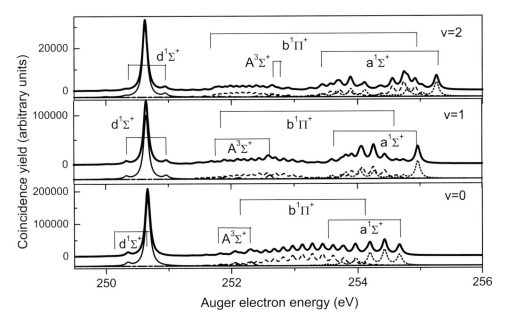

Figure 1.16 Simulation of the APECS spectra generated in the photoionization states of the $C(1s^{-1})$ $v = 0$, 1 and 2 states. The potential energy curves used in the simulations are taken from [118]. In each panel the separate contributions of the CO^{2+} states are reported: $a^1\Sigma^+$ (dotted line), $b^1\Pi$ (dashed line), $A^3\Sigma^+$ (dashed-dotted line), and $d^1\Sigma^+$ (thin full line).

the same Auger energy. The theoretical calculation, convoluted with an asymmetric lineshape, which takes into account PCI effects and the response function of the analyzers, gives a satisfactory representation of the experimental results (Fig. 1.15). The change in shape and dispersion can be rationalized in terms of the potential energy curves of the intermediate cation and final dication states. The dications are all metastable states and have a local minimum capable of supporting vibrational and rotational levels. The minimum in the potential energy curve of the $a^1\Sigma^+$, $b^1\Pi$ and $A^3\Sigma^+$ dication states occur at C–O internuclear distances [118] definitely larger than the one of the $C(1s^{-1})$ intermediate state, while the internuclear distance of the $d^1\Sigma^+$ state is very close to the one of the intermediate state [118]. Therefore, in the latter case vertical transitions are strongly favoured, as clearly seen in Figs. 1.15 and 1.16. The shift of the minimum in the potential curve of the other dication states leads to the population of higher vibrational states (Fig. 1.16). This has also been confirmed by a high-resolution APECS measurement by Ulrich *et al.* [117].

These results clearly show that the Auger spectrum is strongly dependent on the vibrational internal degree of freedom of the intermediate state, and only the selectivity of APECS allows us to observe this effect.

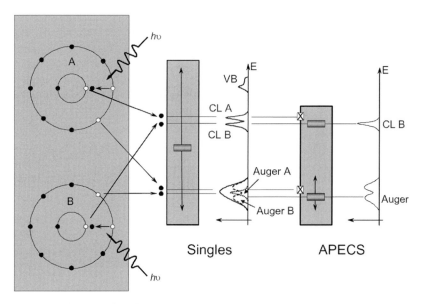

Figure 1.17 Schematic diagram comparing conventional, or 'singles', photo-emission spectroscopy to APECS. Singles spectroscopy uses one electron energy analyzer and is unable to resolve the Auger spectra associated with the decay of core holes of atoms A and B. In APECS, by using two electron energy analyzers (one fixed on the core level of interest and the other scanning the Auger spectrum), one can separate the contributions from each atom to the Auger spectrum [112].

1.3.2 Auger photoelectron coincidence spectroscopy (APECS) applied to solids

Also for solids, the most complete way to study two-hole final states created via an indirect process is to detect, coincident in time, the Auger electron–photoelectron pair and to analyze the two partners in energy (APECS) and in ejection angle (AR-APECS). The general theoretical background for these experiments is independent from the aggregation state and what has already been discussed in the case of atoms and molecules can be extended to solids, except for the description of the initial and final states where suitable solid-state wave functions have to be used.

The schematic diagram of an APECS experiment is shown in Fig. 1.17. It illustrates the high degree of discrimination introduced by this class of coincidence experiment in the analysis of the photoelectron and Auger electron spectra.

To elucidate the main advantages expected from this methodology, let us consider a solid composed by two different types of atoms, A and B in Fig. 1.17. This is a suitable model to represent any change in the energy of the core levels of A and B caused by different species of atoms, by atoms of the same species in different sites or in different chemical states, or even by the spin–orbit interaction in the same atom. In conventional photoelectron spectroscopy, the entire kinetic energy range of the photoemitted electrons is recorded and analyzed. It contains the valence band, the core levels and the Auger bands (Fig. 1.17). The valence spectrum contains information about the local electronic structure of atoms A and B, but in practice the width of the valence band, typically from 1 to 10 eV, makes it difficult to isolate these contributions. The core-valence-valence, CVV, Auger spectrum is in principle a probe of local valence electronic structure and is very sensitive to electron correlation. In practice, the broadness of the spectrum, which makes the spectral features from each atom overlapping in energy, hampers the exploitation of this potentiality. Conversely, the intrinsic width of the core levels, typically a few tenths of an electron volt, is sharp enough to allow us to resolve experimentally the energy shifts that result from differences in the local valence electronic structure. APECS correlates in time well-separated core level spectra to Auger transitions, hence it isolates the coincident Auger spectrum originating from a specific site. As illustrated in the rightmost panel of Fig. 1.17, APECS measurements are performed using (at least) two electron energy analyzers. One analyzer, for example, is set at the kinetic energy of core level B while the other is swept through the CVV Auger spectrum. By detecting only Auger electrons in time coincidence with photoelectrons, only events associated with decay of core level B will contribute to the coincidence Auger spectrum. Vice versa, if the photoemission analyzer is fixed at the energy of emission from core level A only events associated with decay of core level A will contribute to the coincidence Auger spectrum. APECS can then be used to investigate the electronic structure at a particular site in a solid and monitor how that site changes as the solid is perturbed. It can also be used to measure a coincidence photoemission spectrum where an analyzer is tuned on a feature of interest in the Auger spectrum and the other analyzer scans the photoemission spectrum.

In a ground-breaking experiment performed in the late seventies [1], the two final electrons of the copper L-shell ionization process were correlated in time, thus demonstrating the feasibility of APECS and its capability to disentangle features of an Auger spectrum that otherwise overlap in energy. The Cu $L_{23}M_{45}M_{45}$ Auger spectrum was measured in coincidence with the $2p_{\frac{1}{2}}$ and $2p_{\frac{3}{2}}$ photoelectrons and the Coster–Kronig-preceded Auger lines were unambiguously separated from the main $L_{23}M_{45}M_{45}$ Auger transition. This experiment also made clear that the background due to multiple inelastic scattering is largely suppressed in an APECS spectrum as compared to conventional Auger and photoelectron spectra. Shortly after this first experiment an APECS theory was developed [127, 128]. According to this early theory [128], APECS brings little extra information on solids when the core-hole intrinsic energy width is small compared to the energy interval over which the local density of states is distributed. Under these conditions the excitation associated with the core-hole creation propagates away well before the Auger decay takes place [129, 130]. Consequently, the APECS spectrum does not contain more information than individual photoelectron and Auger spectra. Nonetheless, following experimental investigations have clearly shown the unique capability of APECS to disentangle signals originating from different sites within the solid from overlapping spectral features in a variety of different materials, ranging from metals [131–137] to semiconductors [133, 138] and oxides [139, 140]. APECS experiments on the $M_{45}N_{45}N_{45}$ of Pd have shown the capability of this spectroscopy to single out different steps of a complex Auger cascade process [141] in which large widths and energy gains have been observed in the MVV spectra. This provides strong evidence that valence holes created in previous cascade steps participate dynamically in later cascade steps [142].

APECS has also been demonstrated to be more surface sensitive than conventional electron spectroscopies [135, 139, 143–145]. This peculiarity has been modeled [146] by a Monte Carlo calculation, which accounts for the elastic and inelastic multiple scattering suffered by the electron pair within the solid. According to this model, APECS is proposed as a suitable tool for quantitative surface analysis [147] and demonstrated to be capable of performing non-destructive depth profiling with extreme surface sensitivity and with a few angstroms resolution [148]. A recent investigation of the topmost layer of a clean Si(100)-2x1 surface [149] has demonstrated that the escape depth for a specific APECS event is limited to 1.2 Å, while the APECS Si $2p$ photoline in Si(111)-7x7 [150] can discriminate rest atoms from pedestal atoms of the surface dimers and suggests that the former are negatively charged while the latter are positively charged. Surface sensitivity has been further demonstrated by experiments on alloys [151]. A further capability unique to APECS concerns the possibility of performing core photoemission with an energy resolution limited by the experimental resolution rather than by the lifetime broadening [152–154]. It results from the energy conservation applied to the electron pair rather than to the photoelectron and Auger electron individually. In these experiments, the coincidence photoline was narrower than the non-coincidence one and the energy position of the peak maximum displayed a linear dispersion versus the energy selected for the partner Auger electron. In Fig. 1.18 a clear example of the narrowing and the shift in energy that affects the $3p_{\frac{3}{2}}$ photoline measured in an APECS experiment on Cu(111) [154] is reported. Due to the short lifetime of the core hole, the photoline profile undergoes a Lorentzian broadening. On the contrary, when photoelectron and Auger electron are detected correlated in time the uncertainty affecting their kinetic energy cannot exceed the line broadening of the final double-hole valence state [155], which is much narrower than the core-hole

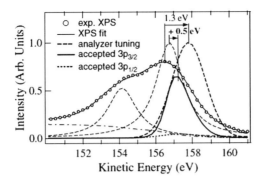

Figure 1.18 APECS $3p$ photoemission spectrum measured on Cu(111) (open circles) superimposed with the best fit (thin solid line) with two Lorentzian components (thin dashed lines) and a Shirley-type integral background (dash-dotted lines). By multiplying each of the two Lorentzian peaks by the analyzer transmission curve (thick dashed line), one obtains the curves representing the accepted $3p_{\frac{3}{2}}$ photoelectrons (thick line) and the negligible amount of accepted $3p_{\frac{1}{2}}$ photoelectrons (thick dotted line). The former one results shifted by about $+0.5$ eV above the corresponding Lorentzian peak [154].

one. The possibility offered by APECS of performing XPS with energy resolution not limited by the photoline natural width, whenever the intermediate core-hole state is short lived, was already foreseen in the early days of this technique [156]. APECS discrimination power is made even more incisive by selecting the Auger electron–photoelectron pair in emission angle. Already, the first angular-resolved experiments [157, 158] showed that a moderate angular resolution amounted to observing Auger decays originating from core-hole states whose alignment can be controlled by varying the mutual directions of light polarization and momentum vectors of the two electrons. This capability, unique to AR-APECS, allows, through propensity rules, to tune selectively on the individual magnetic sublevels of the final two-hole state [159]. The experiment can be performed in two different modes. In the first mode, angular scanning, the analyzers are tuned on a fixed energy and the coincidence yield is measured as a function of the ejection angles [112, 158]. In the second mode, energy scanning, the analyzers are set at a fixed angle and the coincidence rate is recorded as a function of the kinetic energy of either the photoelectron or the Auger electron [154, 160]. In measurements from solids, the angular scans are to be compared to conventional (non-coincidence) photoelectron diffraction (PED) or Auger diffraction (AED) spectra. In the second case, the coincidence spectra are to be compared with conventional Auger and photoelectron spectra. In both cases, the AR-APECS cross section must take into account that both the photoelectron and the Auger wave functions locally generated at the core-hole site (source wave functions) are diffracted by the crystal lattice [161]. At high electron energy (> 300 eV), diffraction will be dominated by small-angle forward scattering. At lower energies, the different sublevels will influence the coincidence angular pattern, but will leave almost unaffected the correlation between the quantum numbers and ejection angles [154]. The extra information gathered in the two coincidence modes is discussed in the following by the help of a few case studies. The use of linearly polarized light simplifies the interpretation of an AR-APECS experiment as the dipole selection rules that govern the core-hole ionization reduce to $\Delta l = \pm 1$ and $\Delta m_l = 0$, where l is the quantum number associated with the orbital angular momentum and m_l is the quantum number of its projection along the quantization axis. It is to be noted that the m_l of the core hole is identical to the photoelectron one and also enters into the selection rules of the Auger process that can be written as [156, 158, 162–164] :

$$
\begin{aligned}
l_C + l_1 + k &= \text{even} \\
l_A + l_2 + k &= \text{even} \\
\beta = m_1 - m_C &= m_A - m_2 \\
|l_C - k| \leq l_1 &\leq l_C + k \\
|l_A - k| \leq l_2 &\leq l_A + k,
\end{aligned}
\tag{1.10}
$$

where l_C, l_A, l_1, and l_2 are the orbital quantum numbers of the core hole, the emitted Auger electron, and the two holes in the solid characteristic of the Auger final state, respectively. Similarly, m_C, m_A, m_1, and m_2 are the magnetic quantum numbers of the core hole, the Auger electron and the two final holes, respectively. k and β are the orbital and magnetic quantum numbers related to the expansion of the Coulomb interaction in spherical harmonics [162–164]. To the first-order approximation, $k = \beta = 0$. The key feature of AR-APECS is that the coincidence angular distribution of the Auger electron and photoelectron is linked to the magnetic quantum numbers m_1 and m_2 of the two final holes [112, 159]. This can be better explained by an example. Let us consider photoemission from an $l = 1$ core level, such as the Cu $3p$. Since $l_C = 1$, the dipole selection rule states that the final-state photoelectron will have either $l = 0$ (i.e., an s-wave) or $l = 2$ (a d-wave) and m_l values equal either to 1 or 0. The photoelectron partial waves having different m_l values will correspondingly have different angular distributions. It is easy to realize that whenever the photoelectron is collected in the direction of the light polarization axis (briefly is aligned with the polarization), only photoelectrons, and consequently core states, with $m_l = 0$ are detected. The following Auger decay, such as the $M_3 M_{45} M_{45}$ in Cu, has been demonstrated to be mostly dominated by f-wave symmetry (i.e., $l_A = 3$) [165, 166], while the two final holes are localized in d levels (i.e., $l_1 = l_2 = 2$). As a consequence, the two-hole final state exhibits a d^8 multiplet structure dominated by the ^1G configuration.

If the Auger electron is also detected aligned with the light polarization axis only waves with $m_A = 0$ are allowed and the third of relations (1.10) implies that $m_1 = m_2$. As already stated, $l_1 = l_2$, hence final states with parallel spins are forbidden by Pauli's principle. In this case, the ^1G component should be largely enhanced with respect to the triplet components. If we were dealing with isolated atoms our discussion of AR-APECS could end at this point. But the atom is part of a crystal and the diffraction of the atomic Auger and photoelectron wave functions (source wave functions) by the surrounding atoms must be accounted for. AR-APECS angular distribution depends on both orbital and magnetic quantum numbers, l and m_l, of the source wave functions [167]. Furthermore, the introduction of the quantization axis constituted by the light polarization allows the definition of non-statistical contributions to the Auger yield from the different sublevels of atomic core even in non-magnetic materials [165]. As a consequence, angular resolution will add to APECS the capability to act as a 'band-pass filter' for l and m_l quantum numbers. This is elucidated by the simulations reported in Fig. 1.19 relative to the diffraction patterns of a Cu $M_{23}M_{45}M_{45}$ Auger electron and to the associated $3p$ photoelectron from the Cu(111) surface. The calculation has been performed within the multiple scattering approach to diffraction [168], assuming in both cases excitation with p-polarized photons and a polar scanning along the crystal direction [112]. In Fig. 1.19, together with the total angular distribution are reported the individual contributions of the different partial wave components of the source wave ($m_l = 0, 1, 2$). It is evident that for both Auger electron and photoelectron, different polar angles will select quite different weights for the individual partial waves. In essence, even in the presence of diffraction, at the polar angle $0°$, i.e., in aligned geometry, the $m_l = 0$ component is overwhelmingly larger than the others, thus realizing a sharp filtering action with respect to the magnetic sublevels of the core-hole state. In summary, diffraction from the crystal lattice does not forbid the interpretation of the Auger photoelectron angular distribution as the angular distribution of Auger processes generated by a polarized core hole whose polarization is determined by the angular resolution of the photoelectron [154].

Therefore, the effective atomic alignment determined by detecting the photoelectron in a specified direction will determine different angular distributions for the different multiplet components of the coincident Auger electron spectrum. As a consequence, the coincident Auger lineshape will be in general different from the non-coincident one and in particular will change upon changing the AR-APECS kinematics. This effect, named the dichroic effect in AR-APECS (DEAR-APECS) has been discovered on a Sn/Ge(111) surface by Gotter *et al.* [169] and confirmed on Cu [154] and Co/Cu [170].

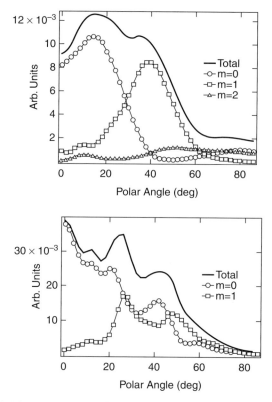

Figure 1.19 Multiple-scattering calculation of diffraction, MSCD, calculated angular distributions of the Cu M_3VV Auger (top panel) and M_3 photoelectron (bottom panel) along the [112] crystallographic direction of the Cu(111) surface (see text for details). Different symbols show the contribution of individual m_l source wave components to the total cross section [112].

This finding has sparked an interest in applying this selectivity in magnetic materials. Indeed, in magnetic systems, where exchange and correlation play a fundamental role, the Auger lineshape is often broad, doesn't change upon crossing the temperatures of magnetic transitions and doesn't allow for extracting the information on correlations that is potentially contained in Auger spectra. The DEAR-APECS has the ability to disentangle different multiplet components of the Auger spectrum, especially high-spin versus low-spin components, thus allowing us to investigate closely the electronic and magnetic structure of a solid – even upon crossing the temperature of a magnetic transition. Lately, Gotter *et al.* [160] have reported results of an AR-APECS experiment used to probe directly the local magnetic properties of an anti-ferromagnetic thin film of CoO/Ag(001). The experiment was performed with the sample at temperatures above and below the Néel temperature of the CoO film. The dramatic differences in the coincident Auger lineshape obtained at these two temperatures are understood in terms of dissolution of the local magnetic order, and provide an upper bound for the transition temperature of the CoO surface.

A panorama of what we learn from APECS experiments has been given by several review papers [112, 135, 157, 171–175] that also briefly summarize the theoretical models developed to describe APECS events. Two different approaches have been used in describing core-hole creation and the following autoionizing decay. The two-step model treats the core hole as a well-defined real state and factors out the Auger event from the core photoionization. The other approach, the one-step model, describes the core hole by virtual intermediate states spanning over the full series of electron excited states, including the continuum [175, 176]. In the latter model, the Coulomb operator responsible for the Auger decay operates over all particles in the target, and both final electrons are involved and their pair wave function has to fulfill the correct boundary conditions. A simple model for core-valence-valence APECS from open band solids that accounts for both screened and unscreened core holes has also been proposed by Marini and Cini [177].

1.3.3 Interference and coherence effects in indirect double photoionization

The picture of indirect DPI consisting of two incoherent successive processes no longer holds if the energy of the incident photon is such as to produce photoelectrons with a kinetic energy close to that of one of the Auger electrons. In such a case, electron exchange plays an important role and the two outgoing electrons cannot be specified as photoelectron or Auger electron. Therefore, one has not only to describe DPI within a one-step model [178], but also account for electron exchange explicitly [98]. Then, strong interference effects are expected to appear in the angular and energy distributions of the outgoing electrons. Moreover, when two electrons with the same or similar kinetic energy are ejected into the continuum, their Coulomb interaction may also play a role and alter the energy and angular distribution. Therefore this effect, known as post-collisional interaction, PCI [179], has to be considered and combined with the ones due to the indistinguishability of the electrons. Depending on the total spin and parity of the electron pair and on the direction of each electron with respect to the polarization axis of the incident photon, the interference due to the indistinguishability of the electrons is either constructive or destructive [98]. The experimental challenge is to achieve an energy resolution comparable to or better than the linewidth of the intermediate state to ensure the indistinguishability of the two electrons.

A case that has attracted substantial attention is the indirect double ionization of Ne via the formation of $Ne^+ 2s2p^5(^3P^0)nl$ states, which mainly decay non-radiatively to the $Ne^{2+}(2s^22p^4\ ^1D^e)$ continuum [92, 180]. An example is represented by the following process:

$$h\nu + Ne \rightarrow Ne^+ \left[2s2p^5 \left(^3P^0\right) 3p \left(^2S^e\right)\right] + e_{ph}\left(E_{ph},\ l = 1\right) \rightarrow$$

$$Ne^{2+} \left[2s^22p^4 \left(^1D^e\right)\right] + e_{ph}(E_{ph}, l = 1) + e_{Auger}\left(E_{Auger} = 13.24\,eV, l = 2\right),$$

$$(1.11)$$

where the natural width of the intermediate state is 155 ± 25 meV [101].

In process (1.11), each photoelectron is related to only one Auger electron and in the LS coupling scheme and dipole approximation the wave function of the photoelectron is an εp partial wave and the one of the Auger electron is an εd partial wave. This choice of the indirect process greatly simplifies the theoretical description and the amplitude of the indirect DPI process reduces to:

$$T_m = \frac{a(E_b, E_a)Y_{10}(\Omega_a)Y_{2m}(\Omega_b)}{E_b - E_{Auger} + i\Gamma/2} + \frac{a(E_a, E_b)Y_{10}(\Omega_b)Y_{2m}(\Omega_a)}{E_a - E_{Auger} + i\Gamma/2}, \qquad (1.12)$$

where Ω_a and Ω_b indicate the angle of emission of the two electrons, E_a and E_b are their kinetic energies and the function $a(E_a, E_b)$ represents the slowly varying radial part of the amplitude, which can be assumed to be constant near the resonance. The Y_{10} and Y_{2m} spherical harmonics are associated with the εp and εd partial waves of the photoelectron and Auger electron, respectively. In the first term of Eq. (1.12), 'electron a' plays the role of the photoelectron and 'electron b' is produced in the Auger decay, whereas, in the second term, the two electrons have exchanged the roles. The TDCS can be calculated as an incoherent summation of the amplitudes T_m over the unobserved magnetic quantum levels of the final ionic state.

Schaphorst et al. [101] followed the evolution of the TDCS shape for the condition of equal energy sharing ($E_a = E_b$) across the energy region of the resonance. They observed a noticeable increase of the coincidence intensity at $E_a = E_b = E_{Auger} = 13.24$ eV in agreement with previous observations on resonance-affected double photoionization. However, the indirect process is expected to yield maxima in the TDCSs for parallel and antiparallel emission along the electric field vector [94]. This latter feature was not observed in the experiments, but the measurements at higher energy resolution clearly displayed the presence of a dip for emission of the two electrons in opposite directions. In different sets of experiments, Rioual et al. [102, 104] also observed a similar dip in the coincidence energy spectra measured at a relative angle $\theta_{ab} = 180°$ and $E_a = E_b = E_{Auger}$, while the dip disappeared when $\theta_{ab} = 60°$ (Fig. 1.20).

The observation of the dip is a consequence of the indistinguishability of the two electrons in Eq. (1.12), which forbids antiparallel emission of the two electrons when they form a singlet state and possess opposite parity as in the studied

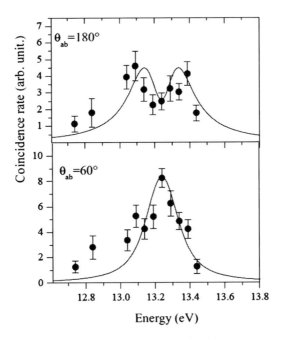

Figure 1.20 Auger electron–photoelectron coincidence spectrum at $h\nu = 92.21$ eV for $\theta_{ab} = 180°$ ($\theta_a = 30°$ and $\theta_b = 210°$) at the top and for $\theta_{ab} = 60°$ ($\theta_a = 30°$ and $\theta_b = 330°$) at the bottom. The coincidence spectrum has been collected at fixed photon energy, varying the kinetic energies E_1 and E_2 of the two electrons according the relationship $h\nu - \varepsilon^{2+}\left[\mathrm{Ne}^{2+}\left(^1D^e\right)\right] = E_a + E_b$ [102, 104].

case. The formula predicts a destructive interference that should result in a vanishing TDCS at $\theta_{ab} = 180°$. The finite values observed in the works by Schaphorst *et al.* [101] and Rioual *et al.* [102, 104] and displayed in Fig. 1.20 are due to the finite experimental resolution. These experimental findings have been subsequently well reproduced by theoretical *ab initio* calculations [181].

It is interesting to observe that in the direct DPI the back-to-back emission of two electrons in the equal energy-sharing conditions is forbidden when the electron pair is of an unfavoured type [182, 183]. The observation of a destructive interference in the indirect process leading to the $\mathrm{Ne}^{2+}(2p^4\,^1D^e)$ final state allows us to establish a link with the direct process leading to the same final state. Indeed, due to the unfavoured symmetry of the three possible states ($^1P^O$, $^1D^O$ and $^1F^O$) of the electron pair, the direct DPI is characterized by a node for the back-to-back emission. This link has been clearly proven by a series of experiments in Xe, where the indirect process leading to final states of different symmetries has been studied and characterized [106]. The destructive/constructive interference in the indirect DPI and the node/antinode in the direct DPI for back-to-back emission find the

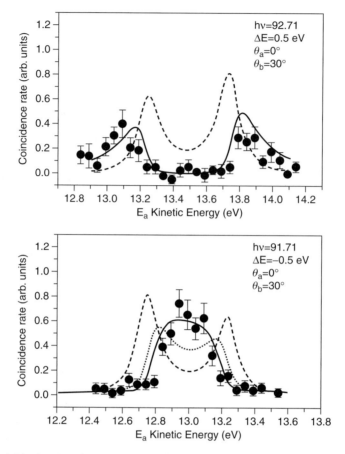

Figure 1.21 Auger electron coincidence photoelectron spectrum at $h\nu = 92.71$ eV, i.e., $\Delta E = 0.5$ eV, and $\theta_{ab} = 30°$ on the top and $h\nu = 91.71$ eV, i.e., $\Delta E = -0.5$ eV, on the bottom. The experiment is compared with the calculations without PCI effects (dashed line), with PCI effects without interference effects (dotted line) and with both effects (full line) [102, 104].

same physical origin in the anti-symmetrization of the wave function of the electron pair in the final state.

Let us now consider the combined effects of interference and PCI in the TDCS. Due to the inherent difficulties of coincidence experiments the investigation of these effects has been restricted to specific conditions [102, 104, 108, 109]. In Fig. 1.21 the results of Rioual *et al.* [102, 104] relative to the process (1.10) are shown. The measurements have been performed at $\Delta E = E_{ph} - E_{Auger} = +0.5$ and -0.5 eV, respectively, with the two detectors placed at a mutual angle $\theta_{ab} = 30°$. At $\Delta E = +0.5$ eV two structures, each with an asymmetric lineshape and shifted by about 100 meV from their nominal values, appear in the experimental spectrum.

The calculation, which includes PCI effects (dotted lines in Fig. 1.21) accounts very well for energy positions and shapes of the two peaks. The inclusion in the model of the interference term produces no effect, as shown by the full line that is superimposed on the dotted one. At $\Delta E = -0.5$ eV the experimental data exhibit a relatively narrow structure centred at about 13 eV. Here both interference and PCI play an important role, as shown by the calculations that include only the PCI (dotted line) and both the effects (full line). The increase in intensity in the energy region between the nominal energies of the photoelectron and Auger electron can be understood as follows. The PCI shifts the energy distribution of the photoelectrons and Auger electrons in the region between their nominal values. Then, in the overlapping part of the resulting distribution the two electrons become indistinguishable and suffer an interference effect. At this relative angle the interference is constructive and produces the structure observed. Scherer *et al.* [108, 109] have made similar observations in the case of $4d_{\frac{5}{2}}$ photoionization and $N_5 O_{23} O_{23}$ ($^1S_0$) Auger decay of Xe.

It is important to stress that the interference discussed in this section is a one-channel effect, due to the indistinguishability of the two electrons. In other words, the two interfering amplitudes reflect the fact that either of the observed electrons can play the role of the photoelectron or of the Auger electron, with the other electron playing the complementary role. Different sorts of interference effect can occur when the DPI process proceeds to the same final state via two different channels. In such a case, interference relies upon the existence of two resonances and therefore on two different pathways, as in the case of the Young two-slits experiment in optics. These phenomena have been studied in Kr and Ar by de Gouw *et al.* [184, 185] in non-coincidence experiments and in Kr by Rioual *et al.* [186] in electron–electron coincidence experiments.

Interference effects are also expected to appear in solids when the creation of double valence hole states is considered as a one-step process mediated by electron correlation in the initial and final state. The intermediate core-hole state is a virtual one and the two final electrons, as discussed in Section 1.3.2 for the case of Cu(111) [154], will continuously share the energy exceeding the valence double ionization threshold. In other words, rather than considering two distinct events, Auger decay and core ionization, the process is considered a core resonant double photoemission, which proceeds via a coherent process ending in an electron pair that continuously share the available energy. A striking manifestation for appropriateness of the core resonant double photoemission model has been provided by an experiment on Cu(001), where electron pairs over a wide range of energy were detected [111]. In Fig. 1.22, the measured yield of electron pairs is reported as a function of the difference of the two-electron kinetic energies. It is evident that the continuously distributed yield resonantly increases in the vicinity

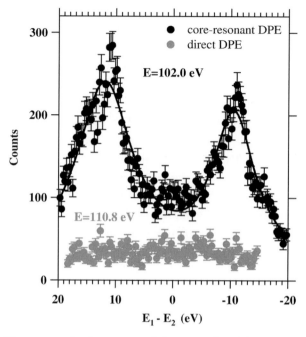

Figure 1.22 The energy-sharing curves of electron pairs with sum energies of $E_1 + E_2 = \varepsilon \pm 0.9$ eV for $E = 102$ eV (upper experimental data) and $E = 110.8$ eV (lower experimental data) obtained by integrating the pair intensity in energy regions corresponding to electron pairs emitted by the core-resonant DPI process, resulting in a $^1G\ M_{45}M_{45}$ final-state configuration. The latter corresponds to pairs emitted by direct DPI of electron pairs from the top of the d-band [111].

of the 'Auger energy' and the 'photoelectron energy'. At resonant energies, the double-hole valence state is created starting from an identical initial state and proceeding through two different channels: the core resonant and the direct one. It is then conceivable that quantum interference effects should show up in the measured cross section. Such interferences have been foreseen in an atomic case [100] and are to be expected when the timescale of relaxation of the resonantly excited core-hole state is much longer than that of the core-hole decay. Experimental investigations [187, 188] have not yet provided definite evidence of such effects.

1.4 Conclusions

The evolution of a specific class of break-up experiments, i.e., the **1-photon IN 2-electrons OUT** has been reviewed. These experiments are a unique tool that provides a direct insight into the role of electron–electron correlation in both initial and final states of the fragmentation reaction.

So far, most of the experiments have been performed on gaseous targets and in an energy regime where the long-range Coulomb interactions among the three final charged reaction products, the two electrons in the presence of the residual doubly charged ion, determine the dynamics of the process. For energy very close to the double ionization threshold, double escape of the electrons is found to be very unlikely, as each of them tend to be attracted back to the nucleus and the system has to maintain itself along the Wannier ridge in the critical phase of double escape.

At intermediate energies, the measurements of the multiply differential cross section and their interpretation have become a fertile playground for both experimentalists and theoreticians. This is because the dynamics of the process are determined by the three-body Coulomb interaction and by the interplay imposed by the symmetry determined by the absorption of the radiation and the electron–electron interaction.

At even higher energies, the electron–electron interaction in the initial state dominates the behaviour of the cross section, thus making these experiments viable to investigate correlations in the bound state.

An interesting outcome is that the aforementioned schematization in three energy regimes is found to be valid for atoms, molecules and solids, irrespective of the aggregation state.

For energies large enough to access inner valence and core ionization, the double valence hole final states can be created either directly (DPI) or via an Auger decay of the hole (APECS). Measuring the fully differential cross section for such processes has shown that the two channels combine either coherently or incoherently depending on the lifetime of the intermediate core hole. Incoherent APECS processes allow us to disentangle site and state selected features in molecular Auger bands and, through the angle selection of the photoelectron, to study Auger spectra originating from spin polarized states in solids. This latter issue is of fundamental interest in physics when applied to magnetic targets as it addresses the atomic origin of the magnetic properties in condensed matter. In other words, AR-APECS provides a unique, spin-sensitive way to probe the local electronic structure of solids.

For short-lived core-hole states the two channels combine coherently and APECS must be regarded as a core-assisted resonance in the valence DPI cross section. It is interesting to note that in these coincidence experiments the tunability of the incident radiation allows us to tune 'coherence/incoherence' effects. These effects are the formal analogue of the quantum optics experiments devoted to the study of mesoscopic superposition of quantum states [189]. This comparison stresses the basic level of detail reached today in electron–electron coincidence spectroscopies.

The main effort in the studies of DPI has been applied to the understanding of processes leading to two holes in the valence. However, nowadays, profiting from the new development of instrumentation and sources, the formation of hollow systems, in which two-core electrons are simultaneously ejected after the absorption of a single photon in synchrotron experiments [190–192] or two independent photons in the short and intense X-Fel pulses [193, 194], has begun to be investigated. These pioneering experiments have shown the potentiality of the molecular double core-hole spectroscopy to provide relevant information on the two-hole repulsion energy, when the two electrons are ejected from the same atomic site, and on the inter-core relaxation energy when the two electrons are ejected from two different sites. The dynamics of the process are, of course, different if the double ionization is generated by a single photon or by two independent photons.

Another way to produce double ionization is to make the system absorb a large number of photons from a laser with *fs* pulses and intensity of TW cm^{-2}. In the case of multiphoton ionization of the He atom with a Ti: sapphire laser ($\lambda \approx$ 800 nm, $E = 1.5$ eV), about 50 photons are needed to produce double ionization. In DPI, the driving force of the process is the electron–electron interaction, while in the case of multiphoton double ionization the physics of the process is different [195] and the basic approaches based on the separation of the interaction with the field and the Coulomb interaction of the two electrons as well as of the selection rules, which greatly contributed to the understanding of the correlated angular patterns of PDI, find scarce application.

References

[1] H. W. Haak, G. Sawatzky and T. D. Thomas, *Phys. Rev. Lett.*, **41**, 1825 (1978).
[2] O. Schwarzkopf, B. Krässig, J. Elmiger and V. Schmidt, *Phys. Rev. Lett.*, **70**, 3008 (1993).
[3] P. Lablanquie, J. Mazeau, L. Andric, P. Selles and A. Huetz, *Phys. Rev. Lett.*, **74**, 2192 (1995).
[4] R. Dörner *et al.*, *Phys. Rev. A*, **57**, 1074 (1998).
[5] L. Avaldi and A. Huetz, *J. Phys. B*, **38**, S861 (2005).
[6] D. B. Thompson *et al.*, *J. Phys. B: At. Mol. Opt. Phys.*, **31**, 222 (1998).
[7] J. H. McGuire *et al.*, *J. Phys. B: At. Mol. Opt. Phys.*, **28**, 913 (1995).
[8] H. Kossmann, V. Schmidt and T. Andersen, *Phys. Rev. Lett.*, **60**, 1266 (1988).
[9] J. M. Bizau and F. J. Wuilleumier, *J. Electron Spectrosc. Relat. Phenom.*, **71**, 205 (1995).
[10] G. H. Wannier, *Phys. Rev.*, **90**, 817 (1953).
[11] R. Wehlitz *et al.*, *Phys. Rev. Lett.*, **67**, 3764 (1991).
[12] J. Viefhaus *et al.*, *J. Phys. B: At. Mol. Opt. Phys.*, **29**, L729 (1996).
[13] J. Viefhaus *et al.*, *Phys. Rev. Lett.*, **77**, 3975 (1996).
[14] F. Maulbetsch and J. S. Briggs, *Phys. Rev. Lett.*, **68**, 2004 (1992).
[15] A. Huetz, P. Selles, D. Waymel and J. Mazeau, *J. Phys. B: At. Mol. Opt. Phys.*, **24**, 1917 (1991).

[16] C. H. Greene, *J. Phys. B: At. Mol. Opt. Phys.*, **20**, L357 (1987).
[17] G. Turri *et al.*, *Phys. Rev. A*, **65**, 034702 (2002).
[18] J. M. Feagin, *J. Phys. B: At. Mol. Opt. Phys.*, **17**, 2433 (1984).
[19] A. S. Kheifets and I. Bray, *Phys. Rev. A*, **65**, 022708 (2002).
[20] A. Huetz and J. Mazeau, *Phys. Rev. Lett.*, **85**, 530 (2000).
[21] G. Dawber *et al.*, *J. Phys. B: At. Mol. Opt. Phys.*, **28**, L271 (1995).
[22] A. Huetz, P. Lablanquie, L. Andric, P. Selles and J. Mazeau, *J. Phys. B: At. Mol. Opt. Phys.*, **27**, L13 (1994).
[23] A. Knapp *et al.*, *Phys. Rev. Lett.*, **89**, 033004 (2002).
[24] V. G. Neudatchin, Yu. F. Smirnov, A. V. Pavlitchenkov and V. C. Levin, *Phys. Lett.*, **64A**, 31 (1977).
[25] P. Bolognesi *et al.*, *J. Phys. B: At. Mol. Opt. Phys.*, **34**, 3193 (2001).
[26] S. Cvejanovic and T. J. Reddish, *J. Phys. B: At. Mol. Opt. Phys.*, **33**, 4691 (2000).
[27] H. Bräuning *et al.*, *J. Phys. B: At. Mol. Opt. Phys.*, **31**, 5149 (1998).
[28] J. Ullrich *et al.*, *J. Phys. B: At. Mol. Opt. Phys.*, **30**, 2917 (1997).
[29] P. Bolognesi *et al.*, *J. Electron. Spectrosc. Relat. Phenom.*, **141**, 105 (2004).
[30] A. Knapp *et al.*, *J. Phys. B: At. Mol. Opt. Phys.*, **38**, 615 (2005).
[31] A. Knapp *et al.*, *J. Phys. B: At. Mol. Opt. Phys.*, **38**, 635 (2005).
[32] J. A. R. Samson, *Phys. Rev. Lett.*, **65**, 2861 (1990).
[33] J. Berakdar and H. Klar, *Phys. Rev. Lett.*, **69**, 1175 (1992).
[34] J. Berakdar, H. Klar, A. Huetz and P. Selles, *J. Phys. B: At. Mol. Opt. Phys.*, **26**, 1463 (1993).
[35] K. Soejima, A. Danjo, K. Okuno and A. Yagishita, *Phys. Rev. Lett.*, **83**, 1546 (1999).
[36] V. Mergel *et al.*, *Phys. Rev. Lett.*, **80**, 5301 (1998).
[37] M. Achler *et al.*, *J. Phys. B: At. Mol. Opt. Phys.*, **34**, 965 (2001).
[38] S. A. Collins *et al.*, *Phys. Rev. A*, **65**, 52717 (2002).
[39] P. Bolognesi *et al.*, *J. Phys. B: At. Mol. Opt. Phys.*, **41**, 051003 (2008).
[40] L. Malegat, P. Selles and A. Huetz, *J. Phys. B: At. Mol. Opt. Phys.*, **30**, 251 (1997).
[41] L. Argenti and R. Colle, *J. Phys. B: At. Mol. Opt. Phys.*, **41**, 245205 (2008).
[42] B. Krässig, in *Correlations, Polarization and Ionization in Atomic Systems*, AIP Conference Proceedings, vol. 604, ed. D. H. Madison and M. Schulz (New York: AIP, 2001), p. 12.
[43] P. Bolognesi *et al.*, *J. Phys. B: At. Mol. Opt. Phys.*, **36**, L247 (2003).
[44] J. S. Briggs and V. Schmidt, *J. Phys. B: At. Mol. Opt. Phys.*, **33**, R1 (2000).
[45] L. Malegat, *Physica Scripta*, **T110**, 83 (2004).
[46] J. Colgan and M. S. Pindzola, *Phys. Rev. A*, **65**, 032729 (2002).
[47] C. W. McCurdy, D. A. Horner, T. N. Rescigno and F. Martin, *Phys. Rev. A*, **69**, 032707 (2004).
[48] P. Selles, L. Malegat and A. K. Kazansky, *Phys. Rev. A*, **65**, 032711 (2002).
[49] P. Selles *et al.*, *Phys. Rev. A*, **69**, 052707 (2004).
[50] P. Selles, L. Malegat and A. K. Kazansky, in *Electron and Photon Impact Ionization and Related Processes*, IOP Conference Series, vol. 183 (Bristol: IOP, 2005), p. 141.
[51] A. Y. Istomin, N. L. Manakov, A. V. Meremiamin and A. F. Starace, *Phys. Rev. Lett.*, **92**, 063002 (2004).
[52] O. Hemmers, R. Guillemin and D. W. Lindle, *Rad. Phys. Chem.*, **70**, 123 (2004) and references therein.
[53] T. J. Reddish, J. P. Wightman, M. A. MacDonald and S. Cvejanovic, *Phys. Rev. Lett.*, **79**, 2438 (1997).
[54] J. P. Wightman, S. Cvejanovic and T. J. Reddish, *J. Phys. B: At. Mol. Opt. Phys.*, **31**, 1753 (1998).

[55] N. Scherer, H. Lörch and V. Schmidt, *J. Phys. B: At. Mol. Opt. Phys.*, **31**, L817 (1998).

[56] J. M. Feagin, *J. Phys. B: At. Mol. Opt. Phys.*, **31**, L729 (1998).

[57] T. J. Reddish and J. M. Feagin, *J. Phys. B: At. Mol. Opt. Phys.*, **32**, 2473 (1999).

[58] T. Weber *et al.*, *Phys. Rev. Lett.*, **92**, 163001 (2004).

[59] F. Maulbetsch and J. S. Briggs, *J. Phys. B: At. Mol. Opt. Phys.*, **28**, 551 (1995).

[60] M. Walter and J. S. Briggs, *Phys. Rev. Lett.*, **85**, 1630 (2000).

[61] S. A. Collins, A. Huetz, T. J. Reddish, D. P. Seccombe and K. Soejima, *Phys. Rev. A*, **64**, 062706 (2001).

[62] D. P. Seccombe *et al.*, *J. Phys. B: At. Mol. Opt. Phys.*, **35**, 3767 (2002).

[63] T. Weber *et al.*, *Nature*, **431**, 43 (2004).

[64] M. Gisselbrecht *et al.*, *Phys. Rev. Lett.*, **96**, 153002 (2006).

[65] T. J. Reddish *et al.*, *Phys. Rev. Lett.*, **100**, 193001 (2008).

[66] A. S. Kheifets, *Phys. Rev. A*, **71**, 022704 (2005).

[67] W. Vanroose, F. Martín, T. N. Rescigno and C. W. McCurdy, *Phys. Rev. A*, **70**, 050703 (2004).

[68] W. Vanroose, F. Martín, T. N. Rescigno and C. W. McCurdy, *Science*, **310**, 1787 (2005).

[69] W. Vanroose, D. A. Horner, F. Martín, T. N. Rescigno and C. W. McCurdy, *Phys. Rev. A*, **74**, 052702 (2006).

[70] D. A. Horner, W. Vanroose, T. N. Rescigno, F. Martin and C. W. McCurdy, *Phys. Rev. Lett.*, **98**, 073001 (2007).

[71] J. Colgan, M. S. Pindzola and F. Robicheaux, *J. Phys. B: At. Mol. Opt. Phys.*, **37**, L337 (2004).

[72] J. Colgan, M. S. Pindzola and F. Robicheaux, *Phys. Rev. Lett.*, **98**, 153001 (2007).

[73] V. G. Neudatchin, N. P. Yudin and V. G. Levin, *Phys. Stat. Sol. B*, **95**, 39 (1979).

[74] Yu. F. Smirnov, A. V. Pavlitchenkov, V. G. Levin and V. G. Neudatchin, *J. Phys. B*, **11**, 3857 (1975).

[75] N. P. Yudin, A. V. Pavlitchenkov and V. G. Neudatchin, *Z. Phys. A*, **320**, 565 (1985).

[76] V. G. Levin, A. V. Neudatchin and Yu. F. Smirnov, *J. Phys. B: At. Mol. Opt. Phys.*, **17**, 1525 (1984).

[77] J. Berakdar, *Phys. Rev. B*, **58**, 9808 (1998).

[78] J. Berakdar, *Phys. Rev. Lett.*, **85**, 665 (2000).

[79] N. Fominykh *et al.*, *Solid State Commun.*, **113**, 4036 (2000).

[80] N. Fominykh, J. Berakdar, J. Henk and P. Bruno, *Phys. Rev. Lett.*, **89**, 086402 (2002).

[81] N. Fominykh and J. Berakhdar, *Surf. Sci.*, **482**, 618 (2001).

[82] N. Fominykh, J. Henk, J. Berakhdar and P. Bruno, *Surf. Sci.*, **507**, 229 (2002).

[83] H. W. Bieste, M. J. Besnard, G. Dujardin, L. Hellner and E. E. Koch, *Phys. Rev. Lett.*, **59**, 1277 (1987).

[84] R. Hermann, S. Samarin, H. Schwabe and J. Kirschner, *Phys. Rev. Lett.*, **81**, 3148 (1998).

[85] R. Hermann, S. Samarin, H. Schwabe and J. Kirschner, *J. de Physique IV*, **9**, 127 (1999).

[86] F. U. Hillebrecht, A. Morozov and J. Kirschner, *Phys. Rev. B*, **71**, 125406 (2005).

[87] S. Krummacher, M. Biermann, M. Neeb, A. Liebsch and W. Eberhardt, *Phys. Rev. B*, **48**, 8424 (1993).

[88] F. O. Schumann, C. Winkler, G. Kerherve and J. Kirschner, *Phys. Rev. B*, **73**, 041404 (2006).

[89] F. O. Schumann, C. Winkler and J. Kirschner, *Phys. Rev. Lett.*, **98**, 257604 (2007).

[90] F. O. Schumann, C. Winkler and J. Kirschner, *Phys. Stat. Solidi.*, **B246**, 1483 (2009).

[91] M. Hattass *et al.*, *Phys. Rev. B*, **77**, 165432 (2008).
[92] U. Becker, R. Wehlitz, O. Hemmers, B. Langer and A. Menzel, *Phys. Rev. Lett.*, **63**, 1054 (1989).
[93] F. Combet-Farnoux, P. Lablanquie, J. Mazeau and A. Huetz, *J. Phys. B: At. Mol. Opt. Phys.*, **33**, 1597 (2000).
[94] B. Kämmerling and V. Schmidt, *Phys. Rev. Lett.*, **67**, 1848 (1991).
[95] N. M. Kabachnik, *J. Phys. B: At. Mol. Opt. Phys.*, **25**, L389 (1992).
[96] P. Bolognesi, A. De Fanis, M. Coreno and L. Avaldi, *Phys. Rev. A*, **70**, 22701 (2004); *Phys. Rev. A*, **72**, 069903(E) (2005).
[97] E. Von Raven, M. Meyer, M. Pahle and B. Sonntag, *J. Electron Spectrosc. Relat. Phenom.*, **52**, 677 (1990).
[98] L. Vegh and J. H. Macek, *Phys. Rev. A*, **50**, 4031 (1994).
[99] L. Vegh, *Phys. Rev. A*, **50**, 4036 (1994).
[100] M. Ohno and L. Sjogren, *J. Phys. B: At. Mol. Opt. Phys.*, **36**, 4519 (2003).
[101] S. J. Schaphorst *et al.*, *J. Phys. B: At. Mol. Opt. Phys.*, **29**, 1901 (1996).
[102] S. Rioual *et al.*, *Phys. Rev. A*, **61**, 044702 (2000).
[103] S. Rioual *et al.*, *Phys. Rev. Lett.*, **86**, 1470 (2001).
[104] S. Rioual *et al.*, *Phys. Rev. A*, **67**, 012706 (2003).
[105] O. Schwarzkopf and V. Schmidt, *J. Phys. B: At. Mol. Opt. Phys.*, **29**, 1877 (1996).
[106] P. Selles, J. Mazeau, P. Lablanquie, L. Malegat and A. Huetz, *J. Phys. B: At. Mol. Opt. Phys.*, **31**, L353 (1998).
[107] J. Viefhaus *et al.*, *Phys. Rev. Lett.*, **80**, 1618 (1998).
[108] N. Scherer, H. Lörch, T. Kerkau and V. Schmidt, *Phys. Rev. Lett.*, **82**, 4615 (1999).
[109] N. Scherer, H. Lörch, T. Kerkau and V. Schmidt, *J. Phys. B: At. Mol. Opt. Phys.*, **34**, L339 (2001).
[110] H. Lörch, N. Scherer, T. Kerkau and V. Schmidt, *J. Phys. B: At. Mol. Opt. Phys.*, **34**, 3963 (2001).
[111] G. A. Van Riessen *et al.*, *J. Phys. Condens. Matter*, **22**, 092201 (2010).
[112] G. Stefani *et al.*, *J. Electron. Spectrosc. Relat. Phenom.*, **141**, 149 (2004).
[113] L. S. Cederbaum, P. Campos, F. Tarantelli and A. Sgamellotti, *J. Chem. Phys.*, **95**, 6634 (1991).
[114] L. S. Cederbaum and F. Tarantelli, *J. Chem. Phys.*, **98**, 9691 (1993).
[115] L. S. Cederbaum and F. Tarantelli, *J. Chem. Phys.*, **99**, 5871 (1993).
[116] P. Bolognesi, R. Püttner and L. Avaldi, *Chem. Phys. Lett.*, **464**, 21 (2008).
[117] V. Ulrich *et al.*, *Phys. Rev. Lett.*, **100**, 143003 (2008).
[118] R. Püttner *et al.*, *Chem. Phys. Lett.*, **445**, 6 (2007).
[119] P. Bolognesi, P. O'Keeffe and L. Avaldi, *J. Phys. Chem. A*, **113**, 15136 (2009).
[120] P. Bolognesi *et al.*, *J. Chem. Phys.*, **134**, 094308 (2011).
[121] P. Bolognesi, M. Coreno, L. Avaldi, L. Storchi and F. Tarantelli, *J. Chem. Phys.*, **125**, 54306 (2006).
[122] W. J. Griffiths *et al.*, *J. Phys. B: At. Mol. Opt. Phys.*, **24**, 4187 (1991).
[123] M. Schmidbauer *et al.*, *J. Chem. Phys.*, **94**, 5299 (1991).
[124] J. Schirmer and A. Barth, *Z. Phys. A.*, **317**, 267 (1984).
[125] F. Tarantelli, A. Sgamellotti and L. S. Cederbaum, *J. Chem. Phys.*, **94**, 523 (1991).
[126] P. van der Straten, R. Mongerstern and A. Niehaus, *Z. Phys. D*, **8**, 35 (1988).
[127] M. Ohno and G. Wendin, *J. Phys. B: At. Mol. Opt. Phys.*, **12**, 1305 (1979).
[128] O. Gunnarson and K. Schonhammer, *Phys. Rev. Lett.*, **46**, 859 (1981).
[129] M. Ohno and G. A. van Riessen, *J. Electron. Spectrosc. Relat. Phenom.*, **128**, 1 (2003).
[130] M. Ohno, *J. Electron. Spectrosc. Relat. Phenom.*, **159**, 73 (2007).

[131] C. P. Lund, S. M. Thurgate and A. B. Wedding, *Phys. Rev. B*, **49**, 11352 (1994).
[132] C. P. Lund, S. M. Thurgate and A. B. Wedding, *Phys. Rev. B*, **55**, 5455 (1997).
[133] S. M. Thurgate and C. P. Lund, *J. Electron. Spectrosc. Relat. Phenom.*, **72**, 289 (1995).
[134] S. M. Thurgate, C. P. Lund, C. A. Craegh and R. P. Craig, *J. Electron. Spectrosc. Relat. Phenom.*, **93**, 209 (1998).
[135] S. M. Thurgate, *J. Electron. Spectrosc. Relat. Phenom.*, **81**, 1 (1996).
[136] C. A. Craegh, S. M. Thurgate, R. P. Craig and C. P. Lund, *Surf. Sci.*, **432**, 297 (1999).
[137] S. M. Thurgate and Z. T. Jiang, *Surf. Sci.*, **466**, L807 (2000).
[138] R. A. Bartynski, E. Jensen, K. Garrison, S. L. Hulbert and M. Weinert, *J. Vac. Sci. Tech.*, **9**, 1907 (1991).
[139] A. K. See *et al.*, *Surf. Sci.*, **383**, L735 (1997).
[140] G. A. Van Riessen, S. M. Thurgate and D. E. Ramaker, *J. Electron. Spectrosc. Relat. Phenom.*, **161**, 150 (2007).
[141] C. A. Craegh and S. M. Thurgate, *J. Electron. Spectrosc. Relat. Phenom.*, **114**, 69 (2001).
[142] R. Sundaramoorthy, A. H. Weiss, S. Hulbert and R. Bartynski, *Phys Rev. Lett.*, **101**, 127601 (2008).
[143] E. Jensen *et al.*, *Phys. Rev. B*, **41**, 12468 (1990).
[144] R. A. Bartynski *et al.*, *Phys. Rev. Lett.*, **68**, 2247 (1992).
[145] A. Liscio *et al.*, *J. Electron. Spectrosc. Relat. Phenom.*, **137**, 505 (2004).
[146] W. S. M. Werner, H. Stori and H. Winter, *Surf. Sci.*, **518**, L569 (2002).
[147] W. S. M. Werner, *Appl. Surf. Sci.*, **235**, 2 (2004).
[148] W. S. M. Werner *et al.*, *Phys. Rev. Lett.*, **94**, 184506 (2005).
[149] T. Kakiuchi *et al.*, *Surf. Sci.*, **604**, L27 (2010).
[150] T. Kakiuchi *et al.*, *Phys. Rev. B*, **83**, 035320 (2011).
[151] D. A. Arena, R. A. Bartynski, R. A. Nayak, A. H. Weiss and S. L. Hulbert, *Phys. Rev. B*, **63**, 155102 (2001).
[152] E. Jensen, R. A. Bartynski, S. L. Hulbert, D. E. Johnson and R. Garret, *Phys. Rev. Lett.*, **62**, 71 (1989).
[153] Z. T. Jiang, M. Ohno, S. M. Thurgate and G. van Riessen, *J. Electron. Spectrosc. Relat. Phenom.*, **143**, 33 (2005).
[154] R. Gotter *et al.*, *Phys. Rev. B*, **79**, 075108 (2009).
[155] M. Ohno, *J. Electron. Spectrosc. Relat. Phenom.*, **149**, 1 (2005).
[156] G. A. Sawatzky, in *Auger Electron Spectroscopy*, ed. C. L. Briant and R. P. Messmer, (Boston: Academic Press, 1988), pp. 167–243.
[157] G. Stefani, S. Iacobucci, A. Ruocco and R. Gotter, *J. Electron. Spectrosc. Relat. Phenom.*, **127**, 1 (2002).
[158] R. Gotter *et al.*, *Phys. Rev. B*, **67**, 033303 (2003).
[159] G. Stefani *et al.*, in *Correlation Spectroscopy of Surface, Thin Films and Nanostructures*, ed. J. Berekdar and J. Kirschner (Weinheim: Wiley-VCH, 2004), pp. 189–202.
[160] R. Gotter *et al.*, *EPL*, **94**, 37008 (2011).
[161] F. Da Pieve *et al.*, *Phys. Rev. B*, **78**, 035122 (2008).
[162] E. J. McGuire, in *Atomic Inner-Shell Processes*, ed. B. Crasemann (New York: Academic, 1975), pp. 293–330.
[163] P. J. Feibelman, E. J. McGuire, K. C. Pandey and P. Weightman, *Phys. Rev. B*, **15**, 2202 (1977).
[164] G. Cubiotti, A. Laine and P. Wieghtman, *J. Phys.: Condens. Matter*, **1**, 7723 (1989).
[165] D. E. Ramaker, H. Yang and I. U. Idzerda, *J. Electron Spectrosc. Relat. Phenom.*, **68**, 63 (1994).

[166] Y. U. Idzerda, *Surf. Sci. Rev. Lett.*, **4**, 161 (1997).

[167] F. J. G. de Abajo, C. S. Van Hove and C. S. Fadley, *Phys. Rev. B*, **63**, 075404 (2001).

[168] Y. Chen and M. A. Van Hove, *MSCD Package Overview* (Lawrence Berkeley National Laboratory, 1997).

[169] R. Gotter *et al.*, *Phys. Rev. B*, **72**, 235409 (2005).

[170] R. Gotter *et al.*, *J. Electron. Spectrosc. Relat. Phenom.*, **161**, 128 (2007).

[171] J. L. Robins, *Prog. Surf. Sci.*, **48**, 167 (1995).

[172] S. M. Thurgate, *J. Electron. Spectrosc. Relat. Phenom.*, **100**, 161 (1999).

[173] R. A. Bartynski, E. Jensen and S. L. Hulbert, *Physica Scripta T*, **41**, 168 (1992).

[174] R. A. Bartynski, E. Jensen, S. L. Hulbert and C. C. Kao, *Prog. Surf. Sci.*, **53**, 155 (1996).

[175] M. Ohno, *J. Electron. Spectrosc. Relat. Phenom.*, **104**, 109 (1999).

[176] M. Ohno, *Phys. Rev. B*, **58**, 12795 (1998).

[177] A. Marini and M. Cini, *J. Electron. Spectrosc. Relat. Phenom.*, **127**, 17 (2002).

[178] T. Aberg, *Phys. Scripta*, **21**, 495 (1980).

[179] M. Y. Kuchiev and S. A. Sheinerman, *Sov. Phys.-JEPT*, **63**, 986 (1986).

[180] C. Sinanis, G. Aspromallis and C. A. Nicolaides, *J. Phys. B: At. Mol. Opt. Phys.*, **28**, L423 (1995).

[181] S. A. Sheinerman and V. Schmidt, *J. Phys. B: At. Mol. Opt. Phys.*, **32**, 5205 (1999).

[182] C. H. Greene and A. R. P. Rau, *Phys. Rev. Lett.*, **48**, 533 (1982).

[183] C. H. Greene and A. R. P. Rau, *J. Phys. B: At. Mol. Opt. Phys.*, **16**, 99 (1983).

[184] J. A. de Gouw, J. Van Eck, A. C. Peters, J. Van der Weg and H. G. M. Heideman, *Phys. Rev. Lett.*, **71**, 2875 (1993).

[185] J. A. de Gouw, J. Van Eck, A. Q. Wollrab, J. Van der Weg and H. G. M. Heideman, *Phys. Rev. A*, **50**, 4013 (1994).

[186] S. Rioual, B. Rouvellou, A. Huetz and L. Avaldi, *Phys. Rev. Lett.*, **91**, 173001 (2003).

[187] J. Danger *et al.*, *Phys. Rev. B*, **64**, 045110 (2001).

[188] P. Le Févre *et al.*, *Phys. Rev. B*, **69**, 155421 (2004).

[189] S. Haroche, *Phys. Scripta*, **T76**, 159 (1998).

[190] P. Lablanquie *et al.*, *Phys. Rev. Lett.*, **106**, 063003 (2011).

[191] P. Lablanquie *et al.*, *Phys. Rev. Lett.*, **107**, 193004 (2011).

[192] J. H. D. Eland *et al.*, *Phys. Rev. Lett.*, **105**, 213005 (2010).

[193] J. P. Cryan *et al.*, *Phys. Rev. Lett.*, **105**, 083004 (2010).

[194] L. Fang *et al.*, *Phys. Rev. Lett.*, **105**, 083005 (2010).

[195] A. Rudenko *et al.*, *J. Phys. B: At. Mol. Opt. Phys.*, **41**, 081006 (2008).

2

The application of propagating exterior complex scaling to atomic collisions

PHILIP L. BARTLETT AND ANDRIS T. STELBOVICS

2.1 Introduction

The accurate solution of the Schrödinger equation (SE) for electron-impact collisions leading to discrete elastic and inelastic scattering progressed rapidly with the increase in computing power from the 1970s. A review of the principal methods, including second Born, distorted wave, R-matrix, intermediate-energy R-matrix, pseudo-state close coupling and optical model is given in [1]. However, electron impact collisions leading to ionization on even the simplest atom, hydrogen, were by comparison poorly described; significant progress dates only from the early 1990s when Bray and Stelbovics [2] developed a technique called convergent close coupling (CCC). In this approach they used an in-principle complete set of functions to approximate the hydrogenic target states, both bound and continuous, and used the coupled channels formalism to expand the scattering wave function in these discretized states, reducing the solution of the SE to a set of coupled linear equations in a single co-ordinate. The method was tested in a non-trivial model [3] and shown to provide convergent cross sections not only for discrete elastic and inelastic processes but also for the total ionization cross section. Shortly thereafter the method was applied to the full collision problem from atomic hydrogen and one of the major achievements of the method was that it yielded essentially complete agreement with the (then) recent experiment for total ionization cross section [4]. In the following years, the method was applied to other atoms with considerable success; the range of applications of CCC are covered in the review of Bray *et al.* [5]. After the success of the CCC model in calculating the total ionization cross section it was extended to differential ionization cross sections [6]; there, however, it emerged that the energy differential cross sections were not manifestly symmetric about the

Fragmentation Processes: Topics in Atomic and Molecular Physics, ed. Colm T. Whelan. Published by Cambridge University Press. © Cambridge University Press 2013.

equal-energy-sharing configuration of the two ionized electrons as is required by the Pauli principle. This behaviour of the formalism was puzzling in the extreme since the converged total ionization cross section agreed perfectly with experiment. Bray [7] found a way forward by noting that this could be resolved only by proposing a 'step-function' hypothesis, a theoretical basis for which was provided later by Stelbovics [8]. There remained a practical difficulty with implementation of the hypothesis in that it applied only in the limit of a complete set of basis functions. In all calculations a finite set of functions had to be employed and, moreover, the approach to convergence was very slow and the differential cross sections exhibit oscillations near the step. Bray adopted a smoothing ansatz to overcome this characteristic feature of the CCC approach and it is still integral to extracting the differential ionization cross sections.

About the same time, a quite different approach was being proposed to attack the calculation of ionization cross section for atomic hydrogen, namely the exterior complex scaling (ECS) method. This method was developed and adapted to atomic collisions in a series of papers by Rescigno, Baertschy McCurdy and co-workers, e.g., see [9–11]. They sought to compute the ionization amplitude for electron scattering from atomic hydrogen by solving the SE directly, using a rotation of the co-ordinates into the complex plane at sufficiently large distances beyond where the asymptotic form of the scattered wave function was valid. The effect of the rotation was to exponentially dampen the scattered wave to zero along the rotated contour, enabling one to bypass the difficulty of matching to Coulomb-wave boundary conditions for ionization. Their work demonstrated the power of the method, although they continued to refine how to extract the ionization amplitude to obtain higher accuracy. Their topical review [12] discusses their methods.

Bartlett and Stelbovics adopted their basic ECS approach [13] but sought to improve the computational methods so that the method could be extended to the whole range of scattering energies. Of particular interest was the near-threshold ionization region where Wannier [14] had predicted a threshold ionization energy power law different from that with a short-range potentials. However, confirmation of this law by a direct numerical solution of the SE was still lacking at the beginning of the twenty-first century. This article will describe their implementation of ECS, called propagating exterior complex scaling (PECS) and some of the successes that have been achieved with it.

2.2 Introduction to exterior complex scaling

The notion of exterior complex scaling is easily illustrated by means of a simple example.

2.2.1 A one-dimensional example

Consider the S-wave scattering problem for a short-range potential $U(r)$. We may write the radial Schrödinger equation as

$$\left(\frac{d^2}{dr^2} + k^2 \right) \psi^{(+)}(r) = U(r)\psi^{(+)}(r). \tag{2.1}$$

The solution of this equation requires two boundary conditions; the first is $\psi(0) = 0$ and, as usual for collision theory, a second is obtained by prescribing the asymptotic form valid for large r, namely

$$\psi^{(+)}(k, r) \underset{r\to\infty}{\sim} \sin kr + e^{ikr} e^{i\delta} \sin \delta. \tag{2.2}$$

The scattering information is in the phase-shift $\delta = \delta(k)$ that is wholly determined through the coefficient of the outgoing wave. Let us now form a decomposition of $\psi^{(+)}(k, r)$ as the sum of the initial state $\phi = \sin(kr)$ and a part ψ_{sc} that asymptotically contains the outgoing scattered wave and has the form

$$\psi_{\text{sc}} \underset{r\to\infty}{\sim} e^{ikr} e^{i\delta} \sin \delta. \tag{2.3}$$

Now it is possible to express Eq. (2.1) using this decomposition as

$$\left(\frac{d^2}{dr^2} + k^2 - U(r) \right) \psi_{\text{sc}}(r) = U(r)\phi(r). \tag{2.4}$$

In practice, this equation for unknown ψ_{sc} is easily solved numerically for short-range potentials using difference equation techniques and propagating out from $r = 0$ to large enough $r = R$, where the asymptotic form (2.3) is valid given the known right-hand side (RHS).

The ECS method of numerical solution relies on a rotation of the radial coordinate into the complex plane by a fixed angle $0 < \theta < \pi/2$ at a finite distance from the origin (R_0) that is in the asymptotic region, as follows:

$$z(r) \mapsto \begin{cases} r, & r < R_0 \\ R_0 + (r - R_0)e^{i\theta}, & r \geq R_0. \end{cases} \tag{2.5}$$

If we apply this transformation to the outgoing wave e^{ikr}, for $r > R_0$ we obtain

$$e^{ikr} \mapsto e^{ikz(r)} = e^{-k(r-R_0)\sin\theta} e^{ik(R_0+(r-R_0)\cos\theta)} \underset{r\to\infty}{\longrightarrow} 0, \tag{2.6}$$

which demonstrates that the outgoing wave diminishes exponentially beyond R_0 under this transformation. However, there is an important point to note in applying this transformation to Eq. (2.4); on the RHS, $\phi = \sin kr$ contains both incoming and outgoing waves. In contrast to the outgoing component, the incoming wave

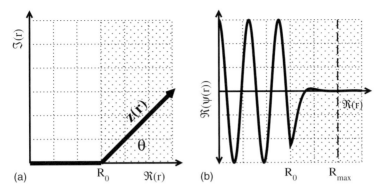

Figure 2.1　(a) Rotation of $z(r)$ by θ radians into the complex plane at the beginning of the complex scaling region (R_0). (b) Example of exterior complex scaling: $\psi(r) = \sin(z(r))$.

diverges exponentially under the transformation. In actual numerical calculations, since this diverging behaviour occurs only for $r > R_0$, by which time the scattering information contained in the phase shift δ is already known, we simply ignore the incoming component of the initial state by applying a smooth, large exponential cut-off to it beyond R_0 that overwhelms the divergence of e^{-ikr} on the rotated contour. An example of the ECS rotation applied to the incoming wave is shown in Fig. 2.1.

In this example, the scattering information, as we have seen, is contained in the coefficient $e^{i\delta} \sin \delta$ of the scattered wave. While this information can be extracted directly using $e^{i\delta} \sin \delta = \lim_{r\to\infty} \psi_{sc} e^{-ikr}$, it is useful to give a more general approach that will form the basis for extracting amplitudes for the full electron–atom collision problems. Consider the integral:

$$I_R(k) = \frac{1}{k} \int_0^R \psi^{(+)}(k, r) \left(\frac{d^2}{dr^2} + k^2 - U(r) \right) \phi(r). \tag{2.7}$$

Making use of Eq. (2.1) it is straightforward to derive

$$I_R(k) = \frac{1}{k} \left(\psi^{(+)}(k, R)\phi'(R) - \phi(R)\psi^{(+)\prime}(k, R) \right) \tag{2.8}$$

and further that

$$I(k) = \lim_{R\to\infty} I_R(k) = e^{i\delta} \sin \delta. \tag{2.9}$$

In practice, the numerical solution would be extracted for sufficiently large R on the grid and the convergence established to a required accuracy, say one per cent.

2.2.2 Numerical method: propagating exterior complex scaling

The numerical approach to solving the scattering equation (2.4) that is favoured by us is to use the Numerov finite difference method that is highly efficient for second-order differential equations that have no first-order derivatives. Denoting $\psi_{sc}(r_i)$ by ψ^i, where $r_i = hi$, $i = 0 \ldots i_{max}$ and where i_{max} is chosen so that $r_{i_{max}} > R_0$ and is sufficiently large so that $\psi^{i_{max}} \approx 0$, on account of the rotation into the complex plane we have

$$\frac{d^2}{dr^2}\psi(r) + \Omega(r) = 0, \tag{2.10}$$

and the Numerov formula relates three successive points of ψ by

$$\psi^{i+1} = 2\psi^i - \psi^{i-1} - \frac{h^2}{12}\left(\Omega^{i+1} + 10\Omega^i + \Omega^{i-1}\right), \tag{2.11}$$

with error term $O(h^6)$. On comparison with (2.4), it follows that we can write the difference equation for ψ^i in the general form:

$$A^i\psi^{i-1} + B^i\psi^i + C^i\psi^{i+1} = F^i \quad i = 1, 2 \ldots i_{max} - 1. \tag{2.12}$$

On the face of it this appears to be a three-term difference equation for the ψ^i, but Poet made the clever observation that it could be reduced to a two-term relation. To see this we make the ansatz

$$\psi^i = D^i\psi^{i+1} + E^i. \tag{2.13}$$

It is then straightforward to apply (2.13) to (2.12) and find that

$$D^i = -\tilde{B}^i C^i \tag{2.14}$$

and

$$E^i = B^i(F^i - A^i E^{i-1}), \tag{2.15}$$

where

$$\tilde{B}^i = (B^i + A^i D^{i-1})^{-1}. \tag{2.16}$$

To initiate the propagation we observe that since $\psi(0) = 0$ we have $\psi^0 = 0$ and so without loss of generality we can set $A^1 = 0$. Then it follows that we have

$$D^1 = -(B^1)^{-1}C^1 \tag{2.17}$$

and

$$E^1 = (B^1)^{-1}F^1, \tag{2.18}$$

which then permits us to calculate all the higher coefficients through the recurrence relations (2.14) through to (2.16). When the exterior complex scaling is applied at $r = R_0$, say beyond $i = i_R$, all the coefficients for $i_R < i < i_{max}$ will become complex. Thus the process of solution is to use the exterior complex scaling boundary condition $\psi^{i_{max}} = 0$ and to propagate backwards to compute the scattering wave function to the asymptotic region where $i < i_R$.

2.3 Application of ECS to electron–hydrogen scattering

In this section we present an outline of the method for electron scattering from atomic hydrogen. We begin by deriving the scattering equations and generalize the discussion of the previous example.

For electron–hydrogen scattering, the total electron spin S is a good quantum and the scattering wave function evolves from the initial state of the system. The label i contains all the quantum numbers necessary to fully specify the initial state of the target atom and incoming electron. As before, we separate the scattering wave function into a sum that isolates the initial state from the outgoing wave function:

$$\Psi_i^{S(+)} = \Phi_i^S + \Psi_{i,sc}^S. \tag{2.19}$$

The Schrödinger equation for the collision becomes

$$(E - H)\,\Psi_{i,sc}^S = (H - E)\,\Phi_i^S, \tag{2.20}$$

where

$$H = -\frac{1}{2}\nabla_0^2 - \frac{1}{2}\nabla_1^2 - \frac{Z}{r_0} - \frac{Z}{r_1} + \frac{1}{|\mathbf{r}_0 - \mathbf{r}_1|}, \tag{2.21}$$

and $Z \geq 1$. Including Z generalizes the discussion of hydrogenic-ion targets. We adopt the static nucleus approximation and use atomic units in the sequel. The incident wave function on the RHS is the appropriately antisymmetrized co-ordinate space function, comprising the product of an incoming Coulomb wave of charge[1] $-Z + 1$ and the initial-state wave function of the ionic target with charge $Z - 1$:

$$\Phi_i^{S(-)}(\mathbf{r}_0, \mathbf{r}_1) = \frac{1}{\sqrt{2}}\big(\phi_c^{(-)}(Z-1; \mathbf{k}_i, \mathbf{r}_0)\phi_i(\mathbf{r}_1) + (-1)^S \phi_c^{(-)}(Z-1; \mathbf{k}_i, \mathbf{r}_1)\phi_i(\mathbf{r}_0)\big). \tag{2.22}$$

The normalization we adopt for the Coulomb wave is that for charge 0: $\phi_c^{(-)}(0; \mathbf{k}, \mathbf{r}) = \exp i\mathbf{k} \cdot \mathbf{r}$.

---

[1] The convention we adopt for describing an attractive potential Coulomb wave corresponding to an attractive charge $-Z$ is to label it as $\phi_c^{(-)}(Z)$.

2.3.1 Extracting scattering amplitudes from surface integrals

Since the ECS method relies on computing the scattering wave function out to sufficiently large distances where its asymptotic form is realized, we can use an elegant method that makes use of surface integrals evaluated in the asymptotic regions where at least one of the electron co-ordinates is very large. We begin by defining and examining the integral

$$I_{\text{fi}}^{S} = \int d\mathbf{r}_0 \int d\mathbf{r}_1 \ \Psi_{\text{i}}^{S(+)}(\mathbf{r}_0, \mathbf{r}_1)(H - E)\Phi_{\text{f}}^{S(-)*}(\mathbf{r}_0, \mathbf{r}_1). \qquad (2.23)$$

Note that this form is essentially a generalization of the one-dimensional example given in Section 1.2. This type of integral has been used extensively in formal development of atomic collision theory, particularly for ionization. The labels i, f used to label the state are used in the sense that they include all momenta $\mathbf{k}_{\text{i,f}}$ needed to describe the state. Using the fact that the scattering function is a solution of the Schrödinger equation, I_{fi}^{S} may be expressed in the following form:

$$I_{\text{fi}}^{S} = \frac{1}{2} \int d\mathbf{r}_0 \int d\mathbf{r}_1 \ \left\{ \Phi_{\text{f}}^{S(-)*}(\nabla_0^2 + \nabla_1^2)\Psi_{\text{i}}^{S(+)} - \Psi_{\text{i}}^{S(+)}(\nabla_0^2 + \nabla_1^2)\Phi_{\text{f}}^{S(-)*} \right\}.$$

$$(2.24)$$

We will now demonstrate that this fundamental form can be used to extract scattering amplitudes for discrete scattering and ionization by applying Green's second theorem,

$$\int_V dV \ \left\{ \Phi\nabla^2\Psi - \Psi\nabla^2\Phi \right\} = \int_S dS \ \left\{ \Phi\frac{\partial}{\partial r}\Psi - \Psi\frac{\partial}{\partial r}\Phi \right\}, \qquad (2.25)$$

where the derivative is along the outwardly drawn normal to the surface S. The form of the final-state wave function with incoming scattered wave depends on the process that one is interested in. For discrete inelastic scattering, the final state is given by Eq. (2.22) and we turn to this case now.

Discrete inelastic scattering

In this situation, both before and after the scattering the target ion is in a bound state and therefore one electron is localized while the other electron recedes to an infinite distance from the scattering centre. So, in deriving the amplitude for this process we choose to apply Green's theorem only to that electron that moves to a large distance, where we can use its asymptotic wave function form. Since we cannot distinguish the electrons, we must separately sum the equal contributions to the amplitude for both cases, where either electron 0 or 1 is the scattered electron. So we need consider only the single case of electron 0 receding to infinite distance and double the result. Thus

$$
I_{\text{fi}}^S \underset{r_0 \to \infty}{\sim} r_0^2 \int d\hat{\boldsymbol{r}}_0 \int d\boldsymbol{r}_1 \left\{ \Phi_{\text{f}}^{S(-)*} \frac{\partial}{\partial r_0} \Psi_{\text{i}}^{S(+)} - \Psi_{\text{i}}^{S(+)} \frac{\partial}{\partial r_0} \Phi_{\text{f}}^{S(-)*} \right\}. \qquad (2.26)
$$

To evaluate the surface integral, we need the asymptotic forms for both the scattering wave function developing from the initial state i and the incoming wave function corresponding to the final state f. We have for the scattered portion of the scattering wave function:

$$
\Psi_{\text{i,sc}}^S(\boldsymbol{r}_0, \boldsymbol{r}_1) \underset{r_0 \to \infty}{\sim} \frac{1}{\sqrt{2}} \sum_{\text{j}} \phi_{\text{j}}(\boldsymbol{r}_1) \frac{\exp\left(\text{i}\left(k_{\text{j}} r_0 + \frac{Z-1}{k_{\text{j}}} \ln(2 k_{\text{j}} r_0)\right)\right)}{r_0} F_{\text{ji}}^S(\hat{\boldsymbol{r}}_0), \quad (2.27)
$$

where $F_{\text{ji}}^S(\hat{\boldsymbol{r}})$ is the scattering amplitude from an initial state i, corresponding to a target in discrete bound state i impacted on by an electron with momentum $\boldsymbol{k}_{\text{i}}$ and leaving the target in a bound state[2] j and scattered electron with momentum $\boldsymbol{k}_{\text{f}} = k_{\text{f}} \hat{\boldsymbol{r}}_0$. Similarly, the initial- and final-state wave functions with incoming Coulomb wave boundary condition $\Phi_{\text{i,f}}^{S(-)}$ in the limit of large r_0 reduce to the form:

$$
\Phi_{\text{i,f}}^{S(-)}(\boldsymbol{r}_0, \boldsymbol{r}_1) \underset{r_0 \to \infty}{\sim} \frac{1}{\sqrt{2}} \phi_c^{(-)}(Z-1; \boldsymbol{k}_{\text{f}}, \boldsymbol{r}_0) \phi_{\text{i,f}}(\boldsymbol{r}_1) \qquad (2.28)
$$

and, further, the incoming Coulomb wave becomes

$$
\phi_c^{(-)}(Z; \boldsymbol{k}, \boldsymbol{r}) \underset{r \to \infty}{\sim} \delta(\hat{\boldsymbol{k}} - \hat{\boldsymbol{r}}) \frac{2\pi}{\text{i}kr} e^{\text{i}kr + \text{i}\frac{Z}{k} \ln(2kr)} + \delta(\hat{\boldsymbol{k}} + \hat{\boldsymbol{r}}) \frac{2\pi}{\text{i}kr} e^{-\text{i}kr - \text{i}\frac{Z}{k} \ln(2kr)}
$$

$$
+ \frac{f_c^*(-\hat{\boldsymbol{k}} \cdot \hat{\boldsymbol{r}})}{r} e^{-\text{i}kr - \text{i}\frac{Z}{k} \ln(2kr)}. \qquad (2.29)
$$

Now inserting the asymptotic forms into the integral (2.26) one finds

$$
I_{\text{fi}}^S = -2\pi F_{\text{fi}}^S. \qquad (2.30)
$$

Ionization

The treatment of ionization scattering again uses the fundamental integral (2.23), except that now the scattered wave function will have an open ionization channel. In this case, both electrons will recede to infinity after the collision. One can still use the asymptotic form of the scattering function (2.27) by replacing the ϕ_{j} discrete hydrogenic states by incoming Coulomb waves $\phi_c^{(-)}(Z-1; \boldsymbol{k}_1, \boldsymbol{r}_1)$ and generalizing the summation over j to an integral over momentum $\boldsymbol{k}_1$. If the total

---

[2] For simplicity we assume the collision takes place at an energy below the ionization threshold. The derived result will still hold but we avoid the necessity to include the ionization channels in the scattered wave.

energy of the collision is E, then we have $E = \frac{1}{2}k_0^2 + \frac{1}{2}k_1^2 = \epsilon_i + \frac{1}{2}k_i^2$. The integral (2.24) now requires a different surface for evaluation due to the fact that both electrons escape to infinite distance. To define the appropriate surface over which to apply Green's second theorem we employ the co-ordinates

$$\rho = \sqrt{r_0^2 + r_1^2}, \quad \alpha = \arctan \frac{r_0}{r_1}, \quad 0 \le \alpha \le \frac{\pi}{2}. \tag{2.31}$$

Then the surface is a hypersphere with surface element

$$dS = \frac{\rho^5}{4} \sin^2 2\alpha \; d\alpha d\hat{r}_0 d\hat{r}_1. \tag{2.32}$$

Applying these co-ordinates to (2.23) and using (2.25) we have

$$I_{fi}^S \underset{\rho \to \infty}{\sim} \frac{\rho^5}{4} \int_0^{\pi/2} d\alpha \sin^2 2\alpha \int d\hat{r}_0 \int d\hat{r}_1 \left(\Phi_f^{S(-)*} \frac{\partial}{\partial \rho} \Psi_i^{S(+)} - \Psi_i^{S(+)} \frac{\partial}{\partial \rho} \Phi_f^{S(-)*} \right). \tag{2.33}$$

This integral may be evaluated using the asymptotic forms for scattering wave function and the final-state wave function that now is an antisymmetrized product of two Coulomb waves with incoming boundary conditions. The evaluation of this integral form is non-trivial. Good discussions of the methods are given both by Peterkop [15] and Rudge [16]. The end result is that in a formal sense the limit does not exist, because there is an oscillatory phase factor $\Delta_{fi}(\rho)$ that has no limiting value unless the two continuum waves comprising the final state satisfy a special screening condition whose details we will not pursue here.[3] Thus for a large but finite ρ we have

$$I_{fi}^S(\rho) \approx -(2\pi)^{\frac{5}{2}} F_{fi}^S \Delta_{fi}(\rho), \tag{2.34}$$

from which it can be seen that the *magnitude* of the amplitude and hence cross section is convergent. In actual calculations we proceed by expanding all the equations in the partial-wave form prior to evaluating this equation for large ρ and forming the partial-wave sum. The calculations indeed confirm that the cross section is convergent. This forms the basis for computed ionization cross sections presented in this chapter.

It is also interesting to note that the choice of the final-state incoming waves is flexible and need not be the one adopted here. However, for numerical work this choice is optimal in the sense that the orthogonality of the continuum Coulomb wave with the discrete excited states manifests itself in improved convergence of the integral form for the amplitude. This has been noted by Peterkop [15] and its usefulness in numerical calculation tested by McCurdy *et al.* [11] and

---

[3] The interested reader is referred to [16] for a full discussion.

Baertschy *et al.* [17]. Before large-scale numerical work was carried out in the past decade using the surface integral methods there had been considerable discussion concerning the fact that unless the final-state wave function formed as a product of two continuum waves obeyed a screening condition (see [16]) the ionization amplitude had a divergent phase. Since no experiments carried out to date have been devised to access phase information about the ionization amplitude it is a moot point. However, it is possible in practical calculations to compute an ionization amplitude with a converged phase. This has been shown in Temkin–Poet model calculations of S-wave scattering by Rescigno *et al.* [18] and Bartlett *et al.* [19].

2.3.2 *Propagating exterior complex scaling considerations*

The Schrödinger equation is reduced by means of partial-wave analysis to a set of coupled second-order differential equations in radial variables r_0 and r_1 for the partial-wave wave function. The equations must be solved for each partial wave. We denote the wave function by $\psi_{il_0l_1}^{LMS\Pi}(r_0, r_1)$ and observe it has the symmetry property

$$\psi_{il_0l_1}^{LMS\Pi}(r_0, r_1) = (-1)^{S+L+l_0+l_1} \psi_{il_1l_0}^{LMS\Pi}(r_1, r_0), \qquad (2.35)$$

where l_0, l_1 are the orbital angular momentum quantum numbers of the electrons, L is the total angular momentum and Π is the parity of the system. For each L, the l_0, l_1 satisfy the usual triangle inequality. The calculations are carried out increasing the maximum value allowed for the l's until convergence in angular momentum is achieved. The principle for solving these equations numerically is similar to the one-dimensional example, except that there are two radial co-ordinates and the Numerov method is generalized to two dimensions. One achieves further computational efficiency by using the symmetry relation (2.35) to store only those grid elements on a triangular grid where $r_0 \leq r_1$. [13] has shown that the propagation method provided in our example is directly applicable to two radial co-ordinates; for the wave function one simply replaces the scalar quantities ψ^i by column vectors $\boldsymbol{\psi}^i$ of length $n_s i$ that store the grid elements for the second electron, where n_s is the number of angular momentum coupled channels included. The length of the column vector is thus proportional to the propagation index. The propagation algorithm of Eqs. (2.12–2.16) is generalized in the sense that the quantities are now matrices. Critically, only the $\tilde{B}^i$ and D^i are dense matrices and the significant time component in computation is in the calculation of the inverse to compute the $\tilde{B}^i$. The algorithm is efficient in that the dimension of the inverse for iteration i is proportional to i; the remaining matrices are banded.[4]

---

[4] For details of the storage requirements, see [13], especially Table 1.

Other refinements can be made to improve the efficiency of the propagations algorithm. For example, it was noted by [20] that the step size could be doubled for larger r without reducing the accuracy, thus reducing computing time. [13] has generalized the algorithm to variable step size; this proves useful in the complex rotation method, where it is necessary to have a finer mesh in the vicinity of the region in co-ordinate space where the co-ordinates are rotated into the complex plane. The time required to compute the wave scattering wave function is of the order $n_s i_{max}^4$; the determining factor for the scaling with time is that required to compute the matrix inversions for the $\tilde{B}^i$. Further enhancements in computational efficiency are possible by use of the technique of 'iterative refinement' as discussed in [21] and an 'energy perturbation method' that is particularly useful for energies near the ionization threshold as mentioned in [13].

2.4 Scattering in electron–hydrogen system

Scattering of electrons from hydrogen atoms has been the prototype atomic collision problem for decades. However, most of the major progress in reconciling theory models based on the non-relativistic Schrödinger equation with experimental data has taken place in the past 20 years, when the computing power became available to enable direct methods based on accurate solution of the SE to be carried out. Here we will be able to mention only a small part of the extensive computational work. For comparison purposes we will principally refer to calculations using the CCC method of Bray and Stelbovics [2, 4] and those of Baertschy, Rescigno, McCurdy and co-workers, who championed in a series of papers[5] the power of the ECS method, especially as applied to ionization of hydrogen.

In order to appreciate how accurately the scattering amplitudes may be computed we show in Fig. 2.2 the convergence rate with increasing hyperradius. It can be seen that for elastic scattering from the ground state, convergence has been already obtained by 20 a.u. For inelastic scattering to the excited final states of hydrogen, the distance required to achieve convergence progressively increases with increasing quantum number. This is typical of these calculations in that convergence is correlated with the mean expectation value of the radius of the state. It is also worth noting that although the cross sections for highly excited states are orders of magnitude smaller than the ground state they can be computed with good accuracy. In the case of ionization there is also excellent convergence for ionization from the ground state by 50 a.u. Nevertheless it is also important to highlight that calculations out to 100 a.u. do require access to powerful computing resources with parallel processing and are time consuming. With the time scaling as mentioned

---

[5] See for example [9], [10] and the topical review [12].

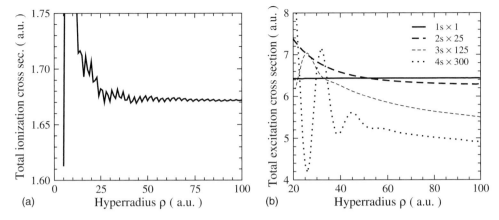

Figure 2.2 Convergence of e–H total cross sections w.r.t. hyperradius ρ for (a) ionization at $E_0 = 27.2$ eV, and (b) discrete final-state scattering at $E_0 = 30.0$ eV.

with the fourth power in number of grid points, extending the calculation to 200 a.u. requires a factor of sixteen increase in computation time.

By and large, discrete scattering processes were already described to very high accuracy by the end of the twentieth century; similarly, total ionization cross sections were described very well for one-electron valence shell atoms (see [1]) over the whole energy range. The PECS method has confirmed these results to very high accuracy. Where the PECS method has made a genuine advance is in near-threshold ionization. There, due to the small cross section, the high precision of the method has permitted us to access what is commonly referred to as the Wannier region. This region, presumed to be less than 1 eV above threshold, was studied by Wannier [14] who predicted a non-linear scaling law for the total ionization cross section. Some results of applying the PECS method to this region are illustrated in Fig. 2.3. They confirm Wannier's prediction for the total ionization cross section.[6] The other physics of this region assumed by Wannier in his reasoning had been that at very low ionization energy the dominant configuration of the ionized electrons was that they would emerge 'back to back', independent of their ratio of energies. This assumption had been questioned by Temkin [22], who proposed that asymmetric energy sharing should dominate. Panel (b) of Fig. 2.3 clearly demonstrates that as threshold is approached the 'back to back' configuration is independent of energy sharing, confirming Wannier's reasoning. The calculations of Bartlett *et al.* [23] also confirm more detailed aspects of the Wannier model applied to partial waves.

---

[6] The oscillations in the PECS curve result from the limitation of a finite hyperradius. They reduce with increasing hyperradius. Nevertheless, with the calculated curve and good statistical methods the Wannier prediction is confirmed to good statistical accuracy.

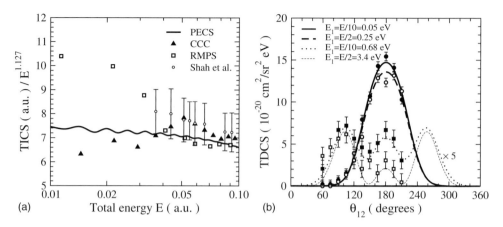

Figure 2.3 (a) Total ionization cross section (TICS) for e–H collisions near threshold ($E = 0.01$–0.10 a.u.) scaled by $E^{1.127}$. CCC and R-matrix with pseudostates (RMPS) calculations, [24], are shown along with the measurements of Shah *et al.* [25]. (b) Fully differential (TDCS) calculations and measurements in the perpendicular plane at various energies and energy-sharing arrangements [26].

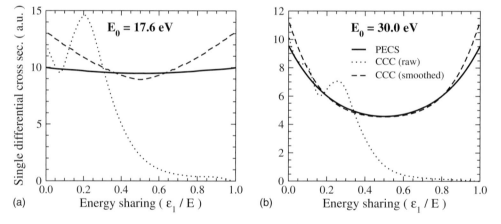

Figure 2.4 Single differential cross section (SDCS) calculations for e–H at $E_0 =$ 17.6 eV and 30.0 eV compared with CCC results [27, 28].

Finally, we conclude our discussion of ECS methods applied to hydrogen targets by illustrating the aspect of the CCC method that incorporates the 'step-function hypothesis' for differential ionization cross sections. In Fig. 2.4 we compare energy differential cross sections between the methods. The ECS and PECS methods produce cross sections symmetric about the equal energy-sharing point; the CCC 'raw' cross sections are not symmetric but do suggest a step-function behaviour as discussed in the introduction. The fitting approach used is reasonably accurate away from threshold but as indicated, near threshold (where the cross sections are small) it becomes less reliable.

2.5 Exterior complex scaling for electron–helium scattering

The electron scattering from helium is a richer problem because the helium spectrum admits metastable doubly excited states and in addition to single ionization, double ionization is also possible. The PECS theory for this system has been described by [29, 30] and we will summarize the essential elements.[7]

The procedure for treating electron–helium scattering closely parallels that of electron–hydrogen scattering in that the SE for the scattering wave function is formally the same as (2.20). The Hamiltonian is given by

$$H = \sum_{i=0}^{2} \left\{ -\frac{1}{2}\nabla_i^2 - \frac{Z}{r_i} + \sum_{j>i}^{2} \frac{1}{|\boldsymbol{r}_i - \boldsymbol{r}_j|} \right\}. \tag{2.36}$$

For ease of presentation we shall introduce some more compact notation; we use $\Psi_{i,s_{12}}^{S(+)}(012) \equiv \Psi_{i,s_{12}}^{S(+)}(\boldsymbol{r}_0, \boldsymbol{r}_1, \boldsymbol{r}_2)$, $\phi_{\boldsymbol{k}}^{(-)}(i) \equiv \phi^{(-)}(\boldsymbol{k}, \boldsymbol{r}_i)$ for one-electron continuum functions and $\phi_i(12) \equiv \phi_{i,s_i}(\boldsymbol{r}_1, \boldsymbol{r}_2)$ for the helium target state with spin s_i.

Then we separate the outgoing scattering wave function into a sum of the initial state and scattered part as in Eq. (2.19). Provided the initial state $\Phi_i^{S(-)}$ is obtained by the formal application of the Pauli principle, then the scattered part of the wave function also has the required symmetry upon solving the Schrödinger equation (2.20). Additionally, the co-ordinate space wave functions require a further spin label s_{12} for the two-body subsystem from which the three electron spin states are formed and must be included in Eq. (2.20). For the three-electron problem the application of the Pauli principle to the co-ordinate-space wave function yields the following more complicated form:

$$\Phi_{i,s_{12}}^{S}(012) = \frac{1}{\sqrt{3}}[\delta_{s_i s_{12}} - \alpha_{01}^{s_i s_{12} S} p_{01} - \alpha_{02}^{s_i s_{12} S} p_{02}]\phi_{\boldsymbol{k}}^{(-)}(0)\phi_i(12), \tag{2.37}$$

where p_{0k} is the operator that interchanges $\boldsymbol{r}_0$ and $\boldsymbol{r}_k$, and

$$\alpha_{01}^{s_i s_{12} S} = -(-1)^{s_i - s_{12}} \hat{s}_i \hat{s}_{12} \begin{Bmatrix} \frac{1}{2} & \frac{1}{2} & s_i \\ S & \frac{1}{2} & s_{12} \end{Bmatrix},$$

$$\alpha_{02}^{s_i s_{12} S} = (-1)^{s_i - s_{12}} \alpha_{01}^{s_i s_{12} S} \tag{2.38}$$

with $\hat{s}_i = \sqrt{2s_i + 1}$.

---

[7] In the sequel for consistency with the electron–hydrogen description we will formulate the theory in terms of space co-ordinates, rather than including spin wave functions as was used in [29, 30].

2.5.1 Extracting scattering amplitudes

The fundamental integral in analogy with Eq. (2.23) is

$$I_{fi}^S = \int dV \sum_{s_{12}} \Psi_{i,s_{12}}^{S(+)}(012)\,(H-E)\,\Phi_{f,s_{12}}^{S(-)*}(012). \qquad (2.39)$$

As before, we convert the integral over the co-ordinates of the three electrons to a surface integral, depending on the type of collision process we are interested in. We consider the three possibilities in turn.

Discrete inelastic scattering

Since the formulation contains explicit anti-symmetrization there are three indistinguishable processes leading to discrete inelastic scattering, namely when either electrons labeled 0,1,2 escape to infinity, leaving the helium target in an excited state. Therefore, it is only necessary to consider the surface integral that leads to the simplest evaluation and multiply the final result by three; from Eq. (2.37) this is clearly when we consider electron 0 escaping to infinity.

We also require the scattering boundary condition for $\Psi_{i,s_{12}}^{S(+)}$ (again for simplicity we assume we are below the ionization thresholds) which is

$$\Psi_{i,s_{12}}^{S(+)}(012) = \frac{1}{\sqrt{3}} \sum_j [\delta_{s_j s_{12}} - \alpha_{01}^{s_j s_{12} S} p_{01} - \alpha_{02}^{s_j s_{12} S} p_{02}] G_{ji}^{S(+)}(0)\phi_j(12), \qquad (2.40)$$

where

$$G_{fi}^{S(+)}(0) \underset{r_0 \to \infty}{\sim} \delta_{fi} \exp i\boldsymbol{k}_i \cdot \boldsymbol{r}_0 + F_{fi}^S(\boldsymbol{k}_f)\frac{\exp^{ik_f r_0}}{r_0}. \qquad (2.41)$$

Applying Green's second theorem and using the asymptotic form (2.40), it is established after some algebra that the relationship (2.30) holds. Thus

$$F_{fi}^S(\hat{\boldsymbol{k}}_f) \underset{r_0 \to \infty}{\sim} -\frac{3}{4\pi}r_0^2 \int d\hat{\boldsymbol{r}}_0 \int d\boldsymbol{r}_1 \int d\boldsymbol{r}_2 \sum_{s_{12}} \left\{ \Phi_{f,s_{12}\boldsymbol{k}_f}^{S(-)*}(012)\frac{\partial}{\partial r_0} \Psi_{i,s_{12}}^{S(+)}(012) \right.$$

$$\left. - \Psi_{i,s_{12}}^{S(+)}(012)\frac{\partial}{\partial r_0} \Phi_{f,s_{12}\boldsymbol{k}_f}^{S(-)*}(012) \right\}. \qquad (2.42)$$

For discrete scattering the general expression can be simplified further by noting that the only contribution to $\Phi_{f,s_{12}}^{S(-)*}$ comes from the Kronecker delta term in (2.40).

Single ionization

Single ionization processes leave one electron in an excited state with the helium ion. Since our surface integral is manifestly symmetrized we can choose any one electron to remain with the helium ion and multiply the result by a factor of three. The convenient choice for evaluation is to choose electrons 0 and 1 as the escaping electrons in the single ionization. Then[8]

$$
F_{\mathrm{fi}}^{S}(\boldsymbol{k}_0, \boldsymbol{k}_1) \underset{\rho \to \infty}{\sim} -\frac{3\,\rho^5}{8(2\pi)^{5/2}} \int_0^{\pi/2} d\alpha \, \sin^2 2\alpha \int d\hat{\boldsymbol{r}}_0 \int d\hat{\boldsymbol{r}}_1 \int dr_2
$$

$$
\times \sum_{s_{12}} \left\{ \Phi_{\mathrm{f},s_{12},\boldsymbol{k}_0,\boldsymbol{k}_1}^{S(-)*}(012) \frac{\partial}{\partial\rho} \Psi_{\mathrm{i},s_{12}}^{S(+)}(012) \right.
$$

$$
\left. - \Psi_{\mathrm{i},s_{12}}^{S(+)}(012) \frac{\partial}{\partial\rho} \Phi_{\mathrm{f},s_{12}\boldsymbol{k}_0,\boldsymbol{k}_1}^{S(-)*}(012) \right\}, \tag{2.43}
$$

where the ρ and α are as defined in Eq. (2.31) and

$$
\Phi_{\mathrm{f},s_{12},\boldsymbol{k}_0,\boldsymbol{k}_1}^{S(-)}(012) \sim \frac{1}{\sqrt{6}} [\delta_{s_{\mathrm{f}}s_{12}} - \alpha_{01}^{s_{\mathrm{f}}s_{12}S} p_{01} - \alpha_{02}^{s_{\mathrm{f}}s_{12}S} p_{02}]
$$

$$
\times [1 + (-1)^{s_{\mathrm{f}}} p_{12}] \phi_{\boldsymbol{k}_0}^{(-)}(0) \phi_{\boldsymbol{k}_1}^{(-)}(1) \phi_\gamma(2). \tag{2.44}
$$

Here, ϕ_γ is a bound state of the helium ion.

Double ionization

At sufficiently high impact energy, double ionization is also possible with the helium atom target. In this case, all three electrons in the final state are asymptotically free. An appropriate co-ordinate system to treat this case is defined by

$$
r_0 = \rho \sin\alpha \cos\beta, \quad r_1 = \rho \cos\alpha \cos\beta, \quad r_2 = \rho \sin\beta,
$$

$$
\rho = \sqrt{r_0^2 + r_1^2 + r_2^2}. \tag{2.45}
$$

The double ionization scattering amplitude is obtained again from the general expression (2.39):

$$
F_{\mathrm{fi}}^{S}(\boldsymbol{k}_0, \boldsymbol{k}_1, \boldsymbol{k}_2) \underset{\rho \to \infty}{\sim} -\frac{\rho^2}{2(2\pi)^4} \int d\hat{\boldsymbol{r}}_0 \int d\hat{\boldsymbol{r}}_1 \int d\hat{\boldsymbol{r}}_2 \int_0^{\pi/2} d\alpha \int_0^{\pi/2} d\beta \, r_0^2 r_1^2 r_2^2 \cos\beta
$$

$$
\times \sum_{s_{12}} \left\{ \Phi_{\mathrm{f},s_{12},\boldsymbol{k}_0,\boldsymbol{k}_1,\boldsymbol{k}_2}^{S(-)*}(012) \frac{\partial}{\partial\rho} \Psi_{\mathrm{i},s_{12}}^{S(+)}(012) - \Psi_{\mathrm{i},s_{12}}^{S(+)}(012) \frac{\partial}{\partial\rho} \Phi_{\mathrm{f},s_{12},\boldsymbol{k}_0,\boldsymbol{k}_1,\boldsymbol{k}_2}^{S(-)*}(012) \right\}. \tag{2.46}
$$

---

[8] We omit the phase factor $\Delta_{\mathrm{fi}}(\rho)$ in the following discussion as it does not contribute to the ionization cross section.

In this case, the final state is a product of three continuum waves defined as

$$\Phi_{f,s_{12}k_0,k_1,k_2}^{S(-)}(012) \sim \frac{1}{\sqrt{6}}[\delta_{s_f s_{12}} - \alpha_{01}^{s_f s_{12} S} p_{01} - \alpha_{02}^{s_f s_{12} S} p_{02}]$$

$$\times [1 + (-1)^{s_f} p_{12}]\phi_{k_0}^{(-)}(0)\phi_{k_1}^{(-)}(1)\phi_{k_2}^{(-)}(2). \qquad (2.47)$$

For calculation purposes, the obvious choice for the continuum waves for each ionized electron is a Coulomb wave with $Z = 2$. Thus far no ECS calculation has been performed for the full electron–helium scattering problem, due to the prohibitive time required to accommodate the four-body nature of the problem. However, the PECS approach has recently been applied to the simplified *s*-wave model for electron–helium scattering. This model admits the complexity of the full four-body for double ionization and has been reported by Bartlett and Stelbovics [29,30].

2.5.2 S-wave model for electron–helium scattering

For a neutral helium-atom target ($Z = 2$), the *s*-wave model Schrödinger becomes

$$(E - H)\psi_{i,s_{12},sc}^{S}(012) = \xi_{i,s_{12}}^{S}(012) \qquad (2.48)$$

and for an initial target in state i

$$\xi_{i,s_{12}}^{S}(012) = \frac{2\sqrt{\pi}}{\sqrt{3}k_i}[\delta_{s_i s_{12}} - \alpha_{01}^{s_i s_{12} S} p_{01} - \alpha_{02}^{s_i s_{12} S} p_{02}]\xi_i(012), \qquad (2.49)$$

where

$$H = -\frac{1}{2}\frac{\partial}{\partial r_0} - \frac{1}{2}\frac{\partial}{\partial r_1} - \frac{1}{2}\frac{\partial}{\partial r_2} - \frac{2}{r_0} - \frac{2}{r_1} - \frac{2}{r_2}$$

$$+ \frac{1}{r_{>01}} + \frac{1}{r_{>02}} + \frac{1}{r_{>12}}, \qquad (2.50)$$

$$\xi_i(abc) = \left(\frac{1}{r_{>ab}} + \frac{1}{r_{>ac}} - \frac{2}{r_a}\right)\sin(k_i r_a)\phi_i(bc), \qquad (2.51)$$

$$r_{>ab} = \max(r_a, r_b) \qquad (2.52)$$

and $\phi_i(r_b, r_c)$ is the radial part of the *s*-wave helium wave function for the initial target state i that is antisymmetrized with respect to electron exchange. The PECS method can once again be applied but now with the generalization that the propagation vector, the scattering wave function, is a three-dimensional array. This means the computational requirement is much greater than for the electron–hydrogen problem.[9]

---

[9] For this reason, [29] developed an improved propagation algorithm for the electron–helium system.

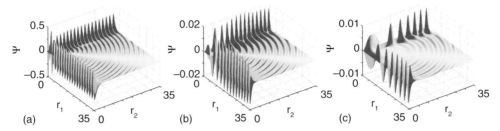

Figure 2.5 Electron–helium *s*-wave triplet scattering wave function (real part) for $E_0 = 150$ eV at slices taken along (a) $r_0 = 1$, (b) $r_0 = 10$ and (c) $r_0 = 35$ a.u. All axes are in atomic units.

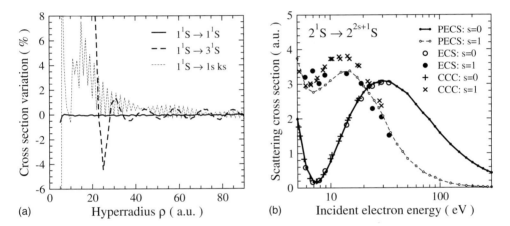

Figure 2.6 (a) Convergence w.r.t. hyperradius (ρ) of various three-body cross sections at $E_0 = 53.9$ eV with a ground-state target, shown as a percentage variation from $\rho \approx 90$ a.u. calculations. (b) PECS excited-state target scattering cross sections ($2^1 S \to 2^{2s+1} S$) compared with ECS [31] and CCC [32, 33] calculations.

In Fig. 2.5 we show an example of an electron–helium scattering wave function computed for a target in the ground state and leading to a final state where $s_{12} = 1$ and the subsystem of electrons 1 and 2 are left in a triplet state. This is the reason the scattering wave function asymptotically has zero amplitude along the the axis $r_1 = r_2$, because both escaping electrons have equal energy and spin.

The convergence with respect to hyperradius for inelastic scattering and single ionization is illustrated in Fig. 2.6(a). As expected for elastic scattering, convergence is obtained by 10 a.u., while inelastic scattering and ionization require propagation out to at least 40 a.u. In Fig. 2.6(b) we show a comparison of our method for scattering from an initial excited target in the $2^1 S$ state with other calculations for the same model. Generally, it appears scattering to the $^3 S$ final state is harder to describe accurately, as evidenced by the comparisons.

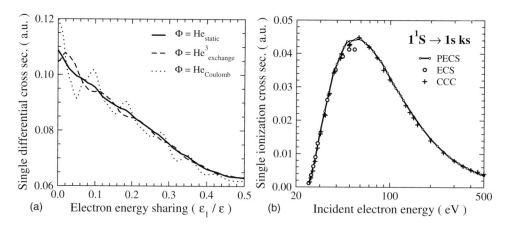

Figure 2.7 (a) Single-ionization s-wave SDCS for $E_0 = 53.9$ eV using helium static (He$_{\text{static}}$), helium triplet static exchange (He$^3_{\text{exchange}}$) and Coulomb ($Z = 1$) waves to approximate the outgoing continuum waves. (b) Total s-wave single-ionization cross section for 20–500 eV incident electron energies. ECS results are from [31] and CCC results are from [32, 33].

Next we show some features of single ionization in Fig. 2.7. Since in single ionization the ejected electron 'sees' a helium ion of charge unity, it seems reasonable to describe this electron by a Coulomb wave, as for the hydrogen target. However, as can be seen for the energy differential ionization cross section there are some oscillations due to this choice. We find that better choices in regard to numerical stability are continuum waves formed from either the helium static or helium triplet static-exchange potential. The total ionization cross section with comparison calculations demonstrate excellent agreement over a wide range of energies. Thus it can be concluded that for three-body aspects of the collision there are few surprises in comparison with electron scattering from hydrogen.

Let us now turn to four-body aspects of the scattering from helium. Consider, first, collisions that leave the helium atom in a doubly excited final state, for example the $2s2s\,^1S$. This state will decay, allowing an electron to escape. Thus this autoionization should be seen as an interference feature in single ionization. That this is the case is demonstrated in Fig. 2.8, where the arrows indicate the feature in the ionization cross section from the doubly excited state.

When considering scattering to doubly excited states it is important to note that they will decay and in the stationary-state formalism this will result in their energies having a small complex component. Therefore, we were careful to ensure that these states were computed accurately using complex rotation. The importance of describing these states accurately is highlighted in the excitation cross section from the ground state to the doubly excited states $2s2s\,^1S$, $2s3s\,^3S$, as indicated in Fig. 2.9. It can be seen that the CCC calculation is quite accurate for the $2s3s\,^{2s+1}S$

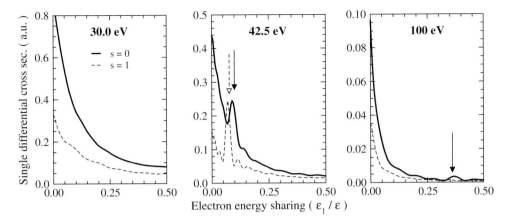

Figure 2.8 SDCS for electron impact single ionization of 2^1S and 2^3S excited-state s-wave helium targets at $E_0 = 30, 42.5$ and 100 eV. The autoionization energy for $2s2s\,^1S \to 1sks$ (34.75 eV) is shown with arrows.

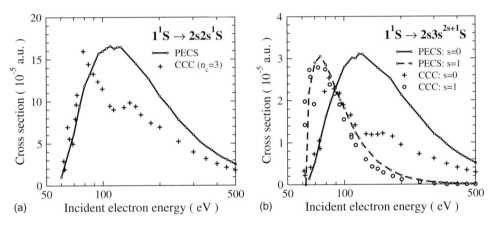

Figure 2.9 Electron-impact double-excitation cross sections of s-wave helium at energies $E_0 = 50$–500 eV. CCC results are from [33].

(panel (b)) but less so for the singlet double-excited states (panels (a) and (b)). This is probably due to the fact that the CCC doubly excited states were computed with a real Laguerre basis and therefore the generated states have no complex energy part. This is a reasonable approximation if the states are very long lived as the triplets are but not so for the singlets.

Finally, we turn to the double-ionization problem from helium. Computationally, this is a most challenging problem, even in this model. The double-ionization cross section is two orders of magnitude smaller and the surface integral we have to evaluate once we have computed the scattering wave function contains three continuum functions, so is an extremely taxing calculation. Since double ionization results in

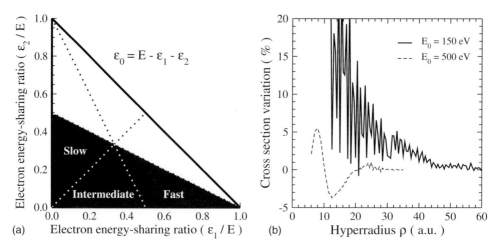

(a)   Electron energy-sharing ratio (ϵ_1 / E)   (b)   Hyperradius ρ (a.u.)

Figure 2.10 (a) Electron energy-sharing lines of symmetry and energy integration regions (shaded) for total and single energy-differential double ionization cross sections. (b) Convergence of s-wave double-ionization cross sections w.r.t. hyperradius (ρ) for $E_0 = 150$ eV and $E_0 = 500$ eV, expressed as a percentage variation from the $\rho = 60$ a.u. and 35 a.u. calculations, respectively.

three identical electrons being ejected, there are six symmetries in momenta of the electrons that one can make use of in evaluating the ionization surface integral. One can always order the momentum and energy of the three electrons from largest to smallest, so for definiteness we choose

$$0 \leq k_2 \leq k_1 \leq k_0 \leq \sqrt{2E},$$

$$0 \leq \epsilon_2 \leq \epsilon_1 \leq \epsilon_0 \leq E$$

$$\epsilon_0 + \epsilon_1 + \epsilon_2 = E. \tag{2.53}$$

Then, with this order, the total double-ionization cross section can be expressed as

$$\sigma_i^S = \int_0^{E/2} d\epsilon_1 \int_0^{\min(\epsilon_1, E-2\epsilon_1)} d\epsilon_2 \frac{d^2\sigma_i^S(k_0, k_1, k_2)}{d\epsilon_1 d\epsilon_2}. \tag{2.54}$$

If we refer to Fig. 2.10(a), the integration limits for the ordering of the momenta in Eq. (2.53) are those over the shaded area labeled 'Intermediate'. Each of the six regions yield the same total cross section for the six possible permutations of the momenta. The region labeled 'Slow' corresponds to $\epsilon_1 \leq \epsilon_2 \leq \epsilon_0$ and 'Fast' to $\epsilon_2 \leq \epsilon_0 \leq \epsilon_1$. These three regions cover the whole range of allowed energy for electron 1, subject to the condition $\epsilon_2 \leq \epsilon_0$. In Fig. 2.10(b) we show the convergence of the integrated double ionization cross section with hyperradius. We are able to achieve converged cross sections to a few per cent for incident electron energies

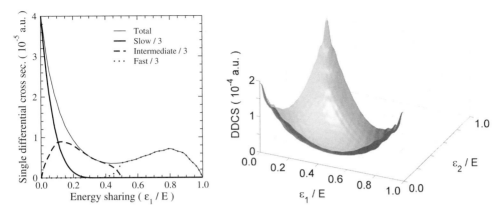

Figure 2.11 Single (left) and double (right) energy-differential cross sections for double ionization of ground-state s-wave helium at $E_0 = 150$ eV.

from 150 eV. As with our previous examples of PECS calculations, one requires propagation to larger hyperradius as the energy decreases.

Now suppose we consider the single differential cross section where only electrons 0 and 2 are detected and with $\epsilon_2 \leq \epsilon_0$. Then we may write:

$$\frac{d\,\sigma_i^S(k_0, k_1, k_2)}{d\epsilon_1} = \frac{1}{3}\int_0^{(E-\epsilon_1)/2} \frac{d^2\,\sigma_i^S(k_0, k_1, k_2)}{d\epsilon_1 d\epsilon_2}d\epsilon_2. \qquad (2.55)$$

In Fig. 2.11, we show the single differential cross section as defined by Eq. (2.55) and its components, as defined by the shaded regions of figure 2.10 for various energies of electron 1 relative to the energies of electrons 0 and 2, subject to $\epsilon_2 \leq \epsilon_0$. It is interesting to see that each of the three regions, 'Slow', 'Intermediate' and 'Fast' provides a dominant contribution to the SDCS over some range of energy sharing. Also presented is the double differential cross section that is peaked whenever one of the electrons has most of the energy available. There is also an exact minimum of zero at the point where all electrons have the same energy; this follows from the Pauli principle, as at least two of the electrons have identical spin and are at the same space co-ordinate. This PECS calculation represents the current state of the art for double ionization and demonstrates that the method can generate converged cross sections, despite the very much smaller cross section compared to single ionization.

2.6 Summary and outlook for the future

We have shown that the propagating exterior scaling method (PECS) is an efficient implementation of the ECS approach and have illustrated its capability for both the electron–hydrogen and electron–helium scattering problems. For electron–hydrogen scattering, the PECS method is sufficiently accurate to allow one to probe the near-threshold region of ionization that previously had been inaccessible to a full numerical solution of the Schrödinger equation.

For electron–helium, the implementation has been applied to the s-wave model that is a subset of the full problem. The method is computationally expensive but is able to deliver converged cross sections for the first time for double ionization. For the full three-electron scattering problem there are no results yet due to the fact that the computational requirements are an order of magnitude greater than for the s-wave model. However, with the expected improvements in computing speeds over the next decade it is almost certain that the full problem will be fully solved.

The combinatorial aspects of the approach make its extension to four–electron problems and beyond unfeasible in the foreseeable future. However, it would be relatively straightforward to adopt an approach used extensively in the field, namely the expansion of the scattering wave function in terms of a basis of the target ion/atom states that leads to a set of coupled channels equations with either two or three active electrons to which the PECS approach can be applied.

References

[1]   I. Bray and A. T. Stelbovics, *Adv. At. Mol. Phys.*, **35**, 209 (1995).
[2]   I. Bray and A. T. Stelbovics, *Phys. Rev. A*, **46**, 6995 (1992).
[3]   I. Bray and A. T. Stelbovics, *Phys. Rev. Lett.*, **69**, 53 (1992).
[4]   I. Bray and A. T. Stelbovics, *Phys. Rev. Lett.*, **70**, 746 (1993).
[5]   I. Bray, D. V. Fursa, A. S. Kheifets and A. T. Stelbovics, *J. Phys. B*, **35**, R117 (2002).
[6]   I. Bray and D. V. Fursa, *Phys. Rev. A*, **54**, 2991 (1996).
[7]   I. Bray, *Phys. Rev. Lett.*, **78**, 4721 (1997).
[8]   A. T. Stelbovics, *Phys. Rev. Lett.*, **83**, 1570 (1999).
[9]   T. N. Rescigno, M. Baertschy, D. Byrum and C. W. McCurdy, *Phys. Rev. A*, **55**, 4253 (1997).
[10] T. N. Rescigno, M. Baertschy, W. A. Isaacs and C. W. McCurdy, *Science*, **286**, 2474 (1999).
[11] C. W. McCurdy, D. A. Horner and T. N. Rescigno, *Phys. Rev. A*, **63**, 022711 (2001).
[12] C. W. McCurdy, M. Baertschy and T. N. Rescigno, *J. Phys. B*, **37**, R137 (2004).
[13] P. L. Bartlett, *J. Phys. B*, **39**, R379 (2006).
[14] G. H. Wannier, *Phys. Rev.*, **90**, 817 (1953).
[15] R. K. Peterkop, *Theory of Ionization of Atoms by Electron Impact* (Boulder, Colorado. Colorado Associated University Press, 1977).
[16] M. R. H. Rudge, *Rev. Mod. Phys*, **40**, 564 (1968).
[17] M. Baertschy, T. N. Rescigno and C. W. McCurdy, *Phys. Rev. A*, **64**, 022709 (2001).
[18] T. N. Rescigno, M. Baertschy and C. W. McCurdy, *Phys. Rev. A*, **68**, 020701 (R) (2003).
[19] P. L. Bartlett *et al.*, *Phys. Rev. A*, **68**, 020702 (R) (2003).
[20] S. Jones and A. T. Stelbovics, *Phys. Rev. A*, **66**, 032717 (2002).
[21] P. L. Bartlett, A. T. Stelbovics and I. Bray, *J. Phys. B*, **37**, L69 (2004).
[22] A. Temkin, *Phys. Rev. Lett.*, **49**, 365 (1982).

[23] P. L. Bartlett, A. T. Stelbovics and I. Bray, *Phys. Rev. A*, **68**, 030701 (R) (2003).

[24] K. Bartschat and I. Bray, *J. Phys. B*, **29**, L577 (1996).

[25] M. B. Shah, D. S. Elliott and H. B. Gilbody, *J. Phys. B*, **20**, 3501 (1987).

[26] J. F. Williams, P. L. Bartlett and A. T. Stelbovics, *Phys. Rev. Lett.*, **96**,123201 (2006).

[27] I. Bray, *Aust. J. Phys.*, **53**, 355 (2000).

[28] I. Bray, Private communication (2004).

[29] P. L. Bartlett and A. T. Stelbovics, *Phys. Rev. A*, **81**, 022715 (2010).

[30] P. L. Bartlett and A. T. Stelbovics, *Phys. Rev. A*, **81**, 022716 (2010).

[31] D. A. Horner, C. W. McCurdy and T. N. Rescigno, *Phys. Rev. A*, **71**, 012701 (2005).

[32] C. Plottke, I. Bray, D. V. Fursa and A. T. Stelbovics, *Phys. Rev. A*, **65**, 032701 (2002).

[33] C. Plottke, P. Nicol, I. Bray, D. V. Fursa and A. T. Stelbovics, *J. Phys. B*, **37**, 3711 (2004).

3

Fragmentation of molecular-ion beams in intense ultrashort laser pulses

ITZIK BEN-ITZHAK

3.1 Introduction

In this chapter we describe molecular fragmentation in intense ultrashort laser pulses, which produce fields comparable to the binding forces in molecules – far from the perturbative regime. The ability to tailor these laser fields holds promise for controlling molecular dynamics. In particular, we focus on molecular-ion beam targets that allow kinematically complete measurements of dissociation. We present the experimental method used to image the molecular fragmentation and illustrate its application in a few recent studies of benchmark and more complex molecules.

Laser molecular science is in the midst of an exciting era when new applications continually emerge, such as molecular imaging [1, 2] and laser control of electron localization – a step toward directing chemical reactions [3]. These applications originate and continue to benefit from basic research on the interaction of intense ultrashort laser pulses with molecules. The hydrogen molecular ion, being a benchmark system, plays a key role in the development of our understanding of the interaction of strong laser fields with molecules (see, e.g., [4, 5] for reviews).

The simple H_2^+ is used in this section to introduce the main molecular fragmentation mechanisms discussed later and to describe the conceptual pictures commonly used to explain these phenomena. We start with the latter by looking at the potential energy curves (PEC) of H_2^+, shown in Fig. 3.1. Given that the energy gap to the $n = 2$ manifold in H_2^+ is much larger than the photon energy at 790 nm, it is sufficient in many cases to consider only the lowest two electronic states. In Fig. 3.1(a) – the photon picture – one represents transitions involving one or more photons by arrows, where 'up' and 'down' denote absorption and emission, respectively. Note that multiphoton transitions are possible due to the strength of the laser field,

Fragmentation Processes: Topics in Atomic and Molecular Physics, ed. Colm T. Whelan. Published by Cambridge University Press. © Cambridge University Press 2013.

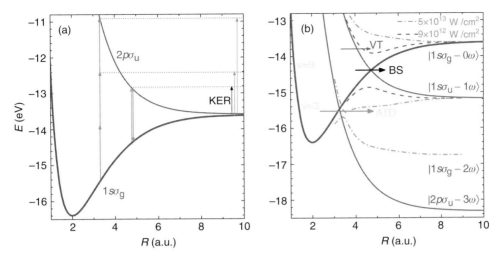

Figure 3.1 A potential-energy curve diagram of H_2^+ showing (a) the lowest two states and (b) the light-dressed diabatic (solid lines) and adiabatic (dash and dash-dot lines) curves. The arrows in panel (a) mark the location where 1- and 3-photon transitions are resonant. The curve crossings between the diabatic PECs in panel (b) occur at these locations, and they become avoided crossings for the adiabatic curves (see text). The main vibrational states undergoing bond softening (BS), vibrational trapping (VT) and above-threshold dissociation (ATD) are marked by arrows in panel (b).

and that dipole selection rules dictate that only odd numbers of photons connect between the lowest $1s\sigma_g$ and $2p\sigma_u$ states of H_2^+.

Another approach used to describe molecular fragmentation in strong laser fields involves the use of light-dressed potentials [4–7], namely a Floquet picture, as shown in Fig. 3.1(b). In this Floquet picture, each of the potential curves is dressed (shifted up or down) by the number of absorbed photons (n), denoted by $-n\omega$, where ω refers to the energy of the photon in a.u. For example, the $1s\sigma_g$ state dressed by two photons is labeled as $|1s\sigma_g - 2\omega\rangle$. Where curves cross one another, transitions from one state to the other are most likely. The transition probability depends on the coupling between these states under the influence of the laser field. Moreover, on such a diagram, the kinetic energy release (KER) upon dissociation can easily be estimated by evaluating the energy difference between the initial state and the final dissociation limit.

The adiabatic Floquet representation, shown also in Fig. 3.1(b), goes a step further by diagonalizing the electronic Hamiltonian including the dipole coupling, producing Born–Oppenheimer potentials that incorporate the laser field. The diabatic crossing between $|1s\sigma_g - 0\omega\rangle$ and $|2p\sigma_u - 1\omega\rangle$ thus becomes an avoided crossing with an energy gap that widens with increasing laser field strength. There-fore, the vibrational states within this energy gap become unbound in the presence of the laser field – a process known as 'bond softening' (BS) [8–10]. This dissoci-ation process is peaked around $v = 9$ in H_2^+ as can be seen on both panels of Fig. 3.1. However, it is easier to see how neighbouring states will undergo BS in the Floquet picture, which we prefer in most cases.

The adiabatic Floquet curve above the one-photon crossing forms a transient bound well at some intermediate intensities, e.g., at 9×10^{12} W/cm^2 in Fig. 3.1(b). Population in high vibrational states is shelved in this laser-induced potential well [10–15]. It is interesting to note that it is harder to visualize this process using the photon picture in Fig. 3.1(a). This 'vibrational trapping' (VT) shelters the population from one-photon bond softening [8–10] until after the pulse, when it may be released back into the bound H_2^+ $1s\sigma_g$ electronic ground state thereby suppressing dissociation of these states. However, calculations including nuclear rotation indicate that vibrational trapping, which leads to stabilization, is strongly inhibited [16–18] (see further discussion in Section 3.3.1).

Multiphoton processes become more likely with increasing laser intensity, and the molecule can dissociate by absorbing more than the minimum number of photons required, resulting in a series of peaks in the KER spectrum separated by one photon energy – this process is known as 'above-threshold dissociation' (ATD) [8–10, 19–21]. The energy gap leading to ATD is clearly visible near the $v = 3$ state for the 5×10^{13} W/cm^2 intensity curves shown in Fig. 3.1(b). From the figure we also see that the dissociating wave packet is most likely to end on the $|1s\sigma_g - 2\omega\rangle$ state by stimulated emission of one photon around $R = 4.7$ a.u., finally resulting in a net two-photon ATD (controlling the ratio of net two- and three-photon ATD is discussed in Section 3.3.1). This ATD process can also be visualized in the photon picture (Fig. 3.1(a)), which shows the near-resonant absorption of three photons around $R = 3.3$ a.u., followed by a near-resonance one-photon emission along the dissociation path around $R = 4.7$ a.u. However, in this picture it is not easy to see why net two-photon ATD is favoured over net three-photon ATD.

Most experimental studies of H_2^+ dissociation in intense ultrashort laser pulses are conducted on commonly available H_2 targets, therefore they require an initial ionization step [4,5]. Probing instead an H_2^+ beam allows one to focus on dissocia-tion over a much wider intensity range, but such experiments are challenging [22].

Pioneering measurements on H_2^+ beams were conducted at the turn of the cen-tury by two groups, Figger and Hänsch [23] and Williams *et al.* [24]. Since then a few additional groups, including our own group, have been conducting intense laser field experiments using atomic- and molecular-ion beam targets [25–36]. In our work [37–39], we introduced the powerful coincidence 3D momentum imag-ing technique, explained in Section 3.2, and applied it in studies of benchmark

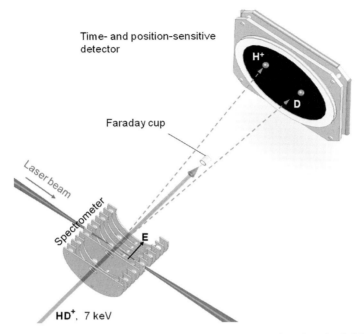

Figure 3.2 Schematic view of our experimental set-up. The electric field of the spectrometer, labeled E, is used to distinguish between fragmentation channels (see text).

molecules, like H_2^+ and H_3^+, described in Section 3.3.1 and Section 3.3.2, respectively. We also extended our studies to more complex molecular ions, a few of which are discussed in Section 3.4.

3.2 Experimental method

To study the fragmentation of molecular-ion beams in intense ultrashort laser pulses we implement a coincidence 3D momentum imaging technique, which takes advantage of the fact that our target ions are moving fast enough to also allow the detection of neutral fragments. The aim is to separate different fragmentation channels from each other and to provide high-resolution and differential data in order to advance our understanding of the molecular dynamics initiated and controlled by the strong laser field. The main components needed to conduct such experiments are (i) a laser system that can provide intense ultrashort pulses, (ii) an ion source and accelerator to provide the ion beams that serve as the target, (iii) a method for crossing the two beams, (iv) an imaging set-up for measuring the reaction products, and finally (v) a method for converting the measured signals to the desired 3D momentum images. The experimental set-up used for these studies is depicted schematically in Fig. 3.2.

3.2.1 Laser

The laser used in our measurements is a Ti:Sapphire system ($\sim$790 nm), providing linearly polarized 28–35 fs (full width at half maximum in intensity), 2 mJ transform-limited pulses at 1–2 kHz. Their bandwidth can be increased in a neon-filled hollow-core fibre, followed by compression to 7–10 fs transform-limited pulses by a set of chirped mirrors [40]. The time envelope of the pulses is characterized by frequency-resolved optical gating (FROG) [41].

The laser beam is focused onto the ion beam target, using an $f = 203$ mm off-axis parabolic mirror, to peak intensities up to about 10^{16} W/cm^2. The peak intensity is determined from the measured laser beam power, together with the temporal and spatial beam profiles. The latter is determined from CCD images of the laser beam profile as a function of the distance from the laser focus. Typically, the laser beam has a Gaussian spatial distribution with a $1/e^2$ diameter of about 20–70 µm and a couple of mm Rayleigh length.

The polarization axis is typically aligned perpendicular to the ion beam direction, but if required it can be rotated by using a half-wave plate. Circular or elliptically polarized light is produced, when needed, using a quarter-wave plate (linear polarization is used in all the figures below).

This 790 nm laser beam can also be frequency doubled to provide higher energy photons. This is accomplished by passing the fundamental beam through a second harmonic generation crystal (beta-barium borate, BBO). In this case the output light from the BBO is separated using a dichromatic beam splitter that transmits the 395 nm light and reflects the 790 nm light.

3.2.2 Ion beam

The molecular ions used in these studies are produced in an electron cyclotron resonance (ECR) ion source and accelerated to about 5–10 keV. The ion beam is selected by a magnetic field according to its charge to momentum ratio, and then collimated to a cross section of about 0.8×0.8 mm^2, with a current of the order of 1 nA. This ion beam propagates through a coaxial electrostatic spectrometer where it is crossed with the laser beam at right angles and is finally directed toward a small Faraday cup, 2 mm in diameter, for normalization purposes and to protect the imaging detector downstream (see Fig. 3.2). It is notable that during the pulse duration the molecular ions move forward less than 50 nm, which is negligible relative to the laser focus, and thus the ions can be considered as stationary during the interaction.

Molecular ions produced in the ion source from a parent neutral molecule by electron impact ionization, which can be treated as a vertical transition, typically have a vibrational population that approximately follows the Franck–Condon factors [42,43]). Some molecular-ion beams contain metastable states, and these states may play a key role in the laser-driven dissociation, as is the case for O_2^+ [44–46]. Moreover, deviations from the Franck–Condon distribution can occur, especially when ionization of higher electronic states, which feed the lower state by cascades, are significant. This is, for example, the case for the metastable $a\,{}^4\Pi_u$ state of O_2^+ [47].

The initial population of molecular ions that are formed in the ion source by more complex reactions, such as H_3^+ created in $H_2^+ + H_2 \rightarrow H_3^+ + H$ reactions [48], may cause an additional challenge when attempting to interpret their laser-induced fragmentation. In the specific example of H_3^+, fortunately, one can rely on the theoretical predictions of the vibrational population in the H_3^+ electronic ground state [48].

3.2.3 Crossing the laser and ion beams

The ion and laser beams cross perpendicular to each other. It is important to note that the laser focus is much smaller than the ion beam width and that the Rayleigh range is bigger than the ion beam thickness. Therefore, averaging over the laser intensity in the interaction volume is simplified as one can neglect the intensity change along the laser beam propagation and assume that the ion beam is infinite when integrating radially.

Crossing the two beams is challenging due to their dimensions, and even more because of the low counting rate, which is mainly limited by the extremely low density of an ion beam target – typically equivalent to a gas target in the 10^{-13}–10^{-12} Torr range – i.e., much thinner than the 10^{-10} Torr residual gas pressure in our set-up. One may wonder how the true signal can be separated from the background counts given that the target of interest is about 2–3 orders of magnitudes less dense. This is accomplished by taking advantage of the high initial velocity of the beam fragments and the coincidence nature of the measurements discussed below.

3.2.4 Coincidence beam-fragment measurements

The fragments from the laser-induced molecular break-up are detected in coincidence by a time- and position-sensitive microchannel plate delay-line detector about one metre downstream from the interaction region, while the ion beam is captured in a miniature Faraday cup as shown in Fig. 3.2. All the time signals are measured relative to a start signal provided by a photodiode detector triggered by the laser pulse. The zero time is defined as the moment the laser pulse crosses the ion beam, and is determined for each experiment by measuring photons arriving

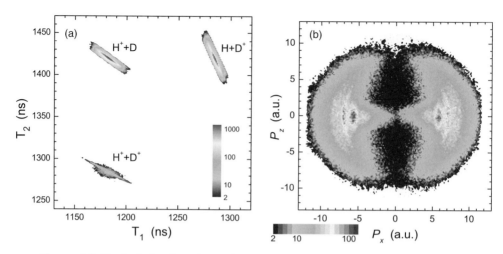

Figure 3.3 Typical density plots of (a) a coincidence time-of-flight for HD^+ fragmentation induced by 33 fs transform-limited pulses with 4×10^{15} W/cm^2 peak intensity, and (b) a momentum image of the $H^+ + D$ dissociation channel in positively chirped 65 fs (from 27 fs transform-limited) laser pulses with 1×10^{14} W/cm^2 peak intensity. The laser polarization is along x, i.e., parallel to the detector plane.

from the interaction region to better than 0.1 ns, therefore enabling the measurement of true times of flight from the interaction to the detector. Finally, this time and position information, recorded event by event, is used to reconstruct the 3D momenta of all fragments.

To experimentally distinguish the reaction channels, the different fragments are separated in time by a constant weak electric field of a coaxial spectrometer, shown in Fig. 3.2. A coincidence time-of-flight spectrum of HD^+ fragmentation by intense ultrashort laser pulses is shown in Fig. 3.3 to demonstrate the clear separation of all break-up channels in this case.

3.2.5 Coincidence 3D momentum imaging of beam fragments

The coincidence 3D momentum imaging of beam fragments method is similar to cold target recoil ion momentum spectroscopy (COLTRIMS) [49,50], in principle, if one thinks of it in the centre-of-mass (CM) frame of reference. The strength of this beam-fragment imaging technique is that it allows one to identify the different dissociation channels experimentally, and through the determination of the momenta of all dissociating fragments one gains the complete kinematic information about the reaction. From the latter, one can evaluate the KER and angular distributions of the reaction products. The separation of the dissociation channels is accomplished, as described earlier, by accelerating the ions along the beam direction.

To understand the basic idea behind the 3D momentum imaging method it is educational to turn off the longitudinal electric field, used to separate the dissociation channels from each other (see Section 3.2.4), and consider first the simpler and more commonly used 'field-free imaging' method (sometimes referred to in literature as 'fast ion beam photofragment spectroscopy' [51–56]. Moreover, we will limit ourselves here to two-body fragmentation, which can easily be extended to three fragments or more following the same approach.

It is convenient to define the z-axis along the beam direction, with the x, y origin in the centre of the beam spot on the imaging detector and the z origin at the laser ion beam crossing. The x and y components of the fragment's momenta are evaluated from the measured position and time of both hits. Specifically, the x-displacements of the first and second fragments hitting the detector are given by

$$x_1 - x_{0_i} = (v_{0x_i} + v'_{1x})t_1 \qquad (3.1)$$

$$x_2 - x_{0_i} = (v_{0x_i} + v'_{2x})t_2 \,, \qquad (3.2)$$

where x_1, t_1 and x_2, t_2 are the measured position and time of the first and second hits, respectively, x_{0_i} is the position where a specific molecule fragmented, v_{0x_i} is its centre-of-mass velocity component, and, finally, v'_{1x} and v'_{2x} are the x velocities of each fragment in the CM system. In addition to the two equations above one can use momentum conservation, explicitly linking the dissociation velocities in the CM system by

$$m_1 v'_{1x} + m_2 v'_{2x} = 0\,, \qquad (3.3)$$

where m_1 and m_2 are the masses of the two fragments.

This set of equations with four unknowns, v'_{1x}, v'_{2x}, v_{0x_i} and x_{0_i}, can be solved once we eliminate one of the unknowns. One possible assumption, which leads to good-quality solutions, is to replace the fragmentation position for each molecule, x_{0_i}, by the average value for the whole ensemble of molecules, $x_0 \equiv \overline{x_{0_i}}$; another is to make a similar substitution for the CM velocity, namely substitute v_{0x_i} with $v_{0x} \equiv \overline{v_{0x_i}}$. The first choice works better when the interaction volume is small while the second works better when the beam divergence is small. In our case, the interaction volume, defined by the overlap of our tightly focused laser beam and the sub-1 mm ion beam, is very small, therefore we selected to replace x_{0_i} by zero and solve for the remaining unknowns.

Specifically, substituting $v'_{2x} = -\frac{m_1}{m_2}v_{1x}$ in Eq. (3.2) and then subtracting it from Eq. (3.1) yields,

$$v'_{1x} = \frac{x_1 - x_2}{t_1 + \beta t_2} + \frac{v_{0x_i}(t_2 - t_1)}{t_1 + \beta t_2}\,, \qquad (3.4)$$

where $\beta \equiv \frac{m_1}{m_2}$ is the mass ratio of the fragments. The second term in the equation above is a small correction due to the beam divergence, which can be evaluated explicitly using the solution for v_{0x_i} given by,

$$v_{0x_i} = \frac{1}{1+\beta}\left[\frac{x_2}{t_2} + \beta\frac{x_1}{t_1}\right]. \tag{3.5}$$

The y velocity component is evaluated following the same procedure, while the z component is evaluated from the times of flight, where we take advantage of the fact that t_1 and t_2 are the true times of flight from the molecular fragmentation, which happens within the ultrashort laser pulse to the imaging detector (see Section 3.2).

The time of flight of each fragment is given by

$$t_1 = \frac{d - z_i}{v_{0z_i} + v'_{1z}} \tag{3.6}$$

$$t_2 = \frac{d - z_i}{v_{0z_i} + v'_{2z}}, \tag{3.7}$$

where d is the distance from the interaction region to the detector, z_i is the location of the dissociation of a specific molecule within the interaction volume, and v_{0z_i} is the beam velocity of each molecule – it is important to note that due to the wide beam-energy spread, the range of v_{0z_i} is much larger than that of the x and y velocity components. As before, momentum conservation provides the third equation,

$$m_1 v'_{1z} + m_2 v'_{2z} = 0. \tag{3.8}$$

Again we have four unknowns, therefore to solve for v'_{1z} and v_{0z_i} we replace z_i with its average value $z_0 \equiv \overline{z_i}$ and get

$$v'_{1z} = \frac{1}{1+\beta}\left(\frac{d-z_0}{t_1} - \frac{d-z_0}{t_2}\right) \tag{3.9}$$

$$v_{0z} = \frac{1}{1+\beta}\left(\frac{\beta(d-z_0)}{t_1} - \frac{d-z_0}{t_2}\right). \tag{3.10}$$

The best value for z_0 is determined by imposing reflection symmetry on the resulting v'_{1z} distribution measured with a laser field having such symmetry, for example, a linearly polarized field perpendicular to the z-axis.

Once all velocity components are evaluated for an individual molecule, we compute the momenta of its fragments given by $p_{1j} = m_1 v'_{1j}$, where $j = x$, y, or z. From the momenta of the fragments we evaluate the KER and the angular distributions shown in the following sections.

The main drawback of the field-free imaging method is that one cannot identify the beam fragments, and therefore one cannot easily distinguish different reaction channels like ionization from dissociation. One can overcome this difficulty by using some known difference between reactions to make such an association, for example, in many cases ionization has been distinguished from dissociation by the expected higher KER. However, in some cases such identification may be misleading, as the KER distributions can overlap. For example, N_2^+ dissociation into $N^+ + N$ through highly excited states yields KER values as high as dissociative ionization to $N^+ + N^+$ [57].

To eliminate this difficulty we have developed imaging methods that introduce a weak electric field in the interaction region, provided by the axial spectrometer, shown in Fig. 3.2. This field accelerates ionic fragments relative to their charge to energy ratio, therefore modifying their time of flight and separating the fragmentation channels from each other, as shown, for example, in Fig. 3.3.

This longitudinal electric field, i.e., $E\hat{z}$, does not affect the analysis of the x and y velocity components, except for minor corrections due to fringe fields and a small misalignment of the beam velocity with respect to the spectrometer field. The discussion of these corrections is too elaborate to include here, so the interested reader is directed to [58, 59], for example.

The z velocity component, in contrast, is strongly affected by the field as the time-of-flight formula for the ions is significantly different than Eq. (3.6). Using SIMION [60] simulations, which reproduce measured times of flight accurately, we established that the model formula given below reproduces the time of flight to better than 0.1 ns, i.e., better than our time resolution.

$$
t_1 = \frac{2d_1}{v_{0z_i}} \frac{1}{\eta_1} \left[\sqrt{(1 + u_{1z})^2 + \eta_1(1 - z_i')} - (1 + u_{1z}) \right]
$$
$$
+ \frac{d_2}{v_{0z_i}} \frac{1}{\sqrt{(1 + u_{1z})^2 + \eta_1(1 - z_i')}} \tag{3.11}
$$

where d_1 and d_2 are the lengths of the electric field and field-free regions, respectively, $z_i' \equiv z_i/d_1$ is a scaled fragmentation location, $u_{1z} \equiv v_{1z}'/v_{0z_i}$ is a scaled dissociation velocity of the first fragment in the CM system, $\eta_1 \equiv 0.8q V_s/\frac{1}{2}m_1 v_{0z_i}^2$ is the ratio of potential and kinetic energies, V_s is the spectrometer voltage, and q is the charge of the ion.

To evaluate v_{1z}' and v_{0z_i} we have to solve a more complex set of equations after making the same assumption as before, replacing z_i with its average value $z_0 \equiv \overline{z_i}$. The equation set for ion–atom break-up includes Eqs. (3.11), (3.7) and (3.8), while the ion–ion break-up includes Eq. (3.11) twice, i.e., for both first and second fragments (with relabeling for the second fragment), and Eq. (3.8). These equations are solved numerically with high precision, or can be approximately solved analytically after expanding Eq. (3.11) to first order in u_{1z}. As the numerical

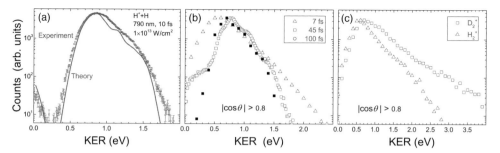

Figure 3.4 (a) Comparison of theoretical predictions of Esry *et al.* (solid line) with our measured KER spectrum (symbols) of H_2^+ dissociation in 10 fs transform-limited pulses centred about 790 nm at a peak intensity of 1×10^{13} W/cm^2. (b) Total dissociation yield as a function of KER for H_2^+, HD$^+$, and D_2^+ in 7 fs transformed-limited pulses at 7.5×10^{15} W/cm^2 scaled to match at their peak. (c) Same as (b) for H_2^+ at different pulse durations of 7, 45 and 100 fs at similar peak intensity 6.5–8.6 $\times 10^{14}$ W/cm^2.

solution converges fast, we favour the more accurate numerical approach when solving for v'_{1_z} and v_{0z_i}.

The extension of both imaging methods described above, namely 'field-free imaging' and 'longitudinal-field imaging', to three fragments or more is straightforward, so we move on to discuss some representative results.

3.3 Benchmark molecules

3.3.1 One electron diatomic molecule – H_2^+

As stated above, the benchmark H_2^+ molecule is a common testing ground for deepening our understanding of molecular response to intense ultrashort laser pulses [4,5]. Despite this intensive research, the complicated multiphoton dynamics of a molecule in a strong laser field continues to unveil fresh surprises even in the simplest H_2^+, as evidenced, for example, by new structure resolved in the laser-induced ionization of this molecule [39, 61]. We have made a concerted effort, together with the theory group of Professor Esry, to study this benchmark system in an attempt to reach agreement between theory and experiment. We have made good progress toward this goal as indicated by the nice match between theory and experiment in Fig. 3.4(a). Our experimental results from this endeavour can be found in [22, 37–39, 45, 62–73], out of which we present a couple of projects to illustrate the physics emerging out of this combined research effort. Specifically, below we briefly describe ATD control and 'vibrational suppression' of H_2^+ by intense ultrashort laser pulses, followed by a brief discussion of future opportunities.

Above-threshold dissociation (ATD)

As mentioned in Section 3.1, ATD involves the absorption of more photons than the minimum required for dissociation. The broad initial vibrational population of H_2^+ obscures clear identification of ATD, as argued by Orr *et al.* [31]. To eliminate this uncertainty, Orr *et al.* trapped an HD^+ beam in an electrostatic storage device [74, 75] until at least 99% of the beam was in the $v = 0$ vibrational state. Then, they crossed it with 40 fs, 800 nm laser pulses and observed $HD^+(v = 0)$ dissociation. Their results at the highest intensity, of about 1×10^{15} W/cm^2, are consistent with the net absorption of 4 or 5 photons, which requires the initial absorption of 5 or 6 photons, clearly involving an ATD process.

The fact that ATD leads to net absorption of fewer photons than required to initiate the dissociation process is typically the case. For example, ATD of $v = 3$ in H_2^+ is initiated at the crossing between the $|1s\sigma_g - 0\omega\rangle$ and $|2p\sigma_u - 3\omega\rangle$ curves (marked by an 'ATD' arrow in Fig. 3.1(b)). However, the adiabatic path, commonly followed by dissociating wave packets, leads to $|1s\sigma_g - 2\omega\rangle$ due to stimulated emission near the crossing between the $|2p\sigma_u - 3\omega\rangle$ and $|1s\sigma_g - 2\omega\rangle$ curves. We estimate the propagation time to the latter crossing, i.e., from $R = 3.3$ a.u. to $R = 4.7$ a.u., to be a few femtoseconds. Therefore, one can switch off, or at least reduce, the stimulated emission rate and by that enhance the net three-photon ATD.

We have accomplished exactly this control scheme by using few cycle laser pulses, 7 and 10 fs long, and have observed an enhancement of high-KER dissociation – the signature of this ATD control [68]. Specifically, we compare the measured KER spectra of H_2^+ dissociation in 7, 45 and 100 fs pulses in Fig. 3.4(b) [68]. The high-KER yield increases for shorter pulses because of the weaker laser field felt by the nuclear wave packet at the second crossing, therefore transitions to the $|1s\sigma_g - 2\omega\rangle$ curve are suppressed [68]. The same result, i.e., control over three-photon ATD, is observed for H_2^+ and D_2^+ dissociation in 10 fs pulses in Fig. 3.4(c). In this case, the high-KER rate is larger for the slower D_2^+ experiencing a weaker laser field at the second crossing [68]. Finally, calculations presented in [68] are in good agreement with the experimental observations. Moreover, they indicate that ATD also happens through the $n = 2$ manifold of H_2^+, which is usually ignored in theoretical treatments.

'Vibrational suppression'

The term 'vibrational suppression' has been used in discussions of vibrational states that have a surprisingly low dissociation probability in a strong laser field. Trapping of high vibrational states in the laser-induced potential well above the crossing between the $|1s\sigma_g - 0\omega\rangle$ and $|2p\sigma_u - 1\omega\rangle$ curves, shown in Fig. 3.1, has been a favoured explanation for the reduced dissociation rate of these states [10–15]. As

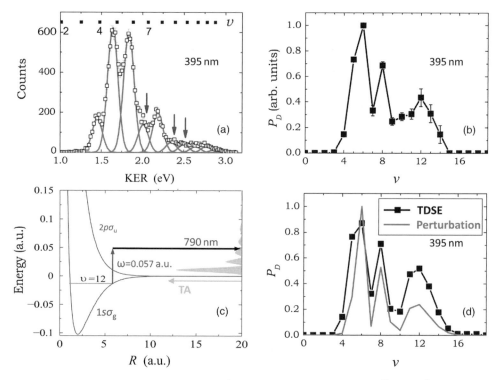

Figure 3.5 (a) KER distribution of H_2^+ dissociation in 40 fs, 3×10^{13} W/cm^2 pulses at 395 nm laser pulses. (b) The dissociation probability of specific vibrational states, P_D derived from the KER spectrum (see text). (c) Schematic transition amplitude (TA) from $1s\sigma_g$ to $2p\sigma_u$ for a given photon energy overlayed on the PEC diagram of H_2^+. (d) Calculated dissociation probability of specific vibrational states, P_D (see text and [72]).

stated in Section 3.1, calculations including nuclear rotation indicate that vibrational trapping, which leads to stabilization, i.e., reduced dissociation probability, is strongly inhibited [16–18].

The measured KER spectra of H_2^+ dissociation in 40 fs, 3×10^{13} W/cm^2 pulses at 790 and 395 nm show clear minima for some of the vibrational states [72]. Specifically, the measured dissociation probability of the $v = 12$ state of H_2^+ seems unusually low at 790 nm [72]. Around the same time as our work in [72], V. S. Prabhudesai, U. Lev, A. Natan and co-workers [76] observed independently this $v = 12$ 'suppressed' dissociation.

The 395 nm data, shown in Fig. 3.5(a,b), has a clearly reduced dissociation probability for a few vibrational states, specifically $v = 7, 9$ and 10. The dissociation probabilities in this figure were evaluated from the KER spectrum in panel (a) by dividing the number of dissociation events associated with each vibrational level by the Franck–Condon population of that vibrational state (see details in [72]).

One may suspect that this reduction in dissociation probability is associated with vibrational trapping, as earlier strong-field work would suggest [4, 5]. However, as discussed in Section 3.1, calculations including nuclear rotation indicate that vibrational trapping is not responsible for this effect. Calculations presented in [72] suggest instead that this surprisingly low dissociation probability is due to small dipole-matrix elements for these specific vibrational levels, i.e., a Cooper minima effect [77]. Essentially, the dipole-matrix elements of the H_2^+ vibrational states oscillate in strength as a function of laser frequency ω for one-photon electronic transitions (see Figure 1 in [72]). In Fig. 3.5(c) we show, for example, the oscillatory behaviour of transitions from the $v = 12$ state. This figure also indicates that excitation of the $v = 12$ state by 790 nm photons lands on a minimum in the transition amplitude as a function of photon energy, therefore leading to its low dissociation probability.

Finally, in Fig. 3.5(d) we show the calculated dissociation probability for 45 fs, 4×10^{12} W/cm^2 pulses at 395 nm (solid squares and line), which match nicely the experimentally determined probabilities in panel (b). Moreover, the normalized perturbation results (solid grey lines) in Fig. 3.5(d) also exhibit the same Cooper minima, thus supporting the assessment that this 'suppression' is not a strong-field phenomena but rather a consequence of the small dipole-matrix elements. The dissociation probability P_D from first-order time-dependent perturbation theory is related to the dipole-matrix elements by $P_D = \int \frac{dP}{dE} dE$, where in the rotating wave approximation (in atomic units)

$$\frac{dP_v}{dE} = \pi \frac{2\ln 2}{(\Delta\omega)^2} I_0 |D_v(E)|^2 \exp\left[-\left(\frac{2\sqrt{\ln 2}(\omega - \omega_{\mathrm{fi}})}{\Delta\omega}\right)^2\right]. \qquad (3.12)$$

In this expression, I_0 is the laser peak intensity, $\omega_{\mathrm{fi}} = E - E_v$, $\Delta\omega$ is the laser bandwidth assuming a Gaussian envelope in time, and $|D_v(E)|$ is the dipole-matrix element (see details in [72]).

Future directions

Our studies of molecular dissociation induced by intense ultrashort laser pulses using molecular-ion beam targets has benefited greatly from the ability to also measure the neutral fragments in coincidence with the routinely measured ions. The main limitation of this method is the very low counting rate caused by the dilute beam target as compared to gas targets. Therefore, it is extremely difficult to conduct pump-probe or two-colour experiments, which have proved very useful in studies of molecular dynamics (see, e.g., [3, 78–81]). Our goal is to conduct, in the near future, such experiments on H_2^+ beams. We have recently taken initial steps toward this goal – hopefully allowing implementation to complex molecular ions next.

3.3.2 Simplest polyatomic molecule – H_3^+

The H_3^+ ion is the simplest polyatomic molecule, therefore one would expect it to be the benchmark system for understanding such molecules in intense ultrashort laser pulses. In contrast to its diatomic counterpart, H_2^+, it cannot be studied by ionizing a neutral parent as H_3 is not a stable molecule. First measurements of H_3^+ fragmentation by intense ultrashort laser pulses have been reported only recently [33, 82], and not because of a lack of earlier efforts – it is just difficult to fragment the H_3^+, as explained below.

The low fragmentation yield of H_3^+ in intense ultrashort laser pulses at 790 nm is mainly due to the high number of photons needed. First, the relatively large energy gap between the ground and first excited potential-energy surfaces sets a minimum of two photons for photofragmentation to occur. Moreover, the vibrational states that may dissociate by absorbing only two photons have hardly any population [48] (see Section 3.2.2). The vibrational states having significant population in H_3^+ require at least four-photon absorption to dissociate, therefore it is not easy to 'break' this molecular ion.

Alexander *et al.* [33] explored the dependence of the H_3^+ dissociation rate on the vibrational population by trapping these ions and measuring their dissociation rate as a function of trapping time. The vibrational population of H_3^+ shifts down with time due to radiative cooling. Therefore, the decrease in dissociation rate as a function of trapping time supports the association of the difficulty to dissociate H_3^+ with its vibrational population.

We reported first measurements of dissociation and ionization of D_3^+ in 7 fs, 790 nm laser pulses with peak intensities in the 10^{14} to 10^{16} W/cm^2 range [82]. Two-body dissociation dominates at the lower intensities, while ionization and Coulomb explosion take over at higher intensities. The coincidence 3D momentum imaging technique, which we implemented for both two- and three-body fragmentation, provides detailed information on the KER and angular distributions in these processes. For example, from the three-body break-up, we have found that D_3^+ is more likely to fragment if the laser field is within the molecular plane (recall that H_3^+ forms an equilateral triangle in its electronic ground state).

Our ongoing studies of H_3^+ and its isotopologues are focused, at present, on (i) the fragmentation dynamics, where we take advantage of the specific fragmentation time of the different isotopologues, as we have done in our ATD studies discussed above (see Section 3.3.1 and [68]), (ii) the competition between two- and three-body dissociation, and (iii) the role of conical intersections present in the excited states of H_3^{2+}, to name a few.

3.4 Complex and/or unique molecular ions

The use of molecular ions as the target for investigation in a strong laser field opens up opportunities for exploring unique molecular systems not available as neutral species. We have already seen one such example in the previous section (Section 3.3.2), namely the benchmark H_3^+ molecular ion, as the H_3 molecule is unstable. Another such example is the simplest two-electron heteronuclear molecule, HeH^+, as the HeH ground state is too fragile (metastable) to be a subject for studies in the gas phase. Metastable molecular ions can also provide unique systems for study as demonstrated below by using CO^{2+} – specifically, only the ground vibrational state of CO^{2+} survives all the way to the interaction region, thus allowing the study of molecular dissociation of a single initial vibrational state.

3.4.1 Vibrationally cold molecular ions – CO^{2+}

As mentioned above, the CO^{2+} molecular-ion target is a unique example for studies of molecular dissociation of a vibrationally cold molecule all the way from weak to strong laser fields.

Similar studies on neutral molecules in the gas phase are limited by the difficulty of detecting neutral fragments, which is circumvented usually by ionizing the molecule or its fragments, therefore limiting studies below the ionization limit. Moreover, in studies of dissociative ionization it is not always easy to distinguish what part of the fragmentation process is associated with the neutral molecule and which with the molecular ion.

Molecular ions are typically formed in ion sources, such as the ECR ion source used in our studies, by fast electron impact ionization of a neutral parent molecule. The creation process involves a vertical transition, which tends to populate the vibrational states of bound electronic states of the molecular ion in a Franck–Condon distribution [42, 43] (though minor deviations are possible [43]). One way to study the fragmentation of molecular ions in their vibrational ground state is to trap these ions and let them cool down to their ground state naturally [42, 74, 75, 83–85]. This is limited, of course, to molecules having a permanent dipole moment. A couple of pioneering measurements employing this approach have been reported recently by Greenwood, Williams and co-workers [31, 33]. They studied multiphoton dissociation of HD^+ [31] and explored the dissociation rate of D_3^+ as a function of the vibrational population [33]. In both cases, the dissociation rate induced by the strong laser field dropped rapidly with vibrational cooling, making the measurements of the dissociation of the $v = 0$ state extremely difficult. Impressively, Orr et al. [31] accomplished these challenging measurements and observed above-threshold dissociation from the ground vibrational state of HD$^+$ by intense 40 fs laser pulses at 800 nm.

We have adopted a different approach to create an electronic and vibrationally cold molecular-ion beam that does not require long-time trapping. Instead, we searched for molecular ions that cool down faster, i.e., on the way from the ion source to the interaction region (about 10 to 20 μs in our set-up). The typical radiative lifetimes of vibrational levels in molecules with a permanent dipole moment are much longer (for example, many milliseconds in HD$^+$ [83,85] and CD$^+$ [86] to 100–1000 μs for H$_3^+$ [84]), therefore we searched for a predissociating molecular ion, such as the CO^{2+} that survives to the interaction region only in the $v = 0$ of the $X\,^3\Pi$ ground state.

As mentioned earlier, the CO^{2+} molecular ion is produced by fast electron impact ionization of CO gas in our ECR ion source in a multitude of electronic states – the lowest lying ones are shown in Fig. 3.6(a). Highly excited electronic states decay rapidly to lower ones, usually even before the molecular ion is extracted from the ion source, thus leaving population only in the two lowest electronic states, specifically the $X\,^3\Pi$ ground state and the $a\,^1\Sigma^+$ first excited state. Note that transitions between these lowest two states are much slower as they belong to different spin multiplets and the transition between them requires a spin flip.

The $X\,^3\Pi$ ground state of CO^{2+} undergoes predissociation through its spin–orbit coupling with the $^3\Sigma^-$ state crossing it around $R = 2.8$ a.u. (see Fig. 3.6) [54, 87–90]. As a result, the lifetime of most vibrational states bound in the $X\,^3\Pi$ potential well are shorter than 1 μs [54, 87–90], i.e., much shorter than the 20 μs flight time from the ion source to the interaction region, therefore only the $v = 0$ of the $X\,^3\Pi$ ground state survives.

The other electronic state that emerges from the ion source, the $a\,^1\Sigma^+$ lowest singlet state, also undergoes predissociation. However, it dissociates in a more complex way, as it is first spin-orbit coupled to the $X\,^3\Pi$ state, which then predissociates to the $^3\Sigma^-$ state as described above [54]. In spite of the two-step process, all the vibrational states bound in the $a\,^1\Sigma^+$ potential also decay before the CO^{2+} arrives to the interaction region, as their lifetimes are shorter than 1 μs [54, 87].

As a result of predissociation in flight to the interaction region, the laser pulses interact only with a well-defined single electronic and vibrational state, the $X\,^3\Pi\,(v = 0)$ of CO^{2+} – the vibrational wave function of which is shown schematically in Fig. 3.6(a). In a way, the CO^{2+} cools down on the way from the ion source by depletion of all the excited states, i.e., as in evaporative cooling.

Given that the laser field couples only electronic states of the same spin multiplet and that only the $X\,^3\Pi\,(v = 0)$ ground state is populated initially we can eliminate

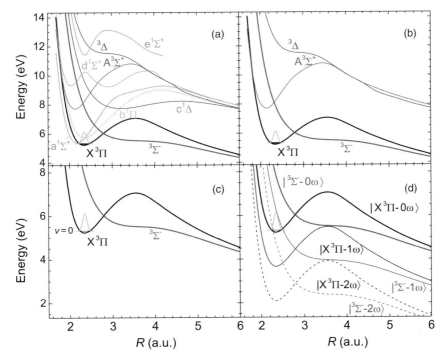

Figure 3.6 (a) Born–Oppenheimer potentials for CO^{2+} calculated by Šedivcová *et al.* [87]. (b) Similar to (a) but only triplet states. (c) Only the lowest two triplet states relevant for the laser intensities used in our studies. (d) The diabatic dressed (Floquet) potentials for the two lowest triplet states. Each curve carries a photon number label. A schematic drawing of the vibrational wave function of the only populated initial state is also shown in all panels.

all the singlet states from the potential-energy diagram, thus simplifying the system one has to deal with (see Fig. 3.6(b)). Moreover, by using moderate laser intensities, i.e., avoiding ionization or excitations to very highly excited states such as the $^3\Delta$, the CO^{2+} dissociation reduces to a two-level problem similar to H_2^+, where only the $1s\sigma_g$ and $2p\sigma_u$ levels are considered. This simplified two-level system is shown in Fig. 3.6(c).

We note that these CO^{2+} states cross each other in the middle of the potential well (at about $R = 2.8$ a.u.) in contrast to merging at the dissociation limit in H_2^+. More interesting is the fact that transitions between these two states of CO^{2+} involve an angular momentum change, $\Delta\Lambda = 1$, while there is no angular momentum change, $\Delta\Lambda = 0$, in H_2^+ transitions. As a result, dipole transitions should peak parallel to the laser polarization for H_2^+ and perpendicular to it for CO^{2+}. It is important to note that perpendicular transitions in molecular fragmentation induced by strong laser fields in the gas phase are seldom observed (see, e.g., [5, 91] for reviews).

To interpret the dissociation in the strong laser field it is convenient to use a dressed-state picture (see Fig. 3.6(d)) as described in the introduction (Section 3.1). We expect one-photon transitions between the $X\,^3\Pi$ and $^3\Sigma^-$ states to dominate the dissociation, as the crossing between the $X\,^3\Pi$ and the $^3\Sigma^- - 1\omega$ states nicely matches the energy of the initial $v = 0$ state making this transition nearly resonant at 790 nm (see Fig. 3.6(d)).

The KER upon dissociation of the $X\,^3\Pi\,(v = 2)$ state of CO^{2+} is known to be 5.65 eV [88]. The energy difference between the $v = 2$ and $v = 0$ in this PEC is evaluated to be 0.349 eV using recent structure calculations [87]. Therefore, the expected KER from the CO^{2+} $v = 0$ ground state is expected to be 5.306 eV. Given that the photon energy used in our measurements was 1.569 eV, one-photon dissociation is expected to have a KER of 6.875 eV, and two-photon dissociation should yield 8.444 eV.

The measured KER spectra for 40 and 7 fs laser pulses with intensities in the 10^{13}–10^{15} W/cm^2 range are shown in Fig. 3.7. As expected, CO^{2+} dissociation is dominated by one-photon transitions from the $X\,^3\Pi\,(v = 0)$ ground state to the $^3\Sigma^-$ state and the resulting KER peaks around the expected value of 6.875 eV (marked by a vertical tick in all panels). This near-resonant one-photon transition (referred to as '1ω') is understood in the dressed-state picture, shown in Fig. 3.6, as leakage of population through the avoided crossing between the $|X\,^3\Pi\,(v = 0) - 0\omega\rangle$ ground state and the $|^3\Sigma^- - 1\omega\rangle$ state – a process known as 'bond softening' [4, 5, 8–10].

A noticeable broadening of this '1ω' peak with increasing intensity of the 40 fs pulses can be seen in Fig. 3.7(a–c). This intensity dependence is masked by the broad bandwidth of the shorter 7 fs pulses (Fig. 3.7(d–f)). It is curious to note that even the narrowest measured one-photon dissociation peak is significantly wider than expected from estimates including the laser bandwidth, experimental resolution, and strong-field effects, which together account for only half the measured width as discussed in [92].

At higher intensities a clear additional KER peak appears around 8.4 eV – spaced by the energy of one IR photon. This peak, marked as '2ω', is a result of dissociation following the absorption of two photons – one extra photon than the minimum needed, which is the signature of above-threshold dissociation (ATD) [4, 5, 8–10]. As suggested in [92], two dissociation pathways contribute to this process. The first is the direct absorption of two photons in a $|X\,^3\Pi\,(v = 0) - 0\omega\rangle \rightarrow |^3\Sigma^- - 2\omega\rangle$ transition (i.e., a transition at the crossing between these two states at the bottom of the $X\,^3\Pi$ potential well; see Fig. 3.6(d)). Alternatively, the CO^{2+} can follow a more complex dissociation path starting at the same one-photon crossing as the dominant '1ω' dissociation process followed by the absorption of another photon along the dissociation path where the transition from the $^3\Sigma^-$ to the $X\,^3\Pi$ is nearly resonant, specifically at the 'crossing' around $R = 3.6$ a.u. shown in Fig. 3.6(d).

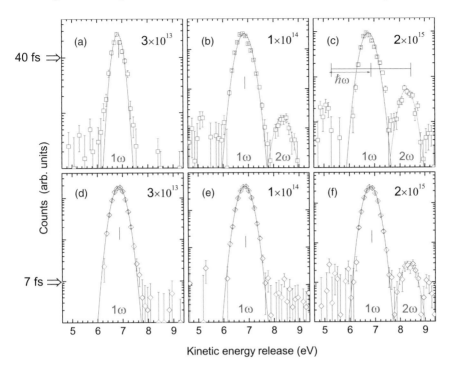

Figure 3.7 Measured KER spectra (symbols) of $CO^{2+} \rightarrow C^+ + O^+$ dissociation induced by intense 790 nm laser pulses with a 40 fs (top) and 7 fs (bottom) pulse duration, respectively. The pulse intensities are marked in W/cm² on each panel. The spectra are arbitrarily scaled to visually match the height of the one-photon dissociation peak. The solid lines show a Gaussian fit to each peak. The ticks in panel (c) mark the expected KER for 0–2 net photon absorption, and the single tick in the other panels marks the position of the dominant one-photon ('1ω') dissociation peak.

This pathway, a net two-photon process, can be written as $|X\,^3\Pi\,(v = 0) - 0\omega\rangle \rightarrow |2\,^1A' - 1\omega\rangle \rightarrow |X\,^3\Pi - 2\omega\rangle$. Clearly this dissociation path requires longer laser pulses to enable the absorption of the second photon after the CO^{2+} stretches to $R = 3.6$ a.u. The relative yield reduction observed for ATD (counts in the '2ω' peak) when using 7 fs pulses as compared with the 40 fs yields suggests that this latter mechanism does contribute significantly for the 40 fs pulses and less for the 7 fs pulses (see [92] for theoretical treatment of these competing mechanisms).

The measured angular distribution of CO^{2+} dissociation into $C^+ + O^+$, shown in Fig. 3.8, is peaked as expected perpendicular to the laser field, that is at $\cos\theta = 0$. This is contrasted in the figure with the angular distribution of the H_2^+ dissociation, which is aligned along the laser polarization as is the case for most molecules dissociating in a strong laser field.

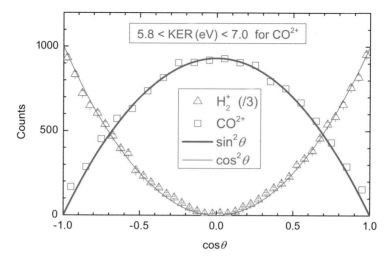

Figure 3.8 Alignment dependence of H_2^+ and CO^{2+} dissociation induced by intense 790 nm, 40 fs, 1×10^{14} W/cm^2 laser pulses. Note that H_2^+ favours dissociation along the laser polarization ($\cos \theta = \pm 1$). In contrast, CO^{2+} favours perpendicular alignment relative to the laser field ($\cos \theta = 0$). The solid lines represent a $\cos^2 \theta$ and $\sin^2 \theta$ fit to the measured H_2^+ and CO^{2+} data, respectively.

The measured angular distributions of H_2^+ and CO^{2+} nicely fit $\cos^2 \theta$ and $\sin^2 \theta$ distributions, respectively, as one would expect for parallel and perpendicular transitions [45, 93]. However, caution is needed when making such assessments as they depend strongly on the initial state of the system (see [92], which shows that the $\sin^2 \theta$ distribution is only an approximate description of the correct distribution resulting from a superposition of different J, m_J initial states).

To summarize, the use of the 'exotic' long-lived metastable CO^{2+} as a target for studies of the interaction of intense ultrashort laser pulses with molecules enabled us to highlight the ATD mechanism and explore the different dissociation pathways involved through their time dependence. This greatly benefited from the fact that the CO^{2+} target was electronically and vibrationally cold.

3.4.2 Vibrationally semi-cold molecular ions – NO^{2+}

The CO^{2+} discussed in the previous section (Section 3.4.1) is not the only molecular ion that can provide some unique features to explore with intense ultrashort laser pulses. Another example we find interesting is the long-lived NO^{2+}. This molecular ion 'cools down' in flight to the interaction region in a similar way as the CO^{2+}. The predissociation of the NO^{2+} electronic ground state, $X\,^2\Sigma^+$, occurs via its spin–orbit coupling with the first excited state, $A\,^2\Pi$, but because the $A\,^2\Pi$ state is

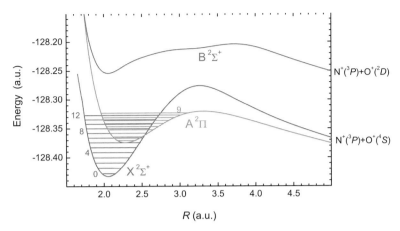

Figure 3.9 Born–Oppenheimer potential-energy curves for the lowest doublet states of NO^{2+} calculated by Baková *et al.* [94]. The vibrational levels shown within the ground state and first excited state were computed using the phase–amplitude method [95]. Note that only $v \leq 12$ states of the NO^{2+} ground state are populated when probed with the laser (see text).

also metastable, only states above $v = 12$ dissociate leaving a truncated vibrational population, shown schematically in Fig. 3.9. The lowest quartet state, $a\,^4\Sigma^+$ (not shown), supports very few vibrational levels, all with very short lifetimes (<20 ps) [94], therefore they decay before exiting the ion source.

At low laser intensities we expect only the lowest two states to play a role in NO^{2+} dissociation. We expect that one 1.569 eV photon should excite most of the populated vibrational levels (for $v \geq 6$) in the $X\,^2\Sigma^+$ potential well to the $A\,^2\Pi$ vibrational continuum, therefore leading to their prompt dissociation. Lower vibrational levels will be excited to the $A\,^2\Pi$ ($v_f < 8$) states, which live long enough to reach the detector before dissociation. One intermediate state, however, the $X\,^2\Sigma^+$ ($v = 5$) state, will be excited to the $A\,^2\Pi$ ($v_f = 8$) or $A\,^2\Pi$ ($v_f = 9$) state, which have lifetimes of 45 ns and 14 ps, respectively, i.e., fast enough to dissociate before detection. Of particular interest are transitions involving the intermediate $A\,^2\Pi$ ($v_f = 8$) state because its lifetime is comparable to the flight time through the field in our spectrometer, and thus we may be able to measure the effect of the strong laser field on its field-free decay rate.

One should note that the one-photon transitions discussed above involve a $\Delta\Lambda = 1$ angular momentum change, therefore they should favour alignment perpendicular to the laser polarization (see Section 3.4.1). In contrast, transitions from the $X\,^2\Sigma^+$ ground state to the highly excited $B\,^2\Sigma^+$ state, which require the absorption of 3 or 4 photons depending on v initial, are expected to be strongly aligned along the laser polarization as $\Delta\Lambda = 0$ in this case. Therefore, we expect dissociation perpendicular to the laser field for low intensities and parallel to the field at high intensities. Preliminary results indicate that these predictions are correct [96], but further work is needed.

3.4.3 *Other complex molecular ions*

The CO^{2+} and NO^{2+} studies, discussed in Sections 3.4.1 and 3.4.2, are good examples of unique molecular systems that one can explore using intense ultrashort laser pulses. Studies of complex molecular-ion beams also benefit from the ability to detect both charged and neutral fragments in coincidence using our imaging technique, and thus provide differential data with high resolution. One recent example of such studies is the measurement of O_2^+ dissociation with vibrational resolution, which allowed us to identify Cooper minima in the kinetic energy release spectra [46]. Moreover, these studies can be extended to polyatomic molecules, more complex than the H_3^+ target presented in Section 3.3.2. For example, we are presently studying the intriguing isomerization process of acetylene cations, $C_2H_2^+$, into vinylidene, i.e., the migration of a hydrogen from one carbon to the other [97–102].

3.5 Summary and outlook

In summary, studies of molecular-ion beams exposed to intense ultrashort laser pulses have been presented, and the unique opportunities they provide have been highlighted. Those include the ability to conduct kinematically complete measurements of dissociation, through the use of the coincidence 3D momentum imaging of beam fragments method, the ability to study the benchmark one-electron H_2^+ molecular ion and the simplest polyatomic molecule, H_3^+. In addition, a couple of other special beam targets, such as the vibrationally cold CO^{2+} have been discussed. Hopefully, this chapter provides the essence of this research effort, as it can by no means represent a complete review of the work done in this field.

In the near future, laser investigations on molecular-ion beam targets can be extended to even more complex systems. Hopefully, the more challenging experiments involving scanning two pulses relative to each other in time, such as pump-probe or two-colour experiments, will become possible with advances in the field.

Acknowledgements   Thanks are due to present and former members of my research group, Dr. Jarlath McKenna, Dr. Pengqian Wang, A. Max Sayler, Bishwanath Gaire, Nora G. Johnson, Mohammad Zohrabi, Utuq Ablikim, Bethany Jochaim and Professor Kevin D. Carnes, for their contributions to the research presented here, and in the case of Mohammad and Kevin for the help in the preparation of this manuscript. Our experimental work benefits greatly from enlightening discussions and the theoretical support of Professor Brett D. Esry and his present and former group members, Fatima Anis, Jianjun Hua, Dustin Ursrey, Dr. Jesus V. Hernández and Dr. Christian Madsen. Thanks are also due to Zenghu Chang and his group members and Charles W. Fehrenbach for assistance with the laser and ion beams, respectively. This work was supported by the Chemical Sciences, Geosciences and Biosciences Division, Office of Basic Energy Sciences, Office of Science, U.S. Department of Energy.

References

[1] J. Itatani *et al.*, *Nature*, **432**, 867 (2004).

[2] M. Meckel *et al.*, *Science*, **320**, 1478 (2008).

[3] M. F. Kling *et al.*, *Science*, **312**, 246 (2006).

[4] A. Giusti-Suzor, F. H. Mies, L. F. DiMauro, E. Charron and B. Yang, *J. Phys. B*, **28**, 309 (1995).

[5] J. H. Posthumus, *Rep. Prog. Phys.*, **67**, 623 (2004).

[6] J. H. Posthumus and J. F. McCann, in *Molecules and Clusters in Intense Laser Fields* ed. J. Posthumus (Cambridge: Cambridge University Press, 2001).

[7] S. I. Chu and D. Telnov, *Phys. Rep.*, **390**, 1 (2004).

[8] P. H. Bucksbaum, A. Zavriyev, H. G. Muller and D. W. Schumacher, *Phys. Rev. Lett.*, **64**, 1883 (1990).

[9] A. Zavriyev, P. H. Bucksbaum, H. G. Muller and D. W. Schumacher, *Phys. Rev. A*, **42**, 5500 (1990).

[10] A. Zavriyev, P. H. Bucksbaum, J. Squier and F. Saline, *Phys. Rev. Lett.*, **70**, 1077 (1993).

[11] A. Giusti-Suzor and F. H. Mies, *Phys. Rev. Lett.*, **68**, 3869 (1992).

[12] A. D. Bandrauk and M. L. Sink, *Chem. Phys. Lett.*, **57**, 569 (1978).

[13] G. Yao and S. I. Chu, *Phys. Rev. A*, **48**, 485 (1993).

[14] E. E. Aubanel, J.-M. Gauthier and A. D. Bandrauk, *Phys. Rev. A*, **48**, 2145 (1993).

[15] L. J. Frasinski *et al.*, *Phys. Rev. Lett.*, **86**, 2541 (2001).

[16] E. E. Aubanel, A. Conjusteau and A. D. Bandrauk, *Phys. Rev. A*, **48**, R4011 (1993).

[17] F. Anis and B. D. Esry, *Phys. Rev. A*, **77**, 033416 (2008).

[18] H. Abou-Rachid, T.-T. Nguyen-Dang and O. Atabek, *J. Chem. Phys.*, **114**, 2197 (2001).

[19] A. Giusti-Suzor, X. He, O. Atabek and F. H. Mies, *Phys. Rev. Lett.*, **64**, 515 (1990).

[20] B. Yang, M. Saeed, L. F. DiMauro, A. Zavriyev and P. H. Bucksbaum, *Phys. Rev. A*, **44**, R1458 (1991).

[21] G. Jolicard and O. Atabek, *Phys. Rev. A*, **46**, 5845 (1992).

[22] I. Ben-Itzhak, in *Progress in Ultrafast Intense Laser Science IV*, Springer Series in Chemical Physics, vol. 91, ed. K. Yamanouchi, A. Becker, R. Li and S. L. Chin (New York: Springer, 2009), p. 67.

[23] K. Sändig, H. Figger and T. W. Hänsch, *Phys. Rev. Lett.*, **85**, 4876 (2000).

[24] I. D. Williams *et al.*, *J. Phys. B*, **33**, 2743 (2000).

[25] A. Assion, T. Baumert, U. Weichmann and G. Gerber, *Phys. Rev. Lett.*, **86**, 5695 (2001).

[26] J. B. Greenwood *et al.*, *Phys. Rev. Lett.*, **88**, 233001 (2002).

[27] D. Pavičić, A. Kiess, T. W. Hänsch and H. Figger, *Euro. Phys. J. D*, **26**, 39 (2003).
[28] W. R. Newell, I. D. Williams and W. A. Bryan, *Euro. Phys. J. D*, **26**, 99 (2003).
[29] D. Pavičić, A. Kiess, T. W. Hänsch and H. Figger, *Phys. Rev. Lett.*, **94**, 163002 (2005).
[30] I. D. Williams *et al.*, *Phys. Rev. Lett.*, **99**, 173002 (2007).
[31] P. A. Orr *et al.*, *Phys. Rev. Lett.*, **98**, 163001 (2007).
[32] A. Kiess, D. Pavičić, T. W. Hänsch and H. Figger, *Phys. Rev. A*, **77**, 053401 (2008).
[33] J. D. Alexander *et al.*, *J. Phys. B*, **42**, 141004 (2009).
[34] V. S. Prabhudesai *et al.*, *Phys. Rev. A*, **81**, 023401 (2010).
[35] F. Baumgartner and H. Helm, *Phys. Rev. Lett.*, **104**, 103002 (2010).
[36] A. Natan *et al.*, submitted.
[37] I. Ben-Itzhak *et al.*, *Phys. Rev. Lett.*, **95**, 073002 (2005).
[38] P. Q. Wang *et al.*, *J. Phys. B*, **38**, L251 (2005).
[39] B. D. Esry, A. M. Sayler, P. Q. Wang, K. D. Carnes and I. Ben-Itzhak, *Phys. Rev. Lett.*, **97**, 013003 (2006).
[40] H. Mashiko *et al.*, *Appl. Phys. Lett.*, **90**, 161114 (2007).
[41] D. J. Kane and R. Trebino, *IEEE J. Quan. Electron.*, **29**, 571 (1993).
[42] Z. Amitay *et al.*, *Phys. Rev. A*, **60**, 3769 (1999).
[43] F. von Busch and G. H. Dunn, *Phys. Rev. A*, **5**, 1726 (1972).
[44] A. Tabch-Fouhaill *et al.*, *J. Chem. Phys.*, **17**, 81 (1976).
[45] A. M. Sayler, P. Q. Wang, K. D. Carnes, B. D. Esry and I. Ben-Itzhak, *Phys. Rev. A*, **75**, 063420 (2007).
[46] M. Zohrabi *et al.*, *Phys. Rev. A*, **83**, 053405 (2011).
[47] W. J. van Der Zande, W. Koot, J. R. Peterson and J. Los, *J. Chem. Phys.*, **126**, 169 (1988).
[48] V. G. Anicich and J. H. Futrell, *Int. J. Mass. Spectrosc. Ion Phys.*, **55**, 189 (1983).
[49] R. Dörner *et al.*, *Phys. Rep.*, **330**, 95 (2000).
[50] J. Ullrich *et al.*, *Rep. Prog. Phys.*, **66**, 1463 (2003).
[51] J. Moseley and J. Durup, *Ann. Rev. Phys. Chern.*, **32**, 53 (1981).
[52] H. Helm, D. P. de Bruijn and J. Los, *Phys. Rev. Lett.*, **53**, 1642 (1984).
[53] H. Helm and P. C. Cosby, *J. Chem. Phys.*, **86**, 6813 (1987).
[54] J. P. Bouhnik *et al.*, *Phys. Rev. A*, **63**, 032509 (2001).
[55] X. Urbain *et al.*, *Phys. Rev. Lett.*, **92**, 163004 (2004).
[56] L. Ph. H. Schmidt *et al.*, *Phys. Rev. Lett.*, **101**, 173202 (2008).
[57] B. Gaire *et al.*, *Phys. Rev. A*, **78**, 033430 (2008).
[58] A. M. Sayler, unpublished PhD thesis, Kansas State University (2008).
[59] B. Gaire, unpublished PhD thesis, Kansas State University (2011).
[60] D. A. Dahl, SIMION 3D Version 8.0, Scientific Instrument Services, Inc., http://simion.com/.
[61] A. Staudte *et al.*, *Phys. Rev. Lett.*, **98**, 073003 (2007).
[62] P. Wang, A. M. Sayler, K. D. Carnes, B. D. Esry and I. Ben-Itzhak, *Opt. Lett.*, **30**, 664 (2005).
[63] I. Ben-Itzhak *et al.*, *Nucl. Instrum. Meth. B*, **233**, 56 (2005).
[64] P. Q. Wang *et al.*, *Phys. Rev. A*, **74**, 043411 (2006).
[65] B. Gaire *et al.*, *Rev. Sci. Instrum.*, **78**, 024503 (2007).
[66] A. M. Sayler, P. Q. Wang, K. D. Carnes and I. Ben-Itzhak, *J. Phys. B*, **40**, 4367 (2007).
[67] I. Ben-Itzhak *et al.*, *J. Phys. Conf. Ser.*, **88**, 012046 (2007).
[68] J. McKenna *et al.*, *Phys. Rev. Lett.*, **100**, 133001 (2008).
[69] I. Ben-Itzhak *et al.*, *Phys. Rev. A*, **78**, 063419 (2008).

[70] J. McKenna *et al.*, *J. Phys. B*, **42**, 21003 (2009).
[71] J. McKenna *et al.*, *Phys. Rev. A*, **80**, 023421 (2009).
[72] J. McKenna *et al.*, *Phys. Rev. Lett.*, **103**, 103006 (2009).
[73] W. Wolff *et al.*, *J. Instrum.*, **5**, 10006 (2010).
[74] D. Zajfman *et al.*, *Phys. Rev. A.*, **55**, R1577 (1997).
[75] M. Dahan *et al.*, *Rev. Sci. Instrum.*, **69**, 76 (1998).
[76] V. S. Prabhudesai *et al.*, private communication (2009).
[77] J. W. Cooper, *Phys. Rev.*, **128**, 681 (1962).
[78] A. S. Alnaser *et al.*, *Phys. Rev. Lett.*, **91**, 163002 (2003).
[79] Th. Ergler *et al.*, *Phys. Rev. Lett.*, **97**, 193001 (2006).
[80] E. Gagnon *et al.*, *Science*, **317**, 1374 (2007).
[81] M. Kremer *et al.*, *Phys. Rev. Lett.*, **103**, 213003 (2006).
[82] J. McKenna *et al.*, *Phys. Rev. Lett.*, **103**, 103004 (2009).
[83] P. Forck *et al.*, *Phys. Rev. Lett.*, **70**, 426 (1993).
[84] M. Larsson *et al.*, *Phys. Rev. Lett.*, **70**, 430 (1993).
[85] T. Tanabe *et al.*, *Phys. Rev. Lett.*, **75**, 1066 (1995).
[86] F. R. Ornellas and F. B. C. Machado, *J. Chem. Phys.*, **84**, 1296 (1986).
[87] T. Šedivcová, P. R. Ždánská, V. Špirko and J. Fišer, *J. Chem. Phys.*, **124**, 214303 (2006); and private communication.
[88] M. Lundqvist, P. Baltzer, D. Edvardsson, L. Karlsson and B. Wannberg, *Phys. Rev. Lett.*, **75**, 1058 (1995).
[89] M. Hochlaf *et al.*, *J. Chem. Phys.*, **207**, 159 (1996).
[90] F. Penent *et al.*, *Phys. Rev. Lett.*, **81**, 3619 (1998).
[91] K. Codling and L. J. Frasinski, *J. Phys. B*, **26**, 783 (1993).
[92] J. McKenna *et al.*, *Phys. Rev. A*, **81**, R061401 (2010).
[93] A. Hishikawa, S. Liu, A. Iwasaki and K. Yamanouchi, *J. Chem. Phys.*, **114**, 9856 (2001).
[94] R. Baková, J. Fišer, T. Šedivcová-Uhlíková and V. Špirko, *J. Chem. Phys.*, **128**, 144301 (2008); and private communication.
[95] E. Y. Sidky and I. Ben-Itzhak, *Phys. Rev. A*, **60**, 3586 (1999).
[96] B. Jochim *et al.*, ICPEAC proceedings in *J. Phys. Conf. Ser.* (2011).
[97] J. Levin *et al.*, *Phys. Rev. Lett.*, **81**, 3347 (1998).
[98] T. Osipov *et al.*, *Phys. Rev. Lett.*, **90**, 233002 (2003).
[99] A. Hishikawa, E. J. Takahashi and A. Matsuda, *Phys. Rev. Lett.*, **97**, 243002 (2006).
[100] A. Hishikawa, A. Matsuda, M. Fushitani and E. J. Takahashi, *Phys. Rev. Lett.*, **99**, 258302 (2007).
[101] Y. H. Jiang *et al.*, *Phys. Rev. Lett.*, **105**, 263002 (2010).
[102] A. Matsuda, M. Fushitani, E. J. Takahashi and A. Hishikawa, *Phys. Chem. Chem. Phys.*, DOI: 10.1039/c0cp02333g (2011).

4

Atoms with one and two active electrons in strong laser fields

I. A. IVANOV AND A. S. KHEIFETS

4.1 Introduction

Recent years have witnessed a remarkable progress in high-power short laser pulse generation. Modern conventional and free-electron laser (FEL) systems provide peak light intensities of the order of 10^{20} W cm^{-2} or above in pulses in femtosecond and sub-femtosecond regimes. The field strength at these intensities is a hundred times the Coulomb field, binding the ground-state electron in the hydrogen atom. These extreme photon densities allow highly non-linear multiphoton processes, such as above-threshold ionization (ATI), high harmonic generation (HHG), laser-induced tunneling, multiple ionization and others, where up to a few hundred photons can be absorbed from the laser field. In parallel with these experimental developments, massive efforts have been undertaken to unveil the precise physical mechanisms behind multiphoton ionization (MPI) and other strong-field ionization phenomena. It was shown convincingly that multiple ionization of atoms by an ultrashort intense laser pulse is a process in which the highly non-linear interaction between the electrons and the external field is closely interrelated with the few-body correlated dynamics [1]. A theoretical description of such processes requires development of new theoretical methods to simultaneously account for the field nonlinearity and the long-ranged Coulomb interaction between the particles.

In this chapter, we review our recent theoretical work in which we develop explicitly time-dependent, non-perturbative methods to treat MPI processes in many-electron atoms. These methods are based on numerical solution of the time-dependent Schrödinger equation (TDSE) for a target atom or molecule in the presence of an electromagnetic and/or static electric field. Projecting this solution onto final field-free target states gives us probabilities and cross sections for various ionization channels.

Fragmentation Processes: Topics in Atomic and Molecular Physics, ed. Colm T. Whelan. Published by Cambridge University Press. © Cambridge University Press 2013.

The chapter is organized as follows. In Section 4.2 we present the key ingredients of our formalism and its numerical implementation. The next sections cover various applications:

Section 4.3: Two-photon double ionization of helium

Section 4.4: DC-assisted double photoionization of He and H$^-$

Section 4.6: ATI and HHG in quasi one- and two-electron atoms

Section 4.5: Strong-field ionization of lithium

Section 4.7: Time delay in atomic photoemission

4.2 Theoretical model

We seek a solution of the time-dependent Schrödinger equation for an atom in the presence of an external electromagnetic (EM) field:

$$i\frac{\partial \Psi}{\partial t} = \left(\hat{H}_{\text{atom}} + \hat{H}_{\text{int}}(t)\right)\Psi, \tag{4.1}$$

where $\hat{H}_{\text{atom}}$ is the field-free Hamiltonian and the operator $\hat{H}_{\text{int}}(t)$ describes interaction of the atom and the EM field. The Hamiltonian of the field-free atomic system can be a completely *ab initio* operator as in the cases of the hydrogen and helium atoms or the hydrogen molecule. For many-electron targets, we apply a frozen core approximation and freeze all the atomic electrons except the valence shell which can contain one or two electrons. This reduces the atomic Hamiltonian to an effective one- or two-body operator. The interaction Hamiltonian $\hat{H}_{\text{int}}(t)$ can be written in various gauges, which are all formally equivalent. We will be using the length and velocity gauges that take the following forms for an n-electron target:

$$\hat{H}_{\text{int}}(t) = \begin{cases} \boldsymbol{E}(t) \cdot \sum_{i=1}^{n} \boldsymbol{r}_i, & \text{length} \\ \boldsymbol{A}(t) \cdot \sum_{i=1}^{n} \hat{\boldsymbol{p}}_i, & \text{velocity.} \end{cases} \tag{4.2}$$

Here the vector potential and the electric field are related via

$$\boldsymbol{A}(t) = -\int_0^t \boldsymbol{E}(\tau)\,d\tau.$$

The time dependence of the electric field is chosen to be

$$\boldsymbol{E}(t) = f(t)\boldsymbol{E}_0 \cos(\omega t + \phi),$$

where ϕ is a carrier envelope phase (CEP), ω is the carrier frequency, and $f(t)$ is an envelope function that is smooth and slowly varying over the interval $(0, T_1)$, thus ensuring that no artificial transient effects are introduced in the calculation. The field is switched off for $t < 0$ and $t > T_1$. In the case of a long pulse $T_1 \gg T = 2\pi/\omega$, neither a precise form of $f(t)$ nor the value of the CEP are important. For short pulses $T_1 \gtrsim T$, the pulse shape and the CEP both have a considerable effect.

We seek a solution of the TDSE on the basis of one- or two-electron wave functions

$$\Psi(r, t) = \sum_{j \equiv \{nlm\} \notin \text{core}} a_j(t) R_{nl}(r) Y_{lm}(\theta, \phi) \tag{4.3}$$

$$\Psi(r_1, r_2, t) = \sum_{j \equiv \{n_1 n_2 l_1 l_2 JM\} \notin \text{core}} a_j(t) \left[1 + \hat{P}_{12} \right] R_{n_1 l_1}(r_1) R_{n_2 l_2}(r_2) |l_1(1) l_2(2) \, JM\rangle \tag{4.4}$$

for atomic systems with one and two valence electrons, respectively. In the one-electron case, the angular dependence is carried by the spherical functions $Y_{lm}(\theta, \phi)$, whereas in the two-electron case it is assumed by the bipolar harmonics $|l_1(1)l_2(2) \, JM\rangle$ [2]. The spatial exchange operator $1 + \hat{P}_{12}$ ensures the proper symmetrization of the wave function (4.4). Convergence of expansions (4.3–4.4) with respect to the angular momenta depends on the nature of the problem at hand and the ionization regime that we consider. It is known that this convergence is generally faster if the velocity gauge is employed for the interaction Hamiltonian [3].

The radial orbitals enetering Eqs. (4.3–4.4) are represented by a square-integrable L^2 basis. This basis can be formed in two different ways. One way is to use a set of B-splines of a certain order ($k = 7$) with the knots located on a sequence of points lying in $[0, R_{\text{max}}]$. Another way is to build a set of positive and negative energy pseudostates of size N, which diagonalizes the target Hamiltonian. A basis-based calculation converges when N is sufficiently large.

As an alternative computational strategy, we may discretize the Hamiltonian and the wave functions on a radial grid of sufficient density and extent R_{max} and seek a direct solution in the form:

$$\Psi(r, t) = \sum_{lm} R_{lm}(r, t) Y_{lm}(\theta, \phi) \tag{4.5}$$

$$\Psi(r_1, r_2, t) = \sum_{l_1 l_2 JM} R_{l_1 l_2 JM}(r_1, r_2, t) |l_1(1) l_2(2) JM\rangle, \tag{4.6}$$

Both grid-based and basis methods of finding the radial solution of the TDSE give us a coupled system of differential equations. The TDSE is solved directly relative to the functions $R_{lm}(r, t)$ or $R_{l_1 l_2 JM}(r_1, r_2, t)$ if we use a grid-based technique, Eqs. (4.5–4.6), or the coefficients of the basis set expansions are obtained if we employ a basis set method, Eqs. (4.3–4.4). In both cases, we can write this system of differential equations in a vector form:

$$i\dot{a} = (H_{\text{atom}} + H_{\text{int}}) \cdot a, \tag{4.7}$$

where H_{atom} and $H_{\mathrm{int}}(t)$ are matrices of the atomic Hamiltonian and the operator of electromagnetic interaction, respectively. A short time propagator for this system can be obtained by using the leading term of the Magnus expansion [4]

$$a(t + \Delta) \approx \exp\left[-i\int_t^{t+\Delta} H(\tau)\,d\tau\right]$$
$$\approx \exp\left[-iH(t + \Delta/2)\Delta\right]a(t) + o(\Delta^2).$$

To compute an exponential of a large matrix efficiently, we employ the so-called Arnoldi–Lanczos method (ALM) [5, 6]. This technique represents the vector $a(t + \Delta)$ as a sum of vectors $H(t)a(t),\ H^2(t)a(t),\ \ldots H^m(t)a(t)$, forming the Krylov subspace. The procedure is unconditionally stable and explicit, which allows us to treat large-scale computational problems efficiently. As an alternative approach, we can approximate the exponential operator in the Magnus expansion by $[1 - iH(t + \Delta/2)\Delta/2][1 + iH(t + \Delta/2)\Delta/2]^{-1}$ with the same accuracy to within the terms of the order of Δ^2. To evaluate the matrix inverse, we employ the so-called matrix iteration method (MIM) [7], which makes use of the fact that the operator in the denominator can be split as: $1 + iH(t + \Delta/2)\Delta/2 = A + B$, where $A = 1 + iH_{\mathrm{atom}}\Delta/2$ and $B = iH_{\mathrm{int}}(t + \Delta/2)\Delta/2$, and expanded into the Neumann series:

$$(A + B)^{-1} = A^{-1} - A^{-1}BA^{-1} + A^{-1}BA^{-1}BA^{-1}\ldots \qquad (4.8)$$

Terms of this expansion are easily computed to any order. On every step, one has to compute only an inverse of the operator A, which can be done fast and efficiently.

The bound atomic or molecular state, which gives the initial condition for this system of differential equations, is obtained by using a relaxation procedure in imaginary time for the grid-based calculation, or is prepared by direct diagonalization of the Hamiltonian in the case of a basis set calculation.

Knowing the solution of the TDSE after the end of the pulse $t = T_1$, we obtain various differential ionization probabilities by projecting this solution on a set of states describing the given ionization channel. This procedure is fairly straightforward in the case of an effective one-electron system where the continuum wave functions can be easily calculated. It is much more difficult in the case of double ionization of the helium atom or hydrogen molecule, where continuous spectra corresponding to single and double ionization may overlap. This issue is addressed in the corresponding sections below.

4.3 Two-photon double ionization of helium

The direct (non-sequential) two-photon double-electron ionization (TPDI) of helium is the simplest and the most fundamental strong-field ionization process with several active electrons, which requires a non-perturbative treatment of the external field as well as a proper account of correlation in the two-electron continuum. Because of the canonical importance of this process, a large number of theoretical methods have been developed and applied to TPDI of He in recent years. These studies allowed us to achieve considerable progress in understanding the qualitative features of the TDPI phenomenon. However, as far as the quantitative description of TDPI is concerned, the available theoretical results paint a somewhat controversial picture. Even though the total TPDI cross section has been measured in He by using the HHG [8, 9] or FEL [10] sources of radiation, the experimental results still remain debatable [11].

In our approach to the TPDI process in He [12], we seek a solution of the two-electron TDSE in the form of the expansion (4.4) in which functions $R_{l_1 l_2 J M}(r_1, r_2, t)$ are expanded on a basis built from the one-electron radial orbitals $\phi_{nl}^N(r)$. The latter are obtained by diagonalizing the He$^+$ Hamiltonian in a Laguerre basis of size N [13]:

$$\langle \phi_{nl}^N | \hat{H}_{\mathrm{He}^+} | \phi_{n'l'}^N \rangle = E_{nl} \delta_{nn'} \delta_{ll'}. \tag{4.9}$$

In the present work, we consider an electric field of the order of 0.1 a.u. corresponding to 3.5×10^{14} W/cm^2 intensity. For this not very high field intensity, we can retain in the expansion (4.4) only the terms with total angular momentum $J = 0$–2. To represent each total angular momentum block, we proceed as follows. For all S, P, D total angular momentum states we let l_1, l_2 vary within the limits 0–3. The total number of pseudostates participating in building the basis states was 20 for each l. To represent $J = 0, 1, 2$ singlet states in expansion (4.4), we used all possible combinations of these pseudostates. Such a choice gave us 840 basis states of S-symmetry, 1200 basis states of P-symmetry and 1430 states of D-symmetry, resulting in a total dimension of the basis equal to 3470. Issues related to the convergence of the calculation with respect to the variations of the composition of the basis set are described in detail in [14].

Initial conditions for the solution of TDSE are determined by solving an eigenvalue problem using a subset of basis functions of the S-symmetry only. This produces the ground-state energy of -2.90330 a.u. We integrate TDSE up to a time T_1 when the external field is switched off. Then we project the solution onto a field-free CCC wave function $\Psi(\boldsymbol{k}_1, \boldsymbol{k}_2)$ representing electron scattering on the He$^+$ ion. Details of the construction of these functions can be found in [15]. This projection gives us a probability distribution function $p(\boldsymbol{k}_1, \boldsymbol{k}_2)$ of finding the helium atom in a field-free two-electron continuum state $\boldsymbol{k}_1, \boldsymbol{k}_2$ at the time $t = T_1$. From this probability, we can compute various differential and total integrated cross

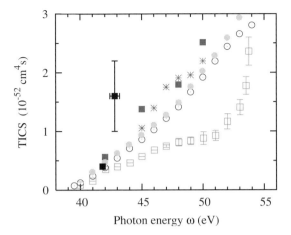

Figure 4.1 Total integrated cross section of TPDI of helium as a function of photon energy. The present TDSE-CCC results (in velocity gauge) are displayed with filled squares. The asterisks display an analogous TDSE calculation with an uncorrelated final state. The literature data are plotted with the following symbols: ECS [16] – open squares, FEDVR-a [17] – open circles, FEDVR-b [18] – filled circles. The experimental data (filled squares) are from [8, 10].

sections (TICS) of TPDI. The latter is computed as

$$\sigma(\omega) = \frac{C}{W} \int p(\mathbf{k}_1, \mathbf{k}_2) \, d\hat{\mathbf{k}}_1 d\hat{\mathbf{k}}_2 dk_1 dk_2. \tag{4.10}$$

Here $W = \int_0^{T_1} E^4(t)dt$, and $C = 12\pi^2 a_0^4 \tau \omega^2 c^{-2}$ is the TPDI constant expressed in terms of the speed of light in atomic units $c \approx 137$, the Bohr radius $a_0 = 0.529 \times 10^{-8}$ cm and the atomic unit of time $\tau = 2.418 \times 10^{-17}$ s.

The TICS results for TPDI of He are presented in Fig. 4.1, along with the most recent calculations reported in the literature: exterior complex scaling (ECS) [16], finite element discrete variable representation (FEDVR-a) [17] and (FEDVR-b) [18] and the experiment [8, 10]. In the same figure, we also plot an alternative set of our results obtained by projecting the solution of the TDSE on the uncorrelated two-electron final state. This state is represented by the product of the two Coulomb waves. Both sets of our TDSE calculations are fairly close. The issue of the final-state correlation in the TPDI process is discussed in detail in [19].

4.4 DC-assisted double photoionization of He and H⁻

Single-photon double ionization (double photoionization – DPI) of helium has been studied extensively over the past decade. Basic mechanisms of this process are now well understood, both qualitatively and quantitatively, with accurate theoretical predictions being confirmed experimentally under a wide range of kinematical conditions [20, 21]. The emphasis in DPI studies is now shifting towards the multiphoton processes in stronger electromagnetic fields or/and more complex atomic and molecular targets where electron correlation may play a more prominent role. In our work [14], we introduced another factor which would complicate the DPI process: a static electric field. We consider the DPI of helium subjected to an external DC field with the strength ranging from a few hundreds to a few tens of the atomic unit. Since the two-electron escape is a 'balancing act' between the inter-electron repulsion and the nucleus drag, the static field may upset this delicate balance or open up new possible two-electron escape routes. This can result in a net decrease or increase of the DPI cross section and changing energy and angular distribution of the photoelectron pair.

The DC-assisted DPI of He is described by the TDSE containing interaction with the external electromagnetic (AC) and static (DC) fields:

$$\hat{H}_{\text{int}}(t) = f(t)(\boldsymbol{r}_1 + \boldsymbol{r}_2) \cdot (\boldsymbol{E}_{\text{AC}} \cos \omega t + \boldsymbol{E}_{\text{DC}}). \tag{4.11}$$

For simplicity, we consider the case when both the AC and DC fields are parallel and controlled by the same smooth switching function $f(t)$, which turns them on and off during one period of the AC field oscillation $T = 2\pi/\omega$ and keeps them constant on the time interval $(T, 4T)$. The total duration of the atom-field interaction is therefore $T_1 = 6T$. Solution of the TDSE is sought on the pseudostate Laguerre basis [13]. The field-free evolution of the two-electron continuum is described by the CCC wave function $\Psi(\boldsymbol{k}_1, \boldsymbol{k}_2)$ representing electron scattering on the He$^+$ ion [15]. By projecting the solution of the TDSE on this function, we obtain a probability distribution function $p(\boldsymbol{k}_1, \boldsymbol{k}_2)$ of finding the helium atom in a field-free state $(\boldsymbol{k}_1, \boldsymbol{k}_2)$ at the time $t = T_1$. The DPI cross section is related to the distribution function $p(\boldsymbol{k}_1, \boldsymbol{k}_2)$ normalized to the field intensity:

$$\sigma(\boldsymbol{k}_1, \boldsymbol{k}_2) = \frac{8\pi\omega}{c} \frac{p(\boldsymbol{k}_1, \boldsymbol{k}_2)}{W}, \tag{4.12}$$

where $W = 2 \int_0^{T_1} E_{\text{AC}}^2(t) dt$ and $c \approx 137$ is the speed of light in atomic units. The total integrated cross section (TICS) is given by:

$$\sigma(\omega) = \frac{1}{2} \int \sigma(\boldsymbol{k}_1, \boldsymbol{k}_2) \, d\hat{\boldsymbol{k}}_1 d\hat{\boldsymbol{k}}_2 dk_1 dk_2. \tag{4.13}$$

In the present work, we consider modestly strong electric fields: the AC field of the order of 0.1 a.u. corresponding to 3.5×10^{14} W/cm^2 intensity, and the DC field not exceeding 0.03 a.u. This allows us to retain terms with total angular momentum

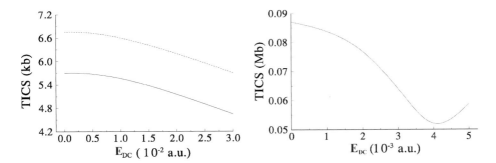

Figure 4.2 Left: TICS of DPI of helium as a function of an external DC field for photon energies of 85 eV (bottom line) and 90 eV (top line). Right: same for H⁻ at the photon energy of 15 eV.

$J = 0 - 2$ in expansion (4.4). To represent each total angular momentum block, we proceed as in the case of the TPDI calculation described in the previous Section 4.3.

On the left panel of Fig. 4.2 we display the TICS results for DPI at photon energies of 85 eV (bottom curve) and 90 eV (top curve) and various applied DC field strengths. For both frequencies, the TICS exhibits a decrease with the DC field. Such a behaviour is not uncommon for two-electron systems in the external DC electric field when there are alternative routes of decay. We documented a similar behaviour of the total (single plus double) photoionization cross section as a function of the external DC field for helium [22].

A much weaker bound system, H⁻, displays a qualitatively different DC field dependence of TICS [23]. This dependence is displayed on the right panel of Fig. 4.2 for the photon energy of 15 eV. As is seen from this figure, the application of even a relatively weak DC field to H⁻ can produce a considerable change in TICS. As a function of the DC field strength, the TICS reaches a minimum at approximately $E_{DC} = 4 \times 10^{-3}$ a.u. and then starts to grow. The growth of TICS indicates that tunneling ionization due to the DC field is becoming a dominant process. The role of electron correlations at this stage is not very significant. Indeed, an ionization probability behaviour similar to Fig. 4.2 can be obtained already in a simple Keldysh-type model of ionization in the presence of both DC and AC fields [23].

4.5 Strong-field ionization of lithium and hydrogen

Various regimes of strong-field ionization can be conveniently categorized by the adiabaticity Keldysh parameter, which relates the frequencies of atomic motion and the laser field, $\gamma = \omega/\omega_{\text{tunnel}}$ [24]. Alternatively, the Keldysh parameter can be expressed in terms of the atomic ionization potential *IP* and the ponderomotive potential U_p, $\gamma = \sqrt{IP/2U_p}$. The MPI regime is characterized by $\gamma \gg 1$, where the characteristic tunneling time of the atomic electron over the Coulomb barrier $\omega_{\text{tunnel}}^{-1}$ is much larger than the timescale ω^{-1} on which electromagnetic field varies considerably. Such a fast ionization process should be described using the quantum-mechanical language of simultaneous absorption of several laser photons. The opposite limit of $\gamma \ll 1$ is reached when the laser field is changing slowly as compared to the characteristic tunneling time. Such a slow adiabatic process can be described quasi-classically using the language of field strength and electron trajectories driven by this field [25].

With increasing field strength and intensity, the width and height of the atomic Coulomb barrier is reduced until it is completely suppressed by the external field. Such a barrier suppression takes place independently of the value of the Keldysh parameter. The lithium atom driven by a femtosecond laser in the near infrared (NIR) spectral range exhibits an unusual example of barrier suppression in the entirely quantum MPI regime. Such a process cannot be treated quasi-classically and analyzed in convenient terms of competing electron trajectories. Instead, a full quantum-mechanical treatment should be given. Because of the large field intensity, such a treatment should be non-perturbative and explicitly time-dependent. Given a large number of field oscillations in the laser pulse and the complexity of the target, an accurate theoretical description of the MPI of Li becomes a challenging task.

In our work [26], we met this challenge by seeking a grid-based solution of the one-electron TDSE by the ALM method, with the local Hamiltonian furnished by an optimized effective potential [27]. We used the velocity gauge to describe the atom-field interaction. The system was enclosed in a box of size $R_{\text{max}} = 2000$ a.u. On the outer boundary of the box the transparent boundary condition was imposed ensuring that edge effects do not appear. The wave function was represented as a series in spherical harmonics (4.3). We included terms with angular momenta up to $\ell_{\text{max}} = 20$ in this expansion. The electron distribution functions were computed by projecting the solution of the TDSE after the end of the pulse on the ingoing distorted waves calculated in the same effective potential.

Results of our calculation are presented in Fig. 4.3 in the form of the photoelectron angular distribution in the polarization plane of laser light. In the momentum distributions shown in this figure, the nominal position of the four-photon line is marked by a dashed semicircle. Four is the minimal number of laser photons $\omega = 1.58$ eV to bridge the ionization potential of the lithium atom $IP = 5.39$ eV. At larger intensities, an additional ring-like structure can be identified due to the five-photon absorption, which is a clear sign of the ATI process. In this process, the photoelectron continues to absorb laser photons when it is already in the ionization continuum. Overall, we see a good agreement between the calculated and experimental spectra in the field intensity range of nearly an order of

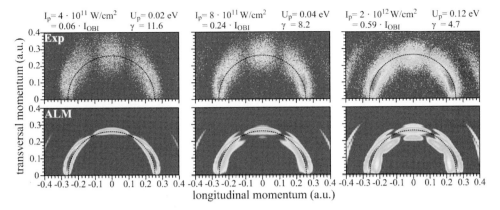

Figure 4.3 Electron momentum distribution parallel (longitudinal) and transversal (perpendicular) to the laser polarization direction. Top row: experimental data obtained by recoil-ion momentum detection. Bottom row: calculated spectra using the ALM model. The laser pulse peak intensity I_p, the ponderomotive potential U_p, and the Keldysh parameter γ are given above each column. The colour scale of the momentum spectra is logarithmic. Dashed semicircle visualizes the nominal position of the four-photon line.

magnitude. Another example of a high non-linearity ionization process is presented by the hydrogen atom driven by an intense NIR laser light [28]. With the intensities reaching 5×10^{14} W/cm^2, the ATI process takes place with as many as 50 photons being readily absorbed from the laser field. Theoretical description of such a process necessitates a complex numerical simulation in order to achieve a quantitative agreement with the experiment, which was documented at the 10% level in the present case.

In our approach to this problem, as in the case of strong-field ionization of lithium, we employed the Arnoldi–Lanczos propagator scheme. The radial orbitals were defined in a box of size $R_{\mathrm{max}} = 2000$ a.u. With the velocity gauge interaction, the terms with $l < \ell_{\mathrm{max}} = 25$ had to be kept in the expansion (4.3) to ensure convergence.

Our results are presented in Fig. 4.4. The left panel displays an example of a calculated ATI spectrum corresponding to the laser peak intensity of 4×10^{14} W/cm^2. Each oscillation on this curve corresponds to a single photon absorption. The theoretical data, averaged over the transverse Gaussian profile of the laser beam, are displayed on the right panel for a range of field intensities in comparison with the experimental data [28]. The full set of experimental data points is simultaneously fitted using only two adjustable parameters: an overall scaling factor and an intensity scaling factor. The former factor accounts for the absolute detector efficiency and the target density, which were not independently measured.

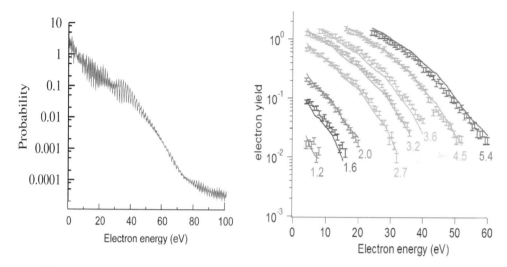

Figure 4.4 The ATI spectrum of atomic hydrogen driven by a 750 nm femtosecond laser. Left: a raw calculated spectrum for laser peak intensity of 4×10^{14} W/cm^2. Right: post-processed theoretical spectra for various peak intensities (solid lines) plotted versus corresponding experimental data (error bars). The laser intensity ranges from 1.2 to 5.4×10^{14} W/cm^2.

The latter factor gives a fit value for the absolute peak intensity, while leaving the relative intensities of the various data runs at their independently measured values. As shown in Fig. 4.2, we obtain agreement at the 10% level between the experimental data and the theoretical prediction over a wide range of electron energies and laser intensities.

4.6 High harmonics generation

High harmonics generation (HHG) is a non-linear atomic process that manifests itself in the appearance of odd-order multiple frequencies in the spectra of an atom placed in an intense electromagnetic field. Many essential features of the HHG process can be explained using the so-called three-step or recollision model [29]. The three steps of this model refer to (i) tunneling ionization of an atomic electron at the moment when the laser field reaches its peak intensity, (ii) acceleration of the photoelectron by the laser field and its return to the parent ion and (iii) recombination with the nucleus and emission of a single HHG photon. The resulting HHG has a typical pattern: the first few quickly decreasing harmonics followed by a plateau ending with a relatively sharp cut-off at the maximum harmonic number, $N_{\text{cut-off}} = (IP + 3.17U_p)/\omega \gg 1$. Up until recently, investigations of the HHG process have been limited to selecting a rare gas atom driven by an NIR laser field to ensure the tunneling ionization regime with the Keldysh parameter $\gamma \simeq 1$. However, a similar HHG regime with $N_{\text{cut-off}} \gg 1$ can be realized in alkaline earth

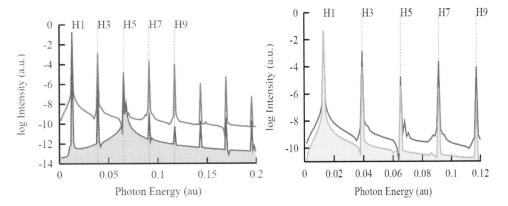

Figure 4.5 Harmonics spectrum of Li driven by a 330 fs laser pulse with the wavelength $\lambda = 3500$ nm. Left panel: the HHG spectrum from the ground $2s$ state at the laser field intensity 8×10^{11} W/cm^2 (shaded area) and the $2p$ excited state at 2×10^{11} W/cm^2 (solid line). Right panel: the $2p$ HHG spectrum with the $n\ell$ excited states manifold included (solid line) or excluded (shaded area) from the expansion (4.3).

atoms, which have a much smaller ionization potential, if they are driven by a mid-infrared laser field. In such systems, a strong resonant enhancement of HHG can be observed [30], which is an important factor for possible applications in tabletop sources of highly coherent XUV radiation.

In our approach to the HHG process in alkaline earth atoms (Li [31], Li and K [32], K and Rb [33]), we adopt a single active electron approximation and seek a solution of the TDSE (4.1) with the Hartree–Fock potential of the frozen atomic core on a one-electron basis (4.3). Once the solution of the TDSE is found, the harmonics spectrum is calculated as the Fourier transform of the time-dependent dipole operator [34]:

$$|d(\omega)|^2 = \left| \frac{1}{t_2 - t_1} \int_{t_1}^{t_2} e^{-i\omega t} d(t) \, dt \right|^2 . \tag{4.14}$$

Here, $d(t) = \langle \Psi(t)|z|\Psi(t) \rangle$ is the expectation value of the dipole operator and $\Psi(t)$ is the solution of the TDSE (4.1). In practical computations, the limits of integration t_1, t_2 are chosen to be large enough to minimize the transient effects. In the calculations presented below, we used $t_1 = 20T$, $t_2 = 30T$, i.e., the last 10 cycles of the pulse duration.

An example of resonantly enhanced HHG spectra is presented on the left panel of Fig. 4.5, which correspond to the Li atom prepared in either the ground $2s$ or excited $2p$ state [31]. In the former case, several orders of magnitude in the HHG intensity can be gained, provided the driving laser field is turned down to avoid the excited state depopulation. To prove the resonant excitation $3p \rightarrow n\ell$ as the main mechanism of the HHG enhancement, we performed several additional computations with the functions $R_{lm}(r, t)$ in expansion (4.3) explicitly orthogonalized to the states of the $n\ell$ manifold. Exclusion of these states causes a sharp drop of the HHG intensity, as is seen on the right panel of Fig. 4.5.

Another way to enhance the HHG spectrum is to extend its cut-off beyond the three-step model limit $N_{\text{cut-off}}$. This can be achieved by tailoring the waveforms of the driving laser pulse. In our recent work [35] we considered an electron moving in a periodic EM field:

$$E(t) = 2\mathrm{Re} \sum_{k=1}^{K} a_k e^{ik\Omega t} , \tag{4.15}$$

where the requirement of the fixed fluency implies that $4 \sum_{k=1}^{K} |a_k|^2 = E_0^2$, with the field strength E_0 related to the intensity of the pure cosine wave form via $I = 3.5 \times 10^{16} E_0^2$. Here the field intensity is measured in W/cm^2 and the field strength is expressed in atomic units.

We perform two calculations of this kind. In the first, we impose an additional restriction that only the terms with odd k-values are to be present in Eq. (4.15). This ensures that the resulting HHG spectrum contains only odd harmonics of the main frequency. In the second calculation, we retain the terms with both odd and even k-values in the expansion (4.15). The first calculation was performed with $K = 7$, while in the second we chose $K = 5$.

Results of these calculations, along with the HHG spectrum from a purely cosine waveform, are shown in Fig. 4.6. As one can see from the figure, the set of the parameters corresponding to only odd harmonics allows us to achieve a 10% gain in the position of the cut-off. This rather moderate increase is due to essentially the same structure of the classical returning electron trajectories as in the case of the cosine wave. The situation is different for the case of even and odd harmonics present in Eq. (4.15). Here a noticeable extension of the cut-off can be clearly seen.

4.7 Time delay in atomic photoionization

Among other spectacular applications of the attosecond streaking technique, it has become possible to determine the time delay between subjecting an atom to a short XUV pulse and subsequent emission of the photoelectron [36, 37]. With these observations, the question as to when does atomic photoionization actually begin can be answered by the experiment. We have studied this problem theoretically by solving the time-dependent Schrödinger equation and carefully examining the time

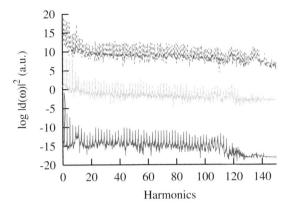

Figure 4.6 Harmonics spectra of Li driven by the laser pulses of different wave-forms: the pure cosine wave (bottom); odd harmonics in Eq. (4.15) with $K = 7$ (middle) dashed line, for convenience of comparison the quantity $\log |d(\omega)|^2 + 10$ is shown; odd and even harmonics in Eq. (4.15) with $K = 5$ (top), the quantity $\log |d(\omega)|^2 + 20$ is shown.

evolution of the photoelectron wave packet. In this way we established the apparent 'time zero' when the photoelectron left the atom. At the same time, we provided a stationary treatment to the photoionization process and connected the observed time delay with the quantum phase of the dipole transition matrix element, the energy dependence of which defines the emission timing. We applied this timing analysis to valence shell single photoionization of neon [38] and DPI of helium [39].

In the case of Ne, we solve a one-electron TDSE (4.1) with the field-free atomic Hamiltonian $\hat{H}_{\text{atom}}$ defined by the parameterized optimized effective potential [27]. The TDSE is solved by radial grid integration using the MIM method. The solution of the TDSE is used to form the wave packet representing the photoelectron ejected from a given shell:

$$\Phi(\mathbf{r}, t) = \sum_L \int a_{kL}(t) \chi_{kL}(\mathbf{r}) e^{-i E_k t} \, dk. \qquad (4.16)$$

Here, $a_{kL}(t) = e^{i E_k t} \langle \chi_{kL} | \Psi(t) \rangle$ are the projection coefficients of the solution of the TDSE on the continuum spectrum of the atom. This solution corresponds to the initial condition $\Psi(t = 0) \rightarrow \Psi_i$, where i indicates the atomic shell to be ionized. The continuum state $\chi_{kL}(\mathbf{r}) = R_{kl}(r) Y_L(\mathbf{r}/r)$ is the product of the radial orbital with the asymptotic

$$R_{kl} \propto \sin \left[kr + \delta_l(k) + 1/k \ln(2kr) - l\pi/2 \right]$$

and the spherical harmonic $Y_L(\mathbf{r}/r)$ with $L \equiv l, m$.

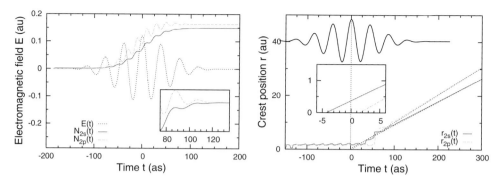

Figure 4.7 Left panel: the norm of the wave packets $N(t)$ (scaled arbitrarily) emitted from the $2s$ and $2p$ sub-shells of Ne is plotted as a function of time with the solid and dashed lines, respectively. The XUV pulse is over-plotted with the black dotted line. In the inset, the norm variation $[N(t) - N(T_1)]/N(T_1)$ is shown on an expanded timescale near the pulse end. Right panel: the crest position of the $2s$ and $2p$ wave packets is shown with the same line styles. The crest position after the pulse end is fitted with the straight line, which corresponds to the free propagation. In the inset, extrapolation of the free propagation inside the atom is shown.

We use two convenient indicators of the evolution of the wave packet. One is the norm given by the integral $N(t) = \sum_L \int dk \, |a_{kL}(t)|^2$. This norm is plotted on the left panel of Fig. 4.7, with the solid and dashed lines for the wave packets originating from the $2s$ and $2p$ sub-shells, respectively. For better clarity, these curves are scaled and over-plotted on the electromagnetic pulse. The figure shows clearly that the evolution of the $2s$ and $2p$ wave packets starts and ends at the same time without any noticeable delay. This is further visualized in the inset where the variation of the norm $[N(t) - N(T_1)]/N(T_1)$ is plotted on an expanded timescale near the driving pulse end. Indeed, the norm starts deviating from zero with the rise of the XUV pulse and reaches its asymptotic value once the interaction with the XUV pulse is over.

Another marker of the wave-packet dynamics is the crest position, defined as a location of the global maximum of the electron density. The latter quantity is truly informative only when the electron is outside the atom and the wave packet is fully formed, having one well-defined global maximum. On the right panel of Fig. 4.7, we show the crest position of the $2s$ and $2p$ wave packets propagating in time. We see that evolution of the norm and the movement of the crest commence at about the same time. The movement of the crest becomes almost linear when the norm reaches its asymptotic value and the wave packet is fully formed. Once fitted with the linear time-dependence $r = k(t - t_0) + r_0$ for large times $t > T_1$ (shown as a dotted straight line) and back propagated inside the atom, the $2s$ wave packet seems to have an earlier start time t_0 than that of the $2p$ wave packet. This difference is magnified in the inset.

We see that at the origin $t_0^{2s} < 0$ and $t_0^{2p} > 0$ are shifted to the opposite direction with respect to the peak of the driving XUV pulse, which sets the start time of the photoionization process. We relate the opposite signs of the time delays with the energy dependence of the corresponding scattering phases $\delta_{l=1}^{2s}$ and $\delta_{l=2}^{2p}$, which is governed by the Levinson–Seaton theorem [40].

In the case of DPI of He, we solve a two-electron TDSE using the ALM method. The field-free solution of the TDSE at $t > T_1$ is used to construct a two-electron wave packet $\Psi_1(\mathbf{r}_1, \mathbf{r}_2, t)$ with the asymptotics corresponding to the given photoelectron momenta $\mathbf{k}_1, \mathbf{k}_2$ $\mathbf{k}_2$:

$$\Psi_1(\mathbf{r}_1, \mathbf{r}_2, t) = \hat{P}_{\mathbf{k}_1, \mathbf{k}_2} \Psi(\mathbf{r}_1, \mathbf{r}_2, t). \tag{4.17}$$

Here, the kernel of the projection operator is constructed as

$$\langle \mathbf{r}_1', \mathbf{r}_2' | \hat{P}_{\mathbf{k}_1, \mathbf{k}_2} | \mathbf{r}_1, \mathbf{r}_2 \rangle = \int_\Omega \Psi_{\mathbf{q}_1}^-(\mathbf{r}_1) \Psi_{\mathbf{q}_2}^-(\mathbf{r}_2) \Psi_{\mathbf{q}_1}^-(\mathbf{r}_1')^* \Psi_{\mathbf{q}_2}^-(\mathbf{r}_2')^* d\mathbf{q}_1 d\mathbf{q}_2, \tag{4.18}$$

where $\Psi_{\mathbf{k}_i}^-(\mathbf{r}_i), i = 1, 2$ are one-electron scattering states with the ingoing boundary condition describing a photoelectron moving in the Coulomb field with $Z = 2$.

The wave packet function $\Psi_1(\mathbf{r}_1, \mathbf{r}_2, t)$ is plugged into the one-electron density function

$$\rho(\mathbf{r}, t) = \int |\Psi_1(\mathbf{r}_1, \mathbf{r}_2, t)|^2 \left[\delta(\mathbf{r} - \mathbf{r}_1) + \delta(\mathbf{r} - \mathbf{r}_2) \right] d\mathbf{r}_1 d\mathbf{r}_2. \tag{4.19}$$

The maxima of this density function are then traced to determine the trajectories of both photoelectrons which, at large distances, can be approximated by

$$r_i(t) - k_i t - r_i'(t) \asymp k_i t_{0i}. \tag{4.20}$$

Here t_{0i} are the time delays and $r_i'(t)$ are known functions which vary logarithmically slow with t.

An example of our timing analysis is illustrated in Fig. 4.8. Here we consider a DPI process in which one photoelectron escapes with energy 32 eV along the z-axis and another with energy 10 eV along the x-axis, thus sharing the excess energy of 42 eV, which corresponds to the photon energy $\omega = 121$ eV. A sequence of snapshots of the one-electron density function (4.19) is taken with an interval of $2T$ and the maxima of the electron density are traced in time. With the known logarithmic function $r_i'(t)$, this procedure defines the trajectories $r_i(t) - r_i'(t)$ for both photoelectrons, which are exhibited on the left panel of Fig. 4.8. The raw data, shown by the points, are fitted with the straight lines $k_z(t - t_0)$ and $k_x(t - t_0)$, which visualize the free propagation. The intersect of these straight lines with the abscissa gives the corresponding time delays t_{0i}. We ran an analogous simulation for other

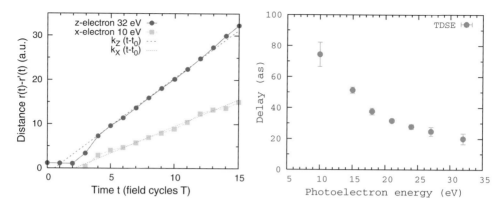

Figure 4.8 Left panel: trajectories of 32 eV and 10 eV photoelectrons propagating along the z and x axes, respectively. The co-ordinates of both photoelectrons, measured with intervals of two field cycles, are shown as dots. The straight lines visualize the free propagating $k_z(t - t_0)$ and $k_x(t - t_0)$. Right panel: time delay as a function of the photoelectron energy.

energy-sharing cases. Thus obtained time-delay data are collated on the right panel of Fig. 4.8. The error bars indicate the uncertainty of the fitting procedure.

The rapid change of the time delay with increase of the photoelectron energy corresponds to the crossover between the two leading mechanisms of DPI: the fast shake-off (SO) and the slow knock-out (KO) processes. The SO mechanism is driven by a fast rearrangement of the atomic core after departure of the primary photoelectron. The KO mechanism involves repeated interaction of the primary photoelectron with the remaining electron bound to the singly charged ion. Thus, future attosecond time-delay measurements on DPI of He can provide information on the absolute quantum phase and elucidate various mechanisms of this strongly correlated ionization process.

References

[1] A. Becker, R. Dörner and R. Moshammer, *J. Phys. B*, **38**, S753 (2005).
[2] D. A. Varshalovich, A. N. Moskalev, and V. K. Khersonskii, *Quantum Theory of Angular Momentum* (Singapore: World Scientific, 1988).
[3] E. Cormier and P. Lambropoulos, *J. Phys. B*, **29**, 1667 (1996).
[4] W. Magnus, *Comm. Pure and Appl. Math.*, **7**, 649 (1954).
[5] T. J. Park and J. C. Light, *J. Chem. Phys.*, **85**, 5870 (1986).
[6] D. Dundas, *Phys. Rev. A*, **65**, 023408 (2002).
[7] M. Nurhuda and F. H. M. Faisal, *Phys. Rev. A*, **60**, 3125 (1999).
[8] H. Hasegawa, E. J. Takahashi, Y. Nabekawa, K. L. Ishikawa and K. Midorikawa, *Phys. Rev. A*, **71**, 023407 (2005).
[9] Y. Nabekawa, H. Hasegawa, E. J. Takahashi and K. Midorikawa, *Phys. Rev. Lett.*, **94**, 043001 (2005).

[10] A. A. Sorokin, M. Wellhöfer, S. V. Bobashev, K. Tiedtke, and M. Richter, *Phys. Rev. A*, **75**, 051402 (2007).

[11] P. Antoine *et al.*, *Phys. Rev. A*, **78**, 023415 (2008).

[12] I. A. Ivanov and A. S. Kheifets, *Phys. Rev. A*, **75**, 033411 (2007).

[13] I. Bray, *Phys. Rev. A*, **49**, 1066 (1994).

[14] I. A. Ivanov and A. S. Kheifets, *Phys. Rev. A*, **74**, 042710 (2006).

[15] I. Bray and A. T. Stelbovics, *Adv. Atom. Mol. Phys.*, **35**, 209 (1995).

[16] D. A. Horner, F. Morales, T. N. Rescigno, F. Martin, and C. W. McCurdy, *Phys. Rev. A*, **76**, 030701(R) (2007).

[17] J. Feist *et al.*, *Phys. Rev. A*, **77**, 043420 (2008).

[18] X. Guan, K. Bartschat and B. I. Schneider, *Phys. Rev. A*, **77**, 043421 (2008).

[19] I. A. Ivanov, A. S. Kheifets and J. Dubau, *Eur. Phys. J. D*, **61**, 563 (2011).

[20] J. S. Briggs and V. Schmidt, *J. Phys. B*, **33**, R1 (2000).

[21] L. Avaldi and A. Huetz, *J. Phys. B*, **38**, S861 (2005).

[22] I. A. Ivanov and A. S. Kheifets, *Eur. Phys. J. D*, **38**, 471 (2006).

[23] I. A. Ivanov and A. S. Kheifets, *Phys. Rev. A*, **75**, 062701 (2007).

[24] L. V. Keldysh, *Sov. Phys. -JETP*, **20**, 1307 (1965).

[25] P. B. Corkum, *Phys. Rev. Lett.*, **71**, 1994 (1993).

[26] M. Schuricke *et al.*, *Phys. Rev. A*, **83**, 023413 (2011).

[27] A. Sarsa, F. J. Gálvez and E. Buendia, *Atomic Data and Nuclear Data Tables*, **88**, 163 (2004).

[28] M. G. Pullen *et al.*, *Opt. Lett.*, **36**, 3660 (2011).

[29] P. B. Corkum, *Phys. Rev. Lett.*, **71**, 1994 (1993).

[30] P. M. Paul *et al.*, *Phys. Rev. Lett.*, **94**, 113906 (2005).

[31] I. A. Ivanov and A. S. Kheifets, *J. Phys. B*, **41**, 115603 (2008).

[32] I. A. Ivanov and A. S. Kheifets, *J. Phys. B*, **42**, 145601 (2009).

[33] I. A. Ivanov and A. S. Kheifets, *Phys. Rev. A*, **79**, 053827 (2009).

[34] J. L. Krause, K. J. Schafer and K. C. Kulander, *Phys. Rev. A*, **45**, 4998 (1992).

[35] I. A. Ivanov and A. S. Kheifets, *Phys. Rev. A*, **80**, 023809 (2009).

[36] M. Schultze *et al.*, *Science*, **328**, 1658 (2010).

[37] K. Klünder *et al.*, *Phys. Rev. Lett.*, **106**, 143002 (2011).

[38] A. S. Kheifets and I. A. Ivanov, *Phys. Rev. Lett.*, **105**, 233002 (2010).

[39] A. S. Kheifets, I. A. Ivanov and I. Bray, *J. Phys. B*, **44**, 101003 (2011).

[40] L. Rosenberg, *Phys. Rev. A*, **52**, 3824 (1995).

5

Experimental aspects of ionization studies by positron and positronium impact

G. LARICCHIA, D. A. COOKE, Á. KÖVÉR AND S. J. BRAWLEY

5.1 Introduction

As well as probing matter–antimatter interactions, positrons (as positive electrons) have been employed to highlight charge and mass effects in the dynamics of collisions, including those resulting in the ionization of atoms and molecules (see, e.g., [1]). Positronium (Ps), the hydrogenic atom formed from the binding of a positron and an electron, is readily produced in the scattering of positrons from matter. Ps is quasi-stable with a lifetime against annihilation dependent upon its spin: ground-state para-Ps ($1\ ^1S_0$) has a lifetime $\tau \simeq 125\,\text{ps}$, whilst ortho-Ps ($1\ ^3S_1$) is considerably longer lived ($\tau \simeq 142\,\text{ns}$). The beam employed for the scattering work discussed in this chapter consists solely of ortho-Ps atoms. In a dense medium, Ps may undergo several cycles of formation and break-up prior to the annihilation of the positron (see, e.g., [2–6]). A quantitative understanding of this cycle is important also for practical applications such as nanodosimetry relating to positron emission tomography (PET) [e.g., 4].

In this chapter, we consider experimental methods employed to investigate positron and positronium impact ionization and fragmentation in collision with atoms and molecules, and associated results. In the case of positrons, an extensive database now exists of integral cross sections for the inert atoms (see, e.g., [5]), less so for molecules (see, e.g., [6]); differential data remain sparse (e.g., [5]). Our focus will be on the latter two topics as well as studies with positronium projectiles.

Fragmentation Processes: Topics in Atomic and Molecular Physics, ed. Colm T. Whelan. Published by Cambridge University Press. © Cambridge University Press 2013.

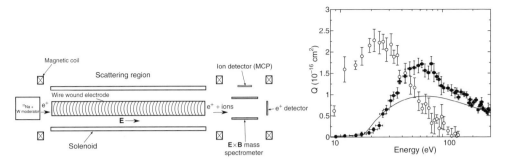

Figure 5.1 Experimental apparatus (left) and results obtained by [7] for $e^+ + H_2$ (right): Q_i^+ ($\bullet$) and $Q_{Ps} = Q_i^t - Q_i^+$ ($\circ$); the line corresponds to electron Q_i^+ [8].

5.2 Integral cross sections for positron impact ionization

The main outcomes of a positron colliding with an atomic or molecular target are summarized below:

$$e^+ + A \rightarrow e^+ + A \qquad \text{Elastic scattering, } Q_{elas} \qquad (5.1)$$

$$e^+ + A \rightarrow A^+ + n\gamma \qquad \text{Annihilation, } Q_{ann} \qquad (5.2)$$

$$e^+ + A \rightarrow A^* + e^+ \qquad \text{Excitation, } Q_{ex} \qquad (5.3)$$

$$e^+ + A \rightarrow A^+ + Ps \qquad \text{Positronium formation, } Q_{Ps} \qquad (5.4)$$

$$e^+ + A \rightarrow A^{+*} + Ps \qquad \text{Ps formation–ionic excitation, } Q_i^{ex/Ps} \qquad (5.5)$$

$$e^+ + A \rightarrow A^+ + e^+ + e^- \qquad \text{Direct ionization, } Q_i^+ \qquad (5.6)$$

$$e^+ + A \rightarrow A^{+*} + e^+ + e^- \qquad \text{Direct ionization–excitation, } Q_i^{ex/+} \qquad (5.7)$$

$$e^+ + A \rightarrow A^{n+} + Ps + (n-1)e^- \qquad \text{Transfer ionization, } Q_{ti} \qquad (5.8)$$

$$e^+ + A \rightarrow A^{n+} + e^+ + ne^- \qquad \text{Multiple ionization, } Q_i^{n+} \qquad (5.9)$$

where n is an integer greater than 1. The inelastic reactions have been listed in approximately ascending order of their energy thresholds; annihilation, of course, is exothermic. For molecules, in addition, there are combinations of the above reactions which result in dissociation into charged and/or neutral fragments.

The earliest studies of positron impact ionization employed energy-loss measurements of the scattered projectile using time-of-flight (TOF) spectroscopy, as in [12], later combined with a retarding field analysis (RFA) by [13]. These investigations were soon followed by several more [14–16], including the first coincidence measurements between positrons and time-correlated ions [7,8]. The experimental set-up and associated results for H_2 are shown in Fig. 5.1, where Q_{Ps} was obtained by subtracting Q_i^+ from Q_i^t, the latter defined as the sum of the cross sections for all processes resulting in ion production, assuming annihilation to be negligible. The absolute scale was set by performing analogous Q_i^+ measurements by

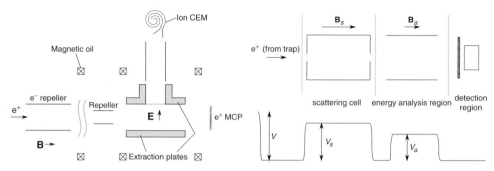

Figure 5.2 Left: parallel plate extraction system used by [9]. The electric field across the interaction region is triggered by detection of a positron. The second repeller is for collection of back-scattered positrons and, additionally, for determining a background by biasing off the slow portion of the beam. Right: schematic of the interaction region of the experimental systems at UCSD [10] and ANU [11]. Detection of the positron is achieved using an annihilation plate/NaI scintillation detector in the UCSD system and a multichannel plate stack in that at ANU. The beam incident energy is set by the voltage difference ($V - V_s$) and the energy of the scattered beam is analyzed by the variable voltage V_a.

electron impact, having verified the convergence of the shape of the cross sections for the two projectiles above 600 eV, in accordance with the Born approximation. Multiple ionization cross sections for atomic targets were later measured [17, 18], the ion charge state distinguished by TOF spectroscopy. The inference that the absence of the positron in the final state signalled Ps formation was first employed by [19] – the approach requiring the efficient confinement of all scattered positrons – and has been adopted more recently by the groups at San Diego (UCSD) [10] and Canberra (ANU) [11]. Both groups employ a modified Penning–Malmberg trap capable of producing pulses of positrons with an energy resolution typically less than ∼50 meV (see Fig. 5.2 (right)) and direct ionization is again associated with events in which the projectile loses an amount of energy equal or greater than the target ionization energy. A magnetic field gradient ensures that most of the positron kinetic energy lies axially in the RFA region.

As part of a drive to clarify the role of the projectile charge and mass in ionization (e.g., [20]), studies of proton and antiproton impact ionization were extended to positrons (e.g., [9, 21]). These employed positron–ion coincidences using a pulsed high-voltage ion-extraction system to ensure efficient detection also of the ions accompanying dissociation. A schematic representation of this apparatus is shown in Fig. 5.2 (left). The positron detection pulse is used to trigger the ion-extraction field, enabling a time-of-flight mass spectrum of the ions to be constructed.

Figure 5.3 shows the gas cell of the UCL magnetic positron beam and a typical block diagram representing a multiple coincidence measurement. A detailed

Table 5.1 *Summary of the cross sections measured in the UCL set-up, classified according to which particles are detected. Note that some are indistinguishable by detection of two particles only, for example, excitation and direct ionization–excitation.* $Q_{Ps(2P)}$ *indicates formation of Ps in the 2P state.*

	e^+	A^+	γ-ray	photon ($h\nu$)
e^+	—	Q_i^+	—	$Q_{ex}/Q_i^{ex/+}$
A^+	Q_i^+	—	Q_{Ps}/Q_{ann}	$Q_{Ps(2P)}/Q_i^{ex}$
γ-ray	—	Q_{Ps}/Q_{ann}	—	$Q_i^{ex/Ps}/Q_i^{ex/ann}$
photon ($h\nu$)	$Q_{ex}/Q_i^{ex/+}$	$Q_{Ps(2P)}/Q_i^{ex}$	$Q_i^{ex/Ps}/Q_i^{ex/ann}$	—

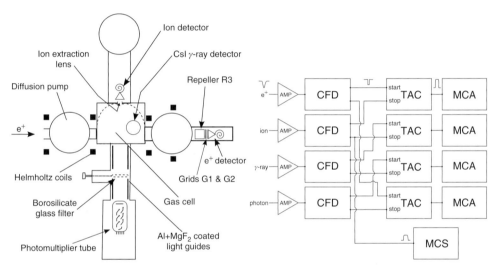

Figure 5.3 Left: schematic of interaction region of the magnetic positron beam at UCL, showing the location of multiple detectors. Right: typical coincidence measurement set-up for this experimental arrangement.

description about the operation of this system is available elsewhere (e.g., [22]). Briefly, the gas cell is surrounded by three detectors: a channel electron multiplier (CEM) mounted behind an extraction lens to detect ions; a photomultiplier for low-energy (200–500 nm) photons and a CsI scintillation crystal coupled to a photodiode for the detection of high-energy photons (γ-rays) produced primarily by the decay of the Ps atom. Another CEM detects (transmitted and scattered) positrons at the end of the beamline. The simultaneous measurement of various cross sections is possible by recording coincidences between resultant particles, as summarized in Table 5.1, the experiment relying mainly on the detection of the resultant ion in coincidence with another product particle. While continuous ion

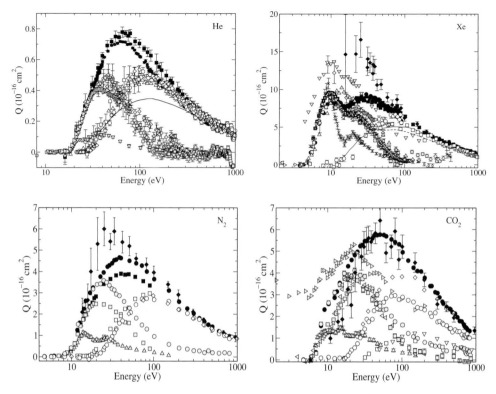

Figure 5.4 In each case, Q_i^t is presented in black points, Q_i^+ in hollow point and Q_{Ps} in grey-filled points. Top left: He. Q_i^t [2] (○), [7] (□); Q_i^+ [25] (○), [16] (□), [21] (◇), [7] (△); Q_{Ps} [26] (○), [2] (□), [7] (◇), [19] (△), [14] (◁), [27] (▽), [28] (▷). Electron measurement of [29] (black line). Top right: Xe. Q_i^t [10] (○), [30] (□), [31] (◇); Q_i^+ from [10] (○), [24] (□); Q_{Ps} [32] (▷), [10] (□), [30] (○), [14] (◇), [27] (×), [33] (upper limit ▽, lower limit △), [3] (◁). Electron data from [29] (Q_i^t solid line, single ionization dashed line). Bottom left: N$_2$. Q_i^t [34] (○), [35] (□), [23] (◇); Q_i^+ [34] (○), [35] (□), [23] (◇); Q_i^{diss} [23] (△); Q_{Ps} from [22] (○), [35] (□), [36] (△). Electron impact data from [37] solid line is total ionization, dashed line and chain curve are non-dissociative and dissociative ionization, respectively. Bottom right: CO$_2$. Q_i^t from [38] (○), [31] (◇); Q_i^+ from [38] (○), [31] (◇); Q_i^{diss} from [38] (□), [23] (▽); Q_{Ps} from [22] (○), [2] (□), [36] (△), upper and lower limits of [39] (◁ and ▷, respectively). Electron impact data from [37], legend as for N$_2$.

extraction with a weak radial electric field provides a high extraction efficiency, its resolution is poor due to the variability in initial position and velocity of the ions. This problem may be avoided by using a uniform stronger electric field, pulsed so as to avoid perturbing the collision energy [23, 24].

Figure 5.4 shows a compendium of positron impact ionization cross sections for He, Xe, CO$_2$ and N$_2$. As may be seen here, discrepancies persist even for the simplest atom. For the more complex targets, there are significant disagreements

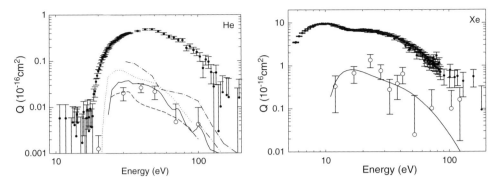

Figure 5.5 Cross section for Ps formation in the $2P$ state $Q_{Ps}(2P)$, experiment (hollow circles) [40] compared with available theories. He: solid curve [41]; dotted line [42]; chain curve [43]; dashed curve [44]; double chain curve [45]. Xe: [46] (solid line). Also shown is Q_{Ps}(all n) (filled circles) [2].

in absolute scale between different studies, even though internal consistency may exist within each set of measurements. Variations in magnitude often arise from differences in the normalization methods and/or reference data chosen in such procedures.

With respect to the cross sections themselves, the principal features of positron impact ionization are illustrated by these examples (see also [5]). In comparison with electrons (shown as lines in Fig. 5.4), positrons have a greater ionizing capability, the enhancement mainly arising from the formation of Ps. A conspicuous difference between atoms and molecules is the non-negligible Ps formation up to energies of several hundred eV. For several targets, there is also a slight excess around its maximum value of the direct ionization cross section by positron over electron impact, as in the case here of He and Xe. This is attributed (e.g., [20]) to target polarization;[1] the reversal of this situation at the lowest energies has been interpreted as arising from 'trajectory effects', whereby light particles are deflected in the field of the target nucleus, leading to an increase in the collision probability for negative projectiles and a corresponding decrease for positive projectiles [1].

Results of recent photon–ion coincidence measurements are shown in Figs. 5.5 and 5.6. In the case of atoms, significant Ps formation in the $2P$ state has been found, as illustrated for He and Xe in Fig. 5.5. Here the experimental cross sections for Ps formation in the $2P$ state ($Q_{Ps(2P)}$) are compared with theories and the total experimental Q_{Ps}(all n) [2, 30], to which $Q_{Ps(2P)}$ contributes maximum values of $(6 \pm 1)\%$ and $(26 \pm 9)\%$ in He and Xe, respectively. For He also shown are the coupled-state calculation of [41], the close-coupling results of [42] and [44], the distorted-wave approximation of [43], and the second Born approach of [45],

---

[1] This feature becomes less significant with increasing Z, owing to the larger static interaction of higher Z targets.

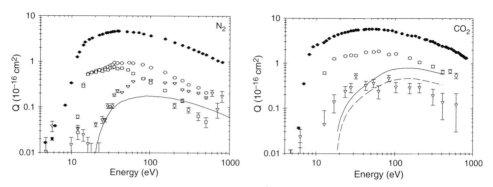

Figure 5.6 Cross sections for ionization–excitation Q_i^{ex} [3]. CO_2: Q_i^{ex} for the $A\ ^2\Pi_u \rightarrow X\ ^2\Pi_g$ (○) and $B\ ^2\Sigma_u^+ \rightarrow X\ ^2\Pi_g$ (▽) transitions in CO_2^+, compared with equivalent electron measurements of [47] (solid and dotted curves, respectively). N_2: Q_i^{ex} for the $B\ ^2\Sigma_u^+ \rightarrow X\ ^2\Sigma_g^+$ transition in N_2^+, showing the total (○) and contributions from direct ionization (▽) and Ps formation (□). Also shown for both targets are Q_i^t (●), as in Fig. 5.4.

the closest agreement with experiment being found with the theory of [41]. For Xe, the measurements are shown together with a distorted-wave Born approximation (DWBA) [46] with which a fair agreement is noted.

In the case of molecules, attempts to measure $Q_{Ps(2P)}$ have thus far been inconclusive or yielded null results [3]. Instead, a considerable probability of Ps formation occurring with ionic excitation has been observed in N_2 and CO_2. With reference to Fig. 5.6 for CO_2, the positron impact cross section for total ionization (i.e., including Ps formation) simultaneous to ionic excitation into the $A\ ^2\Pi_u$ state was found to be enhanced over both the equivalent electron impact cross section and the $B\ ^2\Sigma_u^+$ state. For N_2^+, a summary is shown of ionization–excitation measurements for the $B\ ^2\Sigma_u^+ \rightarrow X\ ^2\Sigma_g^+$ transition [3]. For both targets, the excess of Q_i^{ex} for positron impact over that for electron impact is largely due to Ps formation and has been attributed to accidental resonances between electronically excited states of the neutral molecule and those of the remnant ion produced via Ps formation [3].

5.3 Differential cross sections for positron impact ionization

Several theoretical and experimental results have been published in the past couple of decades for positron–atom collisions investigating the angular distribution of the ejected particles (e$^+$, e$^-$ and Ps), the most advanced investigation among these for probing the collision dynamics being the measurement of fully differential cross sections.

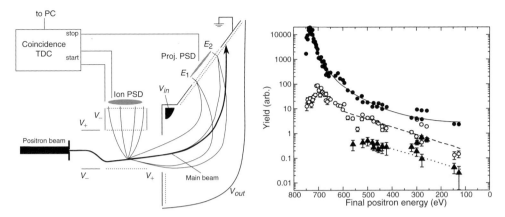

Figure 5.7 Left: schematic of target region and projectile energy analyzer showing main and scattered beam trajectories in the range 0°–17° [54]. Right: single and multiple ionization yields, as a function of scattered positron energy, for 750 eV positron impact on argon. Single (•), double (○) and triple (▲) ionization are shown, together with a DWBA calculation [55] (solid curve). Dashed curves are to guide the eye.

The first doubly-differential measurements to determine the angular and energy distribution of the ejected electrons and scattered positrons from positron–atom collisions were carried out at UCL using a magnetically guided positron beam [48] and later an electrostatic system [49,50]. In the latter case, the energy distribution of the electrons and positrons ejected at 30° and 45° were determined in coincidence with the remnant ion for e^+/e^-–Ar collisions at 100 eV incident energy. This was the first scattering measurement where the ionized electron and scattered projectile (for positron impact) were clearly distinguished, demonstrating that the target electrons are ejected mainly with a low energy while scattered projectiles account for the high end of the final energy spectrum. The measured data were in good agreement with a classical trajectory Monte Carlo (CTMC) calculation [51]. Similar measurements were carried out to determine the angular distribution of 15 eV ejected electrons by 100 eV e^+–Ar collisions [52], and in the 40–150 eV impact region at 60° [53].

More recently, a modified cylindrical electrostatic spectrometer has been developed to measure the energy of the scattered projectiles simultaneously within a certain energy range, and in the 0°–17° angular region, in coincidence with the recoil ions [54]. This system is depicted in Fig. 5.7 (left). They determined the projectile energy-loss spectra for single, double and triple ionization of argon by 750 eV positrons, as shown in Fig. 5.7 (right). The single ionization data are in good agreement with DWBA calculations [55], also shown on Fig. 5.7. It was found that the percentage of double ionization rapidly increases for the first

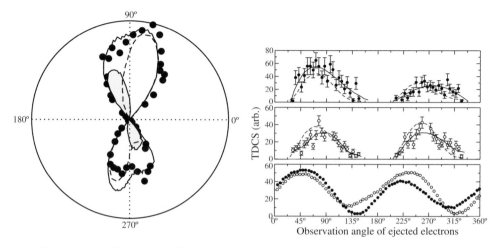

Figure 5.8 Left: triply differential 2.4 eV electron emission yields for single ionization of argon by 500 eV positrons scattered to the 1.5°–4.5° region. The solid and dashed lines and the filled areas are simulated curves for the binary and recoil emission convoluted over experimental conditions [57]. Right: triply differential 5.4 eV electron emission distribution for single ionization of Ar by 200 eV positrons (top) and electrons (middle) compared with a CDW-EIS theory (bottom). The solid curves in the upper two panels are polynomial fits to the data, and the dashed curves are the CDW-EIS theory from the bottom panel convoluted over experimental parameters. The unconvoluted and convoluted theory curves have each been normalized to electron impact experimental fit at 90° [58].

80 eV of energy loss, with the ratio slowly approaching a constant value that is in good agreement with the ratios of total cross sections for photoionization of the *M* shell.

Arcidiacono *et al.* [56] investigated the energy-loss spectra of positrons scattered around 0° from H_2O in coincidence with the remnant ions (H_2O^+, OH^+ and H^+) at 100 eV and 153 eV impact energies. They found that the maxima of the distributions associated with the production of OH^+ and H^+ are about 5–10 times smaller than that for H_2O^+. At both incident energies, a small shoulder was observed at around 28 eV in the energy-loss spectra associated with H_2O^+ production, possibly arising from a shake-up band associated with the $2a_1$ orbital.

A modification to the system of DuBois *et al.* [54] added an additional position-sensitive detector (PSD) to monitor ejected electrons, remnant target ions and forward-scattered projectiles in coincidence [57, 59]. They investigated the correlation between projectile energy loss and scattering angle for single ionization for 500 eV e^+–Ar collisions. Figure 5.8 (left) shows the angular distribution of ejected electrons at a certain projectile electron loss (2.4 eV), the binary and recoil lobes showing the relative importance of these interactions as a function of energy loss [57].

Recently, [58] have compared the fully kinematic data for 200 eV electron and positron impact using identical experimental conditions. Figure 5.8 (right) shows the experimental results, compared with the theoretical values calculated with CDW-EIS theory (convoluted with the experimental parameters). It can be seen that binary events (data for angles less than 180°) are significantly larger for positron impact whereas the recoil peaks (above 180°) have similar intensities.

During the last decade, the UCL group has investigated mainly the particles ejected into the angular region around 0° in order to probe the occurrence of the electron capture to the continuum process (ECC) for positron impact ionization. In this case, the ionized electron, strongly influenced by the positively charged projectile, is captured into the continuum state of the projectile. Theoretically, it can be considered as an extrapolation across the ionization limit of electron capture into highly excited bound states. For ion impact, a sharp peak appears in the doubly differential electron spectrum around 0° at a velocity matching that of the scattered projectile [60, 61]. Due to the positive charge of the positron, a similar final-state interaction occurs; however, due to the equal masses of the positron and electron, the collision dynamics for positron impact are different to those for ion impact. The positron must transfer nearly one half of its kinetic energy to the ionized electron and hence scatters to larger angles compared to ion projectiles. Furthermore, the energy of the ECC peak is at nearly half that of the projectile energy, i.e.,

$$E_{\mathrm{ECC}} = \frac{E_{\mathrm{proj}} - E_{\mathrm{I}}}{2}, \qquad (5.10)$$

where E_{ECC} and E_{proj} are the energy of the ECC peak and the projectile, respectively, and E_{I} is the target ionization energy.

The theoretical interpretation of ECC is that of a real three-body collision process where the interaction among the particles cannot be ignored, particularly in the case of positron impact. Significant interference effects may occur, the description of which has presented a great challenge for theory, from the first pioneering calculation of [62] who used the three-body scattering formalism of [63] through to CTMC [51, 64–67] and quantum mechanical calculations [66, 68–72]. In the latter case, the first Born approximation was extended by multiplying the commonly used plane waves for the two outgoing particles, with three Coulomb distortion factors representing the interaction between the ionized electron and the nucleus, the projectile and the nucleus and between the projectile and the ionized electron, the so-called BBK approximation [73]. More details have been recently summarized in [72]. Most of the calculations predict a cusp-like structure around 0° in the triply differential cross section (TDCS) but only a hump in the doubly differential cross section (DDCS).

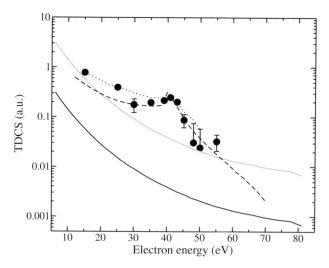

Figure 5.9 The spectra of ejected electrons from 100 eV e^{+}–H_2 collisions around $0°$. The circles are the measured TDCS by [74]; the dashed line corresponds to the calculation by [75] and the dotted line by [76]; the continuous black line is the result of the first Born approximation (FBA); the grey continuous line is the FBA multiplied by 10.

Experimental investigations did not find any structure in the DDCS for the electron spectrum [49, 50, 53], necessitating TDCS measurements for greater sensitivity. In these, coincidences between the scattered positron and the ionized electron, ejected into the same direction, were recorded. A special parallel-plate electron spectrometer (parallel-plate analyzer – PPA) was built to fulfil these requirements [74]. A small, broad peak was found in the ejected electron spectrum at 100 eV positron impact on H_2. This, the first clear experimental evidence for the ECC process by positron impact [74], is shown in Fig. 5.9, together with the theoretical calculations employing BBK final-state wave functions convoluted with the experimental energy and angular resolutions [75, 76], describing the observations well.

A similar measurement was performed for 50 eV impact energy with a modified PPA, extended with time-focusing capabilities [77]: particles entering the analyzer at slightly different angles around $0°$ arrive at the detector at the same time, thus reducing the spread of time of flight and increasing the signal-to-background ratio. A diagram of the analyzer is displayed in Fig. 5.10 (left). Surprisingly, the electron distribution was found to be shifted by 1.6 eV to lower energy, relative to the calculated value of the ECC peak (17.3 eV) [79]. The maximum of the energy distribution of scattered projectiles was also measured [78], showing a similar shift to higher energies. It was speculated that this might indicate a dynamical

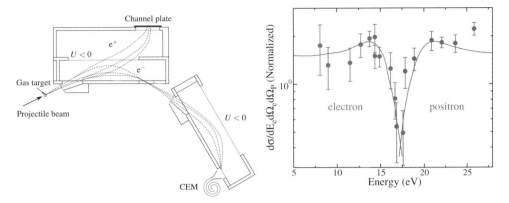

Figure 5.10 Left: cross section of the parallel-plate analyzer with time focusing [77]. Right: the spectra of ejected electrons and scattered positrons from 50 eV e^+–H_2 collisions around $0°$. The circles are the experimental data of [78] and the continuous line from the CTMC calculation of [72].

polarization,[2] as was found with Ps–He collisions by CTMC calculations [80]. Recently another CTMC simulation [72] showed that for electron velocities close to the final positron velocity, the formed dipole is oriented along the direction of the motion of its centre of mass, with the negative charge pointing towards the residual target, which means that its energy is lower than the positron energy. The comparison of their calculation with the experimental data is shown on Fig. 5.10 (right).

As discussed above, the dynamics of the bound-state capture process for positron collisions is quite different to that of protons. The experimental investigation of the differential Ps formation cross section by direct detection of the atom is not easy due to its instability and neutrality, the latter resulting in a low detection efficiency at low incident velocities. A detailed summary of the results may be found in [5, 81]. The most detailed investigation to date was performed by Falke *et al.* [82], extending the angular region from $10°$–$120°$, using three different impact energies (75 eV, 90 eV, 120 eV) and two targets (Ar and Kr). In Fig. 5.11, these data are compared with the theoretical calculations of [83] with which fairly good agreement may be seen, especially for 75 eV impact energy.

Currently, efforts are underway to apply the successful technique, cold target recoil ion momentum spectroscopy (COLTRIMS) [84], to the study of positron impact ionization. COLTRIMS measures the time of flight and angular distribution of the recoil ions (after extraction from the collision region), which gives simultaneous information on its longitudinal and transverse momentum,

---

[2] That is, the outgoing (e^+, e^-) pair is polarized in the electric field of the remnant ion.

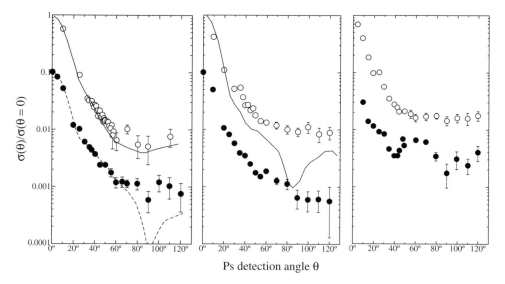

Ps detection angle θ

Figure 5.11   Transfer for Ar (○) and Kr (●) at 75 eV, 90 eV and 120 eV (left to right). The curves represent the theoretical results of [83] combined with the angular resolution of the apparatus. Note that data for Kr has been scaled by a factor of 0.1. Figure adapted from [82].

respectively. Recording the scattered projectiles in coincidence with the outgoing fragments allows a kinematically complete picture of the correlated motion of atomic and molecular break-up processes to be determined. This method has been successfully used with ions, electrons and photons as projectiles ([84, 85] and references therein). The UCL group are in the process of developing a RIMS-based system using positrons as the projectile [86, 87]. The task is not easy because it requires the production of a narrow (1 mm diameter) and parallel monoenergetic positron beam. Owing to the much lower positron intensity in comparison, for example, with electrons or photons projectiles, a fairly high target density is necessary (10^{12}–10^{13} atom cm^{-3}) with a small initial momentum distribution in order to resolve the small momentum transfer (0.3–0.4 a.u.). Such requirements can be achieved by using a supersonic gas jet.

For positron impact, Barrachina *et al.* [88] performed a CTMC calculation of the recoil ion momentum distribution for 100 eV e^+–H collisions. Figure 5.12 shows their results demonstrating the advantage of the method – Ps formation to varying n-states (capture to $1S$, $2S$, etc.) and the ECC process are clearly separated. Recent investigations have been performed by Tökesi [89] and Barrachina *et al.* [90].

Finally, the Heidelberg group, who have extensive experience using electron projectiles [85, 91], have together with the group at ANU [93] coupled their apparatus to the NEPOMUC facility [92] and investigated e^+–He collisions at 80 eV,

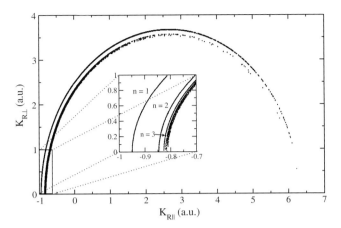

Figure 5.12 CTMC calcultation of the recoil ion momentum for Ps formation in the collision of 100 eV e$^+$–H [88].

measuring triple coincidences and obtaining preliminary results for the longitudinal momentum of the fragments, which was compared with the data measured for electron impact [93]. Further studies are planned.

5.4 Positronium-induced fragmentation

In collisions where its kinetic energy is greater than its binding energy ($E_{Ps} > 6.8/n^2$, n being its principal number), Ps may fragment into its constituent particles and, at sufficiently higher energies, excite or ionize the target, as summarized for He and Xe in Tables 5.2 and 5.3. The first major investigation of Ps fragmentation was made within an impulse approximation in which DDCS and TDCS were calculated for Ps colliding with Xe at 600 eV incident energy [94]. This work probed the emergence of 'footprints' of the constituent electron and positron scattering independently, but coherently, from the target. Experimentally, due to limitations in the Ps beam intensity, investigations have been carried out at lower energies (18–100 eV) by detecting final-state positrons or electrons released from Ps colliding with He and Xe. As discussed below, these studies have found Ps fragmentation to make a sizeable contribution to the Ps total scattering cross section (e.g., [95–97]) in broad agreement with theories (e.g., [98–101]).

At UCL, the beam of tunable Ps atoms is produced by passing a near-monoenergetic beam of positrons through a gas cell (cell 1 in Fig. 5.13). The resultant neutral beam has a kinetic energy which varies according to $E_{Ps} \simeq E_+ - E_I + 6.8$ eV, where E_+ is the positron beam energy and 6.8 eV is the ground-state binding energy of Ps (e.g., [102–104]). Forward-travelling Ps atoms enter cell 2 containing the target gas and positrons (or electrons) ejected from collisions

Table 5.2 *The lowest energy reactions for Ps–He scattering in which a free positron and/or electron are produced in the final state. As well as target elastic (TE) collisions, target inelastic processes (TI, i.e., including target ionization or excitation) are also indicated. Should the Ps negative ion, Ps⁻, be produced, the lifetime of 0.48 ns against annihilation [105] means only an electron remains available for detection.*

	Reaction	Threshold (eV)	Description
(a)	$Ps + He \rightarrow e^+ + e^- + He$	6.8	Ps fragmentation
(b)	$Ps + He \rightarrow Ps^- + He^+$	24.3	Ps⁻ formation
(c)	$Ps + He \rightarrow Ps + e^- + He^+$	24.6	Target ionization
(d)	$Ps + He \rightarrow e^+ + e^- + He^*$	26.6	Ps fragmentation with target excitation
(e)	$Ps + He \rightarrow Ps^* + e^- + He^+$	29.7	Ps excitation with target ionization
(f)	$Ps + He \rightarrow e^+ + 2e^- + He^+$	31.4	Ps fragmentation with target ionization

Table 5.3 *The lowest energy reactions for Ps–Xe scattering in which a free positron and/or electron are produced in the final state.*

	Reaction	Threshold (eV)	Description
(a)	$Ps + Xe \rightarrow e^+ + e^- + Xe$	6.8	Ps fragmentation
(b)	$Ps + Xe \rightarrow Ps^- + Xe^+$	11.8	Ps⁻ formation
(c)	$Ps + Xe \rightarrow Ps + e^- + Xe^+$	12.1	Target ionization
(d)	$Ps + Xe \rightarrow e^+ + e^- + Xe^*$	15.1	Ps fragmentation with target excitation
(e)	$Ps + Xe \rightarrow Ps^* + e^- + Xe^+$	17.2	Ps excitation with target ionization
(f)	$Ps + Xe \rightarrow e^+ + 2e^- + Xe^+$	18.9	Ps fragmentation with target ionization

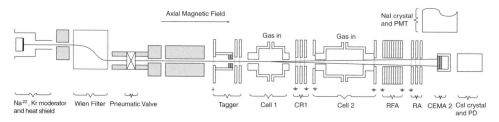

Figure 5.13 Schematic diagram of the Ps beam at UCL. Shown in grey are the collimators. The 'tagger', a timing device consisting of microchannel plates (CEMA 1), a remoderator and accelerating grids, may be removed from the beamline via a UHV manipulator.

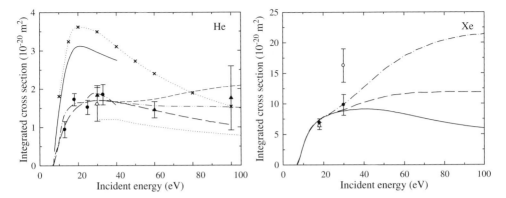

Figure 5.14 Integrated fragmentation cross sections (Q_f^+ and Q_f^-) for Ps collisions with He and Xe. Left: He. Experiment: Q_f^+: (•) [95]; (▲) [96]. Q_f^-: (○) [96]. Theory: $\cdots \times \cdots$, close-coupling approximation and first Born approximation [108]; ———, CTMC [80]; — · — · —, coupled-pseudostate [98]; — — —, TE, impulse approximation for the ejection of positrons [99]; — ·· — ·· —, TI, impulse and first Born approximations (equivalent to Q_f^+) [100]; - - -, TI, impulse and first Born approximations (equivalent to Q_f^-) [101]; $\cdots\cdots\cdots$, Coulomb–Born approximation [106]. Right: Xe. Experiment: (•) Q_f^+ [97]; (○) Q_f^-. Theory: impulse approximation: ———, TE, for the ejection of positrons [101]; — — —, TI, impulse and first Born approximations (equivalent to Q_f^+) [101]; — · — · —, TI, impulse and first Born approximations (equivalent to Q_f^-) [101].

therein are detected downstream with a pair of channel plates (CEMA 2). The signals produced by positron impact may be used in coincidence either with nearby γ-ray detectors (to enhance signal-to-background levels) or with the tagger to generate time-of-flight spectra, discussed further below.

The first experimental study was carried out in collision with He at 13, 18, 25 and 33 eV incident Ps energies [95]. Here, the detection of positrons provided an unambigous signature of Ps fragmentation, comprising reactions (a), (d) and (f) in Table 5.2. A time-of-flight method was used and cross sections integral and differential with respect to the longitudinal energy of the ejected positrons (Q_f^+ and dQ_f^+/dE_ℓ, respectively) were obtained. The tagging system comprises a pair of channel plates (CEMA 1) arranged to detect the secondary electrons emitted from a remoderator. This signal is detected in delayed coincidence with that from CEMA 2. From the time-of-flight spectra, $Q_f^\pm$ was extracted using Eq. (5.11):

$$Q_f^\pm(E_{\mathrm{Ps}}) = \left(\frac{N_\pm}{N_{\mathrm{(scatt)}}}\right) Q_T^{\mathrm{Ps}} S G\left(\frac{\epsilon_{\mathrm{Ps}}}{\epsilon_\pm}\right), \qquad (5.11)$$

where $N_\pm$ is the number of positrons (electrons) ejected in cell 2, $N_{\mathrm{(scatt)}}$ is the number of scattered Ps atoms, S corrects for the in-flight decay of the Ps beam, G corrects for the differences between the solid angle of CEMA 2 and the target

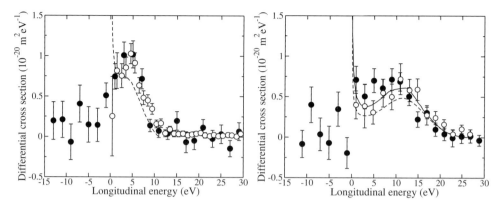

Figure 5.15 Left: dQ_f^+/dE_ℓ for Ps collisions with He and Xe at 18 eV. Xe: (•) [97]. He: (○) [95] multiplied by 4; — — —, TE impulse approximation [99] multiplied by 4. Right: dQ_f^+/dE_ℓ for Ps collisions with He and Xe. Xe: (•) [97] for $E_{Ps} = 30$ eV; ———, TE impulse approximation [100] for $E_{Ps} = 33$ eV. He: (○) [95] for $E_{Ps} = 33$ eV multiplied by 4; — — —, TE impulse approximation [99] for $E_{Ps} = 33$ eV multiplied by 4. There is an additional error of $(-(20–30), +8)\%$ in the He and $(-30, +50)\%$ in the Xe absolute scale due to the uncertainty in determining electron, positron and Ps detection efficiencies.

cell with respect to the production point of the Ps beam, Q_T^{Ps} is the Ps total cross section and $\epsilon_{Ps,\pm}$ are the detection efficiencies for Ps, positrons and electrons. Figure 5.14 (left) shows the results from this work compared with theories. The experimental results rise quickly from 15 to 33 eV, with the contribution of Q_f^+ to Q_T^{Ps} increasing from $\sim 20\%$ to $\sim 40\%$ over this range. The target elastic (TE) 22-pseudostate close-coupled calculation [98], and the impulse approximation [99] are in good agreement with experiment; the results of the Coulomb–Born approach [106] underestimate measured values by a factor of around two, whilst those from the CTMC approach [80] overestimate them by approximately the same factor. The works of [98] and [99] neglected target inelastic (TI) effects having estimated, through a first Born approximation, such a contribution to be very small, a finding later confirmed by Walters *et al.* [100].

The Ps–He measurements for dQ_f^+/dE_ℓ, at 18 eV and 33 eV are shown in Fig. 5.15 (hollow points) [95]. At each energy, a peak is observed at around half the final-state energy, $E_r/2$ (equivalent to 5.6 eV for $E_{Ps} = 18$ eV and 13.1 eV for $E_{Ps} = 33$ eV collision energies) indicating the occurrence of electron loss to the continuum (ELC), a process well documented in the collision of structured projectiles with a target atom/molecule in which the projectile is ionized and drags its daughter electron with a small relative velocity [107]. The observed peaks in dQ_f^+/dE_ℓ were found to be around 1 to 2 eV below $E_r/2$, consistently with particle ejection within a cone of $\sim 20°$ relative to the beam axis and broadly with theories [80, 99], although at least part of the shift arises from the energy resolution of the Ps beam [80].

Following the CTMC [80] and impulse approximation [99] theories in which an asymmetry between the shapes of dQ_f^+/dE_ℓ and dQ_f^-/dE_ℓ was predicted, efforts were made to detect the ejected electrons. An RFA method – in which the energy of the ejected particles would be determined by a varying potential applied to a pair of grids before CEMA 2 (RA in Fig. 5.13) – was developed [96] to counter limitations of the time-of-flight arrangement, such as poor energy resolutions at high energies, small signal-to-background ratios at low energies and low Ps count rates. Additionally, the removal of the tagger increased the Ps beam intensity by an order of magnitude. The scattering cell (cell 2) was floated to a negative potential, in order to accelerate the ejected electrons to energies beyond those typically held by secondary electrons released from various surfaces. Also, the negative potential applied to cell 2 served to reflect background electrons released upstream. As a check on possible systematic effects, cell 2 was maintained at a positive potential when detecting the ejected positrons. A comparison between Q_f^+ measured using the time-of-flight and RFA techniques is shown in Fig. 5.14 (left, black points and triangle) for which a good agreement was found. Also shown is Q_f^- for $E_{Ps} = 30\,eV$ (hollow point). The difference between Q_f^+ and Q_f^- – comprising reactions (b), (c) and (e) in Table 5.2 – corresponds to target ionization events. As Q_f^- and Q_f^+ are similar, there appears to be little evidence of target ionization for $E_{Ps} = 30\,eV$ collisions with He, consistent with theory [100].

More recently, the RFA method has been used to determine Q_f^+, Q_f^- and dQ_f^+/dE_ℓ for Ps collisions on Xe at $E_{Ps} = 18$ and $30\,eV$. Table 5.3 shows the lowest energy reactions to which the experiment is sensitive, and results for Q_f^+ and Q_f^- are shown in Fig. 5.14 along with target elastic and inelastic theories [100, 101]. At low energies, target inelastic effects are negligible according to theoretical calculations, consistent with the near-equality of the experimental Q_f^+ and Q_f^- at 18 eV. Moving from 18 to 30 eV, Q_f^+ increases by a factor of 1.5 – the ratio Q_f^+/Q_T^{Ps} rising from $\sim$20% to $\sim$30% – in agreement with the theories of [100] and [101]. However, at this energy, Q_f^- measurements are almost twice the corresponding theoretical predictions [101] and the experimental Q_f^+, suggesting significant target ionization. Further studies are planned.

5.5 Conclusions and outlook

In this chapter, experimental aspects of studies of ionization/fragmentation of atoms and molecules induced by positron and positronium have been considered. Whilst discrepancies remain, even in the case of simple atomic targets, increasing attention is being directed towards more complex molecular targets (also for their relevance to biomolecular systems) and differential studies to exact greater understanding of the underlying dynamics. The future application of RIMS methods to these studies should greatly enhance the rate of progress in this regard.

Acknowledgements   We are grateful to the Engineering and Physical Research Council (E053521/1 and EP/J003980/1), the Royal Society, the European Union and the Hungarian Scientific Research Fund (K67719 and K7303) for financial support.

References

[1]   K. Paludan *et al.*, *J. Phys. B*, **30**, L581 (1997).
[2]   D. J. Murtagh, M. Szłuińska, J. Moxom, P. Van Reeth and G. Laricchia, *J. Phys. B*, **38**, 3857 (2005).
[3]   D. A. Cooke, D. J. Murtagh and G. Laricchia, *Phys. Rev. Lett.*, **104**, 073201 (2010).
[4]   C. Champion and C. Le Loirec, *Phys. Med. Biol.*, **51**, 1707 (2006).
[5]   G. Laricchia, S. Armitage, Á. Kövér and D. J. Murtagh, in *Advances in Atomic, Molecular, and Optical Physics*, Advances in Atomic Molecular and Optical Physics series, Vol. 56 (The Netherlands: Elsevier 2008), pp. 1–47.
[6]   G. Laricchia, D. A. Cooke and S. J. Brawley, *Radiation Damage in Biomolecular Systems Biological and Medical Physics, Biomedical Engineering*, ed. G. Garcia and M. C. Fuss (Dordrecht: Springer, 2012), pp. 143–153.
[7]   D. Fromme, G. Kruse, W. Raith and G. Sinapius, *Phys. Rev. Lett.*, **57**, 3031 (1986).
[8]   D. Fromme, G. Kruse, W. Raith and G. Sinapius, *J. Phys. B*, **21**, L261 (1988).
[9]   H. Knudsen, L. Brun-Nielsen, M. Charlton and M. R. Poulsen, *J. Phys. B*, **23**, 3955 (1990).
[10]   J. P. Marler, J. P. Sullivan and C. M. Surko, *Phys. Rev. A*, **71**, 022701 (2005).
[11]   J. P. Sullivan, A. Jones, P. Caradonna, C. Makochekanwa and S. Buckman, *Rev. Sci. Instrum.*, **79**, 113105 (2008).
[12]   P. Coleman, J. T. Hutton, D. R. Cook and C. A. Chandler, *Can. J. Phys.*, **60**, 584 (1982).
[13]   O. Sueoka and S. Mori, *J. Phys. Soc. Jpn.*, **53**, 2491 (1984).
[14]   L. M. Diana *et al.*, in *Positron (Electron)–Gas Scattering*, ed. W. E. Kauppila, T. S. Stein and J. M. Wadehra (Singapore: World Scientific, 1986).
[15]   O. Sueoka and S. Mori, *J. Phys. B*, **22**, 963 (1989).
[16]   S. Mori and O. Sueoka, *J. Phys. B*, **27**, 4349 (1994).
[17]   G. Kruse, A. Quermann, W. Raith, G. Sinapius and M. Weber, *J. Phys. B*, **24**, L33 (1991).
[18]   S. Helms, U. Brinkmann, J. Deiwiks, H. Schneider and R. Hippler, *Hyperfine Interactions*, **89**, 395 (1994).
[19]   L. S. Fornari, L. M. Diana and P. G. Coleman, *Phys. Rev. Lett.*, **51**, 2276 (1983).
[20]   H. Knudsen and J. P. Reading, *Phys. Rep.*, **212**, 107 (1992).
[21]   F. M. Jacobsen, N. P. Frandsen, H. Knudsen, U. Mikkelsen and D. M. Schrader, *J. Phys. B*, **28**, 4691 (1995).
[22]   D. A. Cooke, D. J. Murtagh and G. Laricchia, *J. Phys. Conf. Ser.*, **199**, 012006 (2010).
[23]   H. Bluhme *et al.*, *J. Phys. B*, **31**, 4631 (1998).
[24]   V. Kara, K. Paludan, J. Moxom, P. Ashley and G. Laricchia, *J. Phys. B*, **30**, 3933 (1997).
[25]   J. Moxom, P. Ashley and G. Laricchia, *Can. J. Phys.*, **74**, 367 (1996).

[26] N. Overton, R. J. Mills and P. G. Coleman, *J. Phys. B*, **26**, 3951 (1993).

[27] M. Charlton, G. Clark, T. C. Griffith and G. R. Heyland, *J. Phys. B*, **16**, L465 (1983).

[28] P. Caradonna *et al.*, *Phys. Rev. A*, **80**, 032710 (2009).

[29] R. Rejoub, B. G. Lindsay and R. F. Stebbings, *Phys. Rev. A*, **65**, 042713 (2002).

[30] G. Laricchia, P. Van Reeth, M. Szłuińska and J. Moxom, *J. Phys. B*, **35**, 2525 (2002).

[31] H. Bluhme *et al.*, *J. Phys. B*, **32**, 5825 (1999).

[32] M. Szłuińska and G. Laricchia, *Nucl. Instr. and Meth. B*, **221**, 107 (2004).

[33] T. S. Stein *et al.*, *Nucl. Instr. Meth. B*, **143**, 68 (1998).

[34] D. A. Cooke, unpublished. PhD thesis, University College London (2010).

[35] J. P. Marler and C. M. Surko, *Phys. Rev. A*, **72**, 062713 (2005).

[36] T. C. Griffith, *Positron Scattering in Gases* (New York: Plenum, 1984), p. 53.

[37] H. C. Straub, B. G. Lindsay, K. A. Smith and R. F. Stebbings, *J. Chem. Phys.*, **105**, 4015 (1996).

[38] D. A. Cooke, D. J. Murtagh, Á. Kövér and G. Laricchia, *Nucl. Instr. Meth. B*, **266**, 466 (2008).

[39] C. K. Kwan *et al.*, *Nucl. Instr. Meth. B*, **143**, 61 (1998).

[40] D. J. Murtagh, D. A. Cooke and G. Laricchia, *Phys. Rev. Lett.*, **102**, 133202 (2009).

[41] C. P. Campbell, M. T. McAlinden, A. A. Kernoghan and H. R. J. Walters, *Nucl. Instr. Meth. B*, **143**, 41 (1998).

[42] P. Chaudhuri and S. K. Adhikari, *J. Phys. B*, **31**, 3057 (1998).

[43] P. Khan, P. S. Mazumdar and A. S. Ghosh, *Phys. Rev. A*, **31**, 1405 (1985).

[44] R. N. Hewitt, C. J. Noble and B. H. Bransden, *J. Phys. B*, **25**, 557 (1992).

[45] N. K. Sarkar, M. Basu and A. S. Ghosh, *Phys. Rev. A*, **45**, 6887 (1992).

[46] S. Gilmore, J. E. Blackwood and H. R. J. Walters, *Nucl. Instr. Meth. B*, **221**, 129 (2004).

[47] S. Tsurubuchi and T. Iwai, *J. Phys. Soc. Jpn.*, **37**, 1077 (1974).

[48] J. Moxom, G. Laricchia, M. Charlton, G. O. Jones and Á. Kövér, *J. Phys. B*, **25**, L613 (1992).

[49] Á. Kövér, G. Laricchia and M. Charlton, *J. Phys. B*, **26**, L575 (1993).

[50] Á. Kövér, G. Laricchia and M. Charlton, *J. Phys. B*, **27**, 2409 (1994).

[51] R. A. Sparrow and R. E. Olson, *J. Phys. B*, **27**, 2647 (1994).

[52] A. Schmitt, U. Cerny, H. Möller, W. Raith and M. Weber, *Phys. Rev. A*, **49**, R5 (1994).

[53] R. M. Finch, Á. Kövér, M. Charlton and G. Laricchia, *J. Phys. B*, **29**, L667 (1996).

[54] R. DuBois, C. Doudna, C. Lloyd, M. Kahveci, Kh. Khayyat, Y. Zhov and D. H. Madison, *J. Phys. B*, **34**, L783 (2001).

[55] D. H. Madison, *Phys. Rev. A*, **8**, 2449 (1973).

[56] C. Arcidiacono, J. Beale, Z. D. Pešić, Á. Kövér and G. Laricchia, *J. Phys. B*, **42**, 065205 (2009).

[57] O. G. de Lucio, J. Gavin and R. D. DuBois, *Phys. Rev. Lett.*, **97**, 243201 (2006).

[58] O. G. de Lucio, S. Otranto, R. Olson and R. DuBois, *Phys. Rev. Lett.*, **104**, 163201 (2010).

[59] R. DuBois, O. G. de Lucio and J. Gavin, *Nucl. Instr. Meth. B*, **266**, 397 (2008).

[60] M. Lucas and K. G. Harrison, *J. Phys. B*, **5**, L20 (1972).

[61] M. Rodbro and F. D. Andersen, *J. Phys. B*, **12**, 2883 (1979).

[62] P. Mandal, K. Roy and N. C. Sil, *Phys. Rev. A*, **33**, 756 (1986).

[63] L. D. Faddeev, *Zh. Eksp. Teor. Fiz.*, **39**, 1459 (1961).

[64] D. R. Schultz and C. Reinhold, *J. Phys. B*, **23**, L9 (1990).

[65] K. Tőkési and Á. Kövér, *J. Phys. B*, **33**, 3067 (2000).

[66] J. Fiol and R. E. Olson, *J. Phys. B*, **35**, 1173 (2002).

[67] J. Fiol, P. Macri and R. O. Barrachina, *Nucl. Instr. Meth. B*, **267**, 211 (2009).

[68] M. Brauner and J. S. Briggs, *J. Phys. B*, **19**, L325 (1986).
[69] M. Brauner and J. S. Briggs, *J. Phys. B*, **26**, 2451 (1993).
[70] R. O. Barrachina, *Nucl. Instr. Meth. B*, **124**, 198 (1997).
[71] Á. Benedek and R. I. Campeanu, *J. Phys. B*, **40**, 1589 (2007).
[72] J. Fiol and R. O. Barrachina, *J. Phys. B*, **44**, 075205 (2011).
[73] M. Brauner, J. S. Briggs and H. Klar, *J. Phys. B*, **22**, 2265 (1989).
[74] Á. Kövér and G. Laricchia, *Phys. Rev. Lett.*, **80**, 5309 (1998).
[75] J. Berakdar, *Phys. Rev. Lett.*, **81**, 1393 (1998).
[76] J. Fiol, V. D. Rodriguez and R. O. Barrachina, *J. Phys. B*, **34**, 933 (2001).
[77] Á. Kövér and G. Laricchia, *Meas. Sci. Technol.*, **12**, 1875 (2001).
[78] C. Arcidiacono, Á. Kövér and G. Laricchia, *Phys. Rev. Lett.*, **95**, 223202 (2005).
[79] Á. Kövér, K. Paludan and G. Laricchia, *J. Phys. B*, **34**, L219 (2001).
[80] L. Sarkadi, *Phys. Rev. A*, **68**, 032706 (2003).
[81] M. Charlton and J. W. Humberston, *Positron Physics* (Cambridge: Cambridge University Press, 2001).
[82] T. Falke, T. Brandt, O. Kühl, W. Raith and M. Weber, *J. Phys. B*, **30**, 3247 (1997).
[83] M. T. McAlinden and H. R. J. Walters, *Hyperfine Interactions*, **89**, 407 (1994).
[84] R. Dörner *et al.*, *Phys. Rep.*, **330**, 95 (2000).
[85] J. Ullrich *et al.*, *Rep. Prog. Phys.*, **66**, 1463 (2003).
[86] Á. Kövér, D. J. Murtagh, A. I. Williams and G. Laricchia, *J. Phys. Conf. Ser.*, **199**, 012020 (2010).
[87] A. I. Williams, Á. Kövér, D. J. Murtagh and G. Laricchia, *J. Phys. Conf. Ser.*, **199**, 012025 (2010).
[88] R. O. Barrachina, J. Fiol and P. Macri, *Nucl. Instr. Meth. B*, **266**, 402 (2008).
[89] K. Tökesi, *ICPEAC abstracts*, 2011.
[90] R. O. Barrachina, J. Fiol and F. O. Navarrete, *ICPEAC abstracts*, 2011.
[91] M. Dürr *et al.*, *J. Phys. B*, **36**, 4097 (2006).
[92] C. Hugenschmidt *et al.*, *Nucl. Instr. Meth. B*, **593**, 616 (2008).
[93] T. Pflüger *et al.*, *J. Phys. Conf. Ser.*, **262**, 012047 (2011).
[94] J. Ludlow and H. R. J. Walters, *Many-Particle Spectroscopy of Atoms, Molecules, Clusters, and Surfaces* (New York: Kluwer/Plenum, 2001).
[95] S. Armitage, D. E. Leslie, A. J. Garner and G. Laricchia, *Phys. Rev. Lett.*, **89**, 173402 (2002).
[96] S. Armitage, D. E. Leslie, J. Beale and G. Laricchia, *Nucl. Instr. Meth. B*, **247**, 98 (2006).
[97] S. J. Brawley *et al.*, *Nucl. Instr. Meth. B*, **266**, 497 (2008).
[98] J. E. Blackwood, C. P. Campbell, M. T. McAlinden and H. R. J. Walters, *Phys. Rev. A*, **60**, 4454 (1999).
[99] C. Starrett, M. T. McAlinden and H. R. J. Walters, *Phys. Rev. A*, **72**, 012508 (2005).
[100] H. R. J. Walters, C. Starrett and M. T. McAlinden, *Nucl. Instr. Meth. B*, **247**, 111 (2006).
[101] C. Starrett, Mary T. McAlinden and H. R. J. Walters, *Phys. Rev. A*, **77**, 042505 (2008).
[102] A. J. Garner, G. Laricchia and A. Özen, *J. Phys. B*, **29**, 5961 (1996).
[103] D. E. Leslie, S. Armitage and G. Laricchia, *J. Phys. B*, **35**, 4819 (2002).
[104] G. Laricchia, S. Armitage and D. E. Leslie, *Nucl. Instr. Meth. B*, **221**, 60 (2004).
[105] A. P. Mills, Jr., *Phys. Rev. Lett.*, **50**, 671 (1983).
[106] H. Ray, *J. Phys. B*, **35**, 3365 (2002).
[107] G. B. Crooks and M. E. Rudd, *Phys. Rev. Lett.*, **25**, 1599 (1970).
[108] P. K. Biswas and S. K. Adhikari, *Phys. Rev. A*, **59**, 363 (1999).

6

(e,2e) spectroscopy using fragmentation processes

JULIAN LOWER, MASAKAZU YAMAZAKI AND
MASAHIKO TAKAHASHI

6.1 Introduction

The ionization of atoms and molecules by electron impact is of considerable technological and theoretical relevance. From the practical perspective it plays a central role in many atmospheric, industrial and environmental processes. Examples include the physics and chemistry of the upper atmosphere, the operation of discharges and lasers, radiation-induced damage in biological material and plasma etching processes [1–3]. The extent to which such processes can be controlled and/or optimized is limited by our ability to describe the underlying physical mechanisms which drive them. To refine our understanding, new experimental and theoretical results are required. From a broader perspective, the process of electron-impact-induced ionization of atoms and molecules provides an ideal testbed to refine models for the few- and many-body behaviour of identical particles whose interaction is mediated through the Coulomb potential. Moreover, as the ionization process is extremely sensitive to the electronic structure of the target, comparison of experimentally derived ionization cross sections with calculation provides a powerful means to refine models for the target electronic structure. Historically, (e,2e) measurements can be divided into two categories, namely, those whose primary aim is the determination of the target electronic structure and those whose focus is revealing underlying ionization mechanisms. For the former case (so-called electron momentum spectroscopy (EMS) studies), measurements are performed at relatively high impact energies and for roughly equal energies for the two scattered electrons. Under such conditions the ionization mechanism is quite well understood, with the primary electron interacting predominantly with a single bound target electron. From these measurements, electron momentum distributions corresponding to specific values of electron binding energy can be

Fragmentation Processes: Topics in Atomic and Molecular Physics, ed. Colm T. Whelan. Published by Cambridge
University Press. © Cambridge University Press 2013.

derived experimentally (see Section 6.4). For the latter case, measurements are typically performed for lower impact energies and for asymmetric energy sharing between the two scattered electrons. Here the many-body interactions between the continuum electrons and all electrons in the target are enhanced and play a significant role in determining the outcome of the reaction. Both studies are related, to the extent that the quality of the spectroscopic data extracted from EMS measurement depends crucially upon the degree to which the smaller contributions from many-body interactions in the ionization process can be understood and accounted for when comparing experimental results to the predictions of theory.

The most precise information concerning electron-induced ionization is provided by differential cross sections in which the rate of the ionization process is prescribed for specific values of reaction kinematics and quantum states of the target, residual ion and projectile and scattered electrons. For single ionization processes of atoms, the so-called triple differential cross section (TDCS) provides full information, reflecting the ionization rate as a function of the momenta of the projectile electron in the initial state and the two scattered electrons in the final state. Experimentally, TDCSs can be determined in coincidence measurements in which the products, derived from common ionization events, are identified by their correlated arrival times at detectors. For such reactions, however, it is sufficient to measure the momenta of only two of the three charged products (two electrons and the residual ion), with the momentum of the third determined through momentum conservation. In most cases, TDCSs have been determined in so-called (e,2e) measurements [4] where the two scattered electrons are measured in time coincidence. More recently [5], TDCSs have been determined in electron–ion coincidence measurements with the advantage of enabling simultaneous measurement over large regions of momentum space.

For single ionization processes of molecular targets, however, more detailed information regarding the electron-induced ionization is required for a more complete understanding of the (e,2e) reaction, namely, determination of the fivefold differential cross sections (5DCSs) for which the spatial orientation of the target molecule is specified. This is because the separation of nuclei in molecules leads to a strongly non-central scattering potential. In this case, the ionization cross section depends sensitively on the angles that the momentum vectors of the projectile and scattered electrons make with the symmetry axes of the molecule. Electron scattering calculations involving such a reduced-symmetry scattering potential pose a considerable computational challenge; thus (e,2e) measurements in which the alignment or orientation of the molecular axis is determined are highly desirable to test models for molecular ionization and open up the area of stereodynamics of electron–molecule ionizing collision. Moreover, orientation-resolved

EMS enables one to look at three-dimensional patterns of molecular orbitals in momentum space and promotes further momentum space chemistry [6]. As such, orientation-resolved (e,2e) spectroscopy, which can be realized by specifying the orientation of the molecular axis at the time of the ionization, is a method of great capabilities.

The production of ensembles of highly aligned or oriented molecules in the gas phase is, however, very difficult. A train of linearly [7,8] or elliptically polarized [9] laser pulses, which are non-resonant with respect to electronic or vibrational transitions, can be used to achieve a partial degree of molecular orientation. However, the low duty-cycle of the laser pulses renders this technique infeasible for incorporation into crossed-beam experiments involving continuous electron beams. Molecules with a permanent dipole moment can be partially oriented through the use of a supersonic molecular jet and a homogeneous electric field [10] or via state selection with a hexapole field [11]. However, if such fields were employed in a crossed-beam experiment they would perturb the trajectories of projectile electrons and the charged fragments. There is another opportunity for the investigation. If the molecular ion dissociates much faster than it rotates, the direction of fragment-ion departure coincides with the molecular orientation at the moment of the ionization [12]. (e,2e) spectroscopy in conjunction with such axial-recoil fragmentation processes is the method applied for all the orientation-resolved (e,2e) experiments to be subsequently described.

In Section 6.2, the electron-impact-induced dissociative-ionization process is described along with the relationship between the various kinematic variables. A short review of previous experiments and their various experimental approaches is presented. Section 6.3 provides an overview of recent theoretical developments and approaches in the description of orientation-resolved dissociative-ionization processes. Section 6.4 describes the results of high-energy EMS experiments, which demonstrate the ability of orientation-resolved (e,2e) spectroscopy to measure three-dimensional electron momentum densities. Section 6.5 reviews low-energy investigations, whose purpose is to elucidate underlying mechanisms of fragmentation. Finally, Section 6.6 provides a summary of the main results and describes future prospects in the field.

6.2 Background

As previously described, the study of electron-induced dissociative ionization, in the framework of a coincidence experiment, enables orientation-resolved ionization cross sections to be extracted from measurement. In its simplest form (single ionization, two nuclear fragments), the reaction can be written as:

$$e_0^-(\boldsymbol{p}_0) + \mathrm{M} \rightarrow \mathrm{M_a}(\boldsymbol{p}_a) + \mathrm{M_b^+}(\boldsymbol{p}_b) + e_1^-(\boldsymbol{p}_1) + e_2^-(\boldsymbol{p}_2). \qquad (6.1)$$

In this relation, $e_0^-(\boldsymbol{p}_0)$, $e_1^-(\boldsymbol{p}_1)$ and $e_2^-(\boldsymbol{p}_2)$ represent incident and scattered electrons of respective momenta $\boldsymbol{p}_0$, $\boldsymbol{p}_1$ and $\boldsymbol{p}_2$ (energies E_0, E_1 and E_2), M a parent molecule, M_a a fragment molecule/atom and M_b^+ a fragment molecular/atomic ion. Depending on their electronic structure, M_a and M_b^+ may be excited to a range of internal energy states. For the case of molecular hydrogen, the reaction simplifies to:

$$e_0^-(\boldsymbol{p}_0) + H_2 \rightarrow H(n, \boldsymbol{p}_H) + H^+(\boldsymbol{p}_i) + e_1^-(\boldsymbol{p}_1) + e_2^-(\boldsymbol{p}_2). \qquad (6.2)$$

$H(n, \boldsymbol{p}_H)$ represents a hydrogen atom in the electronic state of principal quantum number n, and $\boldsymbol{p}_H$ is its momentum. $H^+(\boldsymbol{p}_i)$ represents a proton of momentum $\boldsymbol{p}_i$ (energy E_i). Given the large mass difference, kinetic energy transfer between the continuum electrons and the hydrogen molecule, hydrogen atom and proton is generally minimal. Thus $\boldsymbol{p}_i \cong -\boldsymbol{p}_H$ and for a known value of $\boldsymbol{p}_0$, determination of $\boldsymbol{p}_1$, $\boldsymbol{p}_2$ and $\boldsymbol{p}_i$ completely determines the reaction kinematics. Furthermore, by invoking energy conservation, the appearance energy $A(n)$ for transitions to the quantum state n of the residual hydrogen atom is determined through the relation:

$$A(n) = \varepsilon_b - 2E_i. \qquad (6.3)$$

Here ε_b is the electron binding energy, defined by the expression $\varepsilon_b = E_0 - E_1 - E_2$, and $2E_i$ accounts for the kinetic energy shared between the proton and the hydrogen atom. This expression allows ionization events to be sorted according to the dissociation limits of the respective transitions with which they are associated. Determination of $\boldsymbol{p}_i$ enables the molecular orientation at the time of ionization to be inferred and orientation-resolved data to be obtained.

Figure 6.1 shows the potential energy diagrams for H_2 and for the four states of H_2^+, the transitions between which are the focus of the ionization measurements described in this chapter. Dissociative ionization of the H_2 molecule may occur through a number of distinct pathways. It can proceed directly via transitions from the $X^1\Sigma_g^+$ electronic ground state of H_2 to the vibrational continuum of the $1s\sigma_g$ ground state of H_2^+. Protons and hydrogen atoms produced through these transitions are released with low values of kinetic energy (typically <1 eV). Alternatively, dissociation may occur by the direct excitation of both target electrons, one to the ionization continuum and the other to an excited state of H_2^+ (all excited states of H_2^+ are dissociative). Dissociative ionization may also occur *indirectly* through transitions to intermediate autoionizing states of H_2.

To date, three groups, located respectively in Sendai, Japan [13–15], Canberra, Australia [16] and Heidelberg, Germany [17,18] have reported (e,2e) measurements on oriented hydrogen molecules. For the Sendai and Canberra groups, dispersive electrostatic electron energy analyzers, terminated by position- and time-sensitive detectors, were employed to measure the energies of scattered electrons. Such

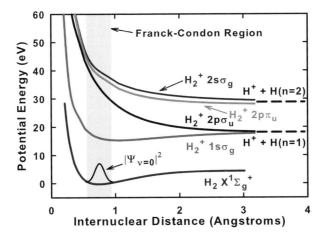

Figure 6.1 Potential energy curves of H_2 and H_2^+ from [37] relevant to transitions discussed in this chapter. Reprinted figure with permission from [16]. Copyright (2010) by the American Physical Society.

analyzers exploit the energy dependence in the degree of deflection electrons experience as they traverse an electric field. Advantages of this approach are that (i) high-energy and momentum resolution can be achieved at high electron energies (a necessary requirement for EMS measurements) and (ii) as a result of only using *position* co-ordinates to determine electron momenta, the arrival *time* of electrons at the detectors can be used to identify (e,2e) pairs. Thus a continuous (non-pulsed) projectile beam can be employed, to the advantage of data-collection efficiency. The disadvantage of the approach is that simultaneous measurement is achieved over only a relatively small volume of electron momentum phase-space. While this is a considerable disadvantage for measurements focusing on (e,2e) reaction dynamics, it is not the case for EMS experiments which, by their very nature, focus only on a small subset of all ionization events (close electron–electron collisions). Thus, for EMS measurements, dispersive electrostatic electron analyzers, with the possibility of continuous electron beam operation, are the ideal choice.

For the Sendai group, measurements were performed under the high-energy EMS conditions and ion momenta were determined by measuring ion drift-times to detectors through a field-free environment. For the Canberra group, a pulsed electric extraction field was employed, enabling the reconstruction of ion momenta for electron emission over the full 4π solid angle. In contrast, the cold target recoil ion momentum spectroscopy (COLTRIMS) technique [19], was employed for the measurements performed by the Heidelberg group. In this technique, the interaction volume, defined by the overlap of electron and molecular beams, is immersed in superposed electric and magnetic fields and the momenta of fragment ions and electrons are determined by their arrival co-ordinates (spatial and temporal) at position- and time-sensitive detectors. This approach has the advantage that, up to some maximum energy values determined by the experimental settings, *both* electrons *and* ions emitted over the 4π solid angle can be simultaneously detected. Being a time-of-flight-based technique, however, a pulsed projectile beam is generally required to determine fragment momenta. In summary, as a result of the low fraction of (e,2e) transitions leading to dissociation of the H_2^+ ion, inefficient detection of electrons and ions and/or the requirement to use a pulsed projectile beam, measurements proved particularly challenging for all three groups and weeks of data time were required.

6.3 Theory

From the theoretical side, there has also been considerable interest in recent years in the topic of molecular effects on electron impact ionization processes, such as interference and orientation effects. For instance, the molecular three-continuum (3C) approximation was applied to the electron impact ionization of H_2 to show that Young's double-slit-type interference should appear in the 5DCSs as a consequence of a coherent sum of effective atomic amplitudes of the ionization transition [20]. The three-body distorted wave (3DW) model [21–23], which used distorted waves to describe all incident and outgoing electrons with averaging over all molecular orientations, reproduced well the differences in TDCSs observed in the perpendicular plane between He and H_2 targets [24]. It has attracted great interest, partly because neutral He and H_2 have an equivalent number of protons and electrons. Furthermore, the model strongly suggested that molecules that have no nuclei at the centre of mass will generally have a minimum for back-to-back scattering and molecules that have a nucleus at the centre of mass will have a maximum, as is found for atoms.

5DCSs from the oriented H_2 molecules at an impact energy of a few tens of eV were explicitly calculated with the time-dependent close-coupling (TDCC) approach [25–27]. The results clearly revealed subtle features in the angular distributions to be observed, which are otherwise washed out by the orientation averaging. Up to now, however, there have been no experiments to be directly compared with the TDCC results; the experimental data for oriented H_2 reported so far are all measured at impact energies above 178 eV. The extension of the TDCC method to the large impact energy conditions is awaited to account for the experimental results, although presently such an extension would require prohibitively large close-coupling calculations due to the need to include a large number of partial waves. A good example of organized efforts of experiment and theory can be seen in a comparison of the experimental 5DCSs for ground-state ionization of oriented

H_2 molecules at $E_0 = 200$ eV with associated calculations using the molecular three-body distorted wave (M3DW) method [17, 18]. It has been found that the calculations successfully reproduce most of the experimental features, though some discrepancies between experiment and theory still remain, especially in the plane perpendicular to the incoming electron beam. In the high-energy EMS conditions, with large energy loss and large momentum transfer, it has been widely believed that the plane wave impulse approximation (PWIA) gives a very good description of the (e,2e) reaction. However, a comparison of the experimental and PWIA results for the transitions to the $2s\sigma_g$ and $2p\sigma_u$ excited states of the oriented H_2^+ molecules also revealed noticeable differences, in particular those in the direction parallel to the molecular axis [13, 14], as will be seen in Section 6.4.

In addition to the above-mentioned studies on the H_2 molecule, molecular orientation effects on electron impact ionization have recently been studied for the water molecule with the first and second Born calculations using an accurate one-centre molecular wave function [28–32]. As such, interest in electron impact ionization of oriented molecules is increasing and broadening out quickly.

6.4 Electron momentum spectroscopy results

EMS refers specifically to (e,2e) spectroscopy near the Bethe ridge [33] at impact energies of the order of 1 keV or higher, where the binary regime, corresponding to a billiard-like collision between the incident and target electrons, plays a leading role in ionization dynamics [15, 34]. The most exciting aspect of this technique is its ability to separately measure the momentum densities for target electrons in different energy levels. Since the electron momentum density is the square modulus of the momentum-space wave function, and since a momentum-space function is the Fourier transform of the position-space function, EMS is expected to serve as an ideal bridge to connect quantum chemical calculations with the actual electronic structures, in particular those relating to the outer, loosely bound electrons that are of central importance in chemical properties such as bonding, reactivity and molecular recognition. For molecules, however, EMS has long been plagued by the fact that the (e,2e) experiments measure averages over all orientations of gaseous targets. The spherical averaging results in enormous loss of versatile information; the intrinsically anisotropic or three-dimensional character of the (e,2e) scattering by molecules is concealed in the one-dimensional momentum distribution or momentum profile. These are the material reasons why what is most wanted in gas-phase EMS is to establish a method to avoid the spherical averaging in order to look at molecular orbitals in the three-dimensional form. The provision of such three-dimensional momentum distributions has the potential to promote important developments in the area of momentum-space chemistry [6].

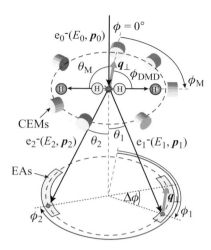

Figure 6.2 Schematic of an electron-electron-ion experimental set-up based on the binary (e,2e) scattering in the symmetric non-coplanar geometry, showing seven channel electron multipliers (CEMs), and a pair of entrance apertures (EAs) of a spherical analyzer followed by a pair of position-sensitive detectors. Reprinted figure with permission from [14]. Copyright (2005) by the American Physical Society.

Recently, an attempt towards the end mentioned above has been made by the Sendai group [13, 14, 35]. Here, EMS experiments have been conducted for ionization–excitation transitions of H_2 that leave the residual ion in the $2s\sigma_g$ and $2p\sigma_u$ excited states of H_2^+, while additionally making use of fragmentation processes of the residual ion H_2^+. Note that as all the transitions to the excited states of H_2 are followed by direct dissociation processes, their axial-recoil fragmentation [12] is unambiguous. Thus, it is possible to measure the electron momentum distribution with respect to the molecular axis or in the three-dimensional form, by investigating vector correlations between the two outgoing electrons and the fragment ion with a triple coincidence technique. The triple coincidence experiments for H_2 were performed using an electron-electron-ion spectrometer shown schematically in Fig. 6.2 (see [13] for details). It employs the symmetric non-coplanar geometry, in which two outgoing electrons having equal energies $(E_1 = E_2)$ and equal scattering angles $(\theta_1 = \theta_2 = 45°)$ with respect to the incident electron momentum (p_0) are detected. Electron impact ionization occurs at an interaction volume where the incident electron beam collides with a gaseous H_2 target. A pair of entrance apertures restricts the electrons leaving the interaction volume so that only those with scattering angle of $\theta = 45°$ and azimuthal angles ϕ_1 and ϕ_2 in the ranges between $70°–110°$ and $250°–290°$ are accepted into a spherical analyzer. The accepted electrons are dispersed, based on their kinetic energies, by the spherical analyzer before being detected by a pair of position-sensitive detectors

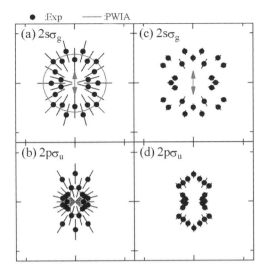

Figure 6.3 Experimental and theoretical 5DCSs of H_2 for the transitions to the $2s\sigma_g$ and $2p\sigma_u$ excited ion states, obtained at an impact energy of 1.2 keV for (a) and (b) and at an impact energy of 2.0 keV for (c) and (d). See text for details. (a) and (b) adapted with permission from [14]. Copyright (2005) by the American Physical Society.

placed behind an exit aperture. On the other hand, fragment ions are detected by the time-of-flight (TOF) technique using seven channel electron multipliers (CEMs) placed in the plane perpendicular to p_0. The azimuthal angular positions (ϕ_M's) of the CEMs are 0, 45°, 90°, 150°, 195°, 240°, and 285°. A retarding electric field is applied in front of each CEM in order to collect only axial-recoil fragments with large kinetic energies. In this experimental set-up, the experiment probes basically the component of the target electron momentum perpendicular to p_0 and hence the angle ϕ_{DMD} of the target electron momentum from the molecular axis can be approximated as ϕ_M, i.e., $\phi_{DMD} \sim \phi_M$ [13, 14].

The first triple coincidence study on H_2 [13, 14] was carried out under experimental conditions where an impact energy of 1.2 keV and a retarding voltage of 2.5 V were employed. In this study, a series of (e,2e) binding energy spectra at each ϕ_M, summed over the momentum range covered, were used to extract contributions from the individual ionization–excitation transitions by a deconvolution procedure. In the data analysis, Franck–Condon transition profiles, calculated using the BCONT program [36] with the relevant potential energy curves [37], were used as fitting curves to reproduce the experiments by summing them with appropriate weight factors. The results for the $2s\sigma_g$ and $2p\sigma_u$ transitions are shown in Fig. 6.3(a) and (b) [14], where they are plotted as a function of ϕ_{DMD} with the molecular axis drawn in the vertical direction. Note that although the measurements were performed at the seven ϕ_M values, the results are presented so as to give the 24 data points, by taking advantage of the inversion symmetry and rotation symmetry of the (e,2e) cross sections about the molecular axis for a homonuclear molecule. Also included in the figures are associated theoretical distributions calculated using the PWIA method with a configuration interaction target H_2 wave function. Here, the experiments and theory are compared by normalizing the $2s\sigma_g$ and $2p\sigma_u$ experimental intensities to the associated PWIA calculations while maintaining the relative intensities of the two transitions.

Although the statistics of the experimental data leave much to be desired, anisotropy of (e,2e) scattering by molecules is clearly seen in Fig. 6.3(a) and (b). Effects of molecular orientation on the (e,2e) scattering amplitudes are evident and, further, the $2s\sigma_g$ and $2p\sigma_u$ experiments are strikingly different from each other. While the former angular distribution is relatively isotropic, the latter shows maxima along the molecular axis. This observation is just what PWIA predicts, which probes excited components of the electron momentum densities of the $2s\sigma_g$ and $2p\sigma_u$ states resulting from electron correlation in the target H_2 wave function.

Another example of the triple coincidence study is given in Fig. 6.3(c) and (d), which has been obtained under experimental conditions where an impact energy of 2.0 keV and a retarding voltage of 0.1 V were employed [35]. Here, extraction of contributions from the individual ionization–excitation transitions was achieved by a different approach. Firstly, a series of spectra at each ϕ_M, which plotted the triple coincidence counts against the kinetic energy of the fragment ion, were constructed. Then, theoretical kinetic energy distributions, obtained by applying the reflection approximation to the Franck–Condon transition profiles, were used as fitting curves to reproduce the experiments. Similar effects of molecular orientation on the (e,2e) scattering amplitudes are also visible in Fig. 6.3(c) and (d).

Nevertheless, one may notice the differences between the experimental and PWIA intensities; the experiments for the $2p\sigma_u$ channel show more significant deviation from theory in the direction of the molecular axis. If this observation is real, it would indicate that the two outgoing electrons escape preferentially so as to leave the ion recoil momentum vector along the molecular axis. The intensity differences should be examined in terms of higher order effects that the first-order PWIA theory does not take into account, because at the impact energies employed it has already been revealed from conventional, spherically averaged EMS measurements on H_2 [38] that the second-order two-step mechanisms give noticeable contributions to the (e,2e) cross sections, in particular for the $2p\sigma_u$ transition. It means that the use of higher impact energies would be required to approach more and more the high-energy limit, where the PWIA theory provides a very good description for two-electron processes of this kind. This is in a sharp contrast with the findings of previous EMS studies [15, 34] that at an impact energy of 1.2 keV PWIA is valid for many of one-electron processes, such as a transition to the ionic ground state.

The comparative studies discussed here [13, 14, 35] can be recognized as a pioneering step to fully take advantage of EMS in the near future, while demonstrating the high reliability of the present two approaches towards quantitatively measuring electron momentum distributions in the three-dimensional form. Indeed, there is ample room for improvements, mainly in collection efficiencies for both the electrons and fragment ion. The latest version of the electron detector arrangement [39], which covers almost completely the available azimuthal angle range (2π) for the symmetric noncoplanar (e,2e) reaction, should be combined with the velocity map imaging technique to detect the fragment ion over the full 4π solid angle [40, 41].

6.5 Low-energy (e,2e) results

Measurements on orientation-resolved electron-impact-induced dissociative ionization at low impact energies were recently performed by the Canberra group [16]. Their study investigated transitions to the $2s\sigma_g$ and $2p\pi_u$ excited states of H_2^+, the same transitions studied previously by the Sendai group [13–15, 35] under high-energy EMS conditions. The apparatus employed two high-efficiency toroidal-sector electrostatic analyzers for momentum analysis of scattered electrons [42] and a pulsed time-of-flight ion spectrometer which collected ions with energies of up to 8 eV over 4π steradians. In this way, cross sections for *all* molecular orientations could be determined, building on the pioneering work of [13–15, 35] in which 'side-on collisions' of the primary electron with the target molecule were treated. However, in contrast to [13–15, 35], results for transitions to the $2p\sigma_u$ state of H_2^+ could not be presented due to a restricted angular capture-range for the higher energy protons associated with that transition. The experimental arrangement had the advantage over a standard COLTRIMS apparatus by allowing independent control of electron and ion energy resolutions and possessed the ability to operate with pulsed projectile beams of high duty-cycle. Nevertheless, in contrast to a COLTRIMS arrangement, it suffered from a low-efficiency collection of scattered electrons inherent to non time-of-flight-based analyzers.

Details of the measurement procedure can be found in [16]. Briefly, electrons emitted within a plane containing the primary electron beam were analyzed in one of two toroidal-sector electrostatic momentum-analyzers, each terminated with a position-and-time-sensitive detector. One analyzer was adjusted to transmit electrons of energy E_1 in the energy range 90 eV $\leq E_1 \leq$ 110 eV over the angular

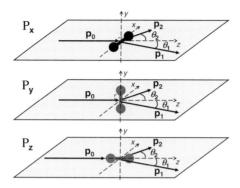

Figure 6.4 Experimental reaction kinematics for Canberra experiments. Electrons of momentum p_0, traveling along the z-axis, ionize an unoriented ensemble of hydrogen molecules. Scattered electrons, emitted in the $x-z$ plane and of respective momenta p_1 and p_2, are detected. For molecules which undergo dissociative ionization, their orientation at the time of ionization is deduced from the momentum p_i of the released proton. The figure shows collisions involving the three orthogonal molecular orientations P_x, P_y and P_z. The projectile-electron energy E_0 is 178 eV and the average energies E_1 and E_2 of the two measured scattered electrons are 100 eV and 40 eV, respectively. Reprinted figure with permission from [16]. Copyright (2010) by the American Physical Society.

range $10° \leq \theta_1 \leq 50°$ on one side of the electron beam. The second analyzer measured electrons of energy E_2 in the energy range 30 eV $\leq E_2 \leq$ 50 eV over the range $30° \leq \theta_2 \leq 70°$ on the other side of the electron beam. Upon detection of each (e,2e) ionization event, identified by the time-correlated arrival of the electrons at the two detectors, an extraction field was applied to the interaction volume for an 8-microsecond period and fragment ions were focused and detected on a third position-and-time-sensitive detector. From measurement of the ion arrival positions and arrival times and by invoking the axial-recoil approximation [12], ion momenta were determined and the molecular orientation at the time of ionization was inferred. The primary electron beam was pulsed with a 30% duty-cycle. Between each electron pulse a 'cleaning cycle' was implemented in which a positive potential was applied to the gas needle through which the target molecular beam eminated. This cycle prevented the build-up of low-energy ions in the interaction region that would otherwise have lead to unacceptably high background levels. Through the determination of electron and ion momenta, partial cross sections describing transitions to the $2s\sigma_g/2p\pi_u$ excited states of H_2^+ could be resolved. Results were presented for the three distinct molecular orientations P_x, P_y and P_z as defined in Fig. 6.4 (see [16] for details), namely for molecules oriented perpendicular to the primary beam direction and within the scattering plane (x orientation), perpendicular to the primary beam and perpendicular to the scattering

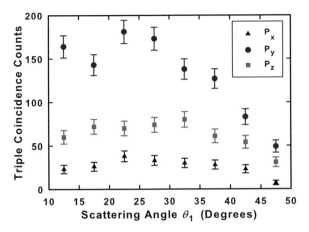

Figure 6.5 Triple-coincidence counts for transitions to the $2s\sigma_g$ and $2p\pi_u$ states of H_2^+ as a function of the fast-electron scattering angle θ_1 for the three molecular orientations $\boldsymbol{P}_x$, $\boldsymbol{P}_y$ and $\boldsymbol{P}_z$. Kinematics as in Fig. 6.4. The data has been averaged over the slow-electron scattering angle θ_2. Reprinted figure with permission from [16]. Copyright (2010) by the American Physical Society.

plane (y orientation) and for molecules oriented along the primary beam direction (z orientation). Due to limited statistics the results were integrated over the scattering angle of the slow scattered electron and over the energy pass-bands of both electron analyzers. In spite of this integration, dramatic orientation-dependent effects remain. Figure 6.5 (see [16] for details) shows the triple-coincidence count rates for transitions to the $2s\sigma_g$ and $2p\pi_u$ states of H_2^+ as a function of the fast-electron scattering angle θ_1. Results correspond to ion emission within 10 degrees of each of the three respective co-ordinate axes. The following observations are made: (i) strong orientation-dependent asymmetries are observed over the full range of projectile-electron scattering angles covered by the measurement; (ii) under the adopted kinematics and for the transitions under investigation, dissociative ionization favours a 'side-on' collision with the molecular axis oriented perpendicular to the scattering plane; (iii) the rate of dissociative ionization depends not only on the angle between the molecular axis and the momentum of the projectile electron, but also on the angle between the molecular axis and the normal to the scattering plane.

More recently, the Heidelberg group [17, 18] presented experimental results for the dissociative ionization of H_2 molecules measured using a COLTRIMS reaction microscope. As mentioned earlier, the great virtue of the COLTRIMS approach is the ability to measure charged reaction fragments simultaneously over a large volume of momentum phase space. Specifically, and in contrast to experimental arrangements employing traditional electrostatic analyzers for electron detection, the technique can access all kinematics at a given electron impact energy, including cases where the incident and scattered electrons are not confined to a common plane [16] or to a cone [13–15, 35]. They employed a 200 eV energy beam of electrons and presented 5DCSs for transitions from $X^1\Sigma_g^+$ electronic ground state of H_2 to the vibrational continuum of the $1s\sigma_g$ ground state of H_2^+. For such transitions, protons are emitted with low energy, typically less than 1 eV. With their chosen electric and magnetic field settings they were able to detect all emitted protons of energy below 1 eV over the complete 4π solid angle. Furthermore, by carefully sorting data according to the energy of the scattered electrons they were able to resolve transitions occurring through direct population of the H_2^+ vibrational continuum from those associated with the excitation and decay of doubly excited states of the H_2 molecule. We note that although the accuracy of the axial-recoil approximation is intrinsically less favourable for low values of ion kinetic energy release [43], the use of a monochromatic and highly collimated molecular beam enabled the Heidelberg group to achieve a determination of the molecular orientation to an uncertainty estimated at less than $\pm 20°$ for kinetic energy releases above 0.13 eV.

In [17], 5DCSs were presented under coplanar kinematics (incident and two scattered electrons confined to a common plane) at an energy of 3.5 ± 2.0 eV for the slow electron and for a projectile scattering angle of $16° \pm 4°$ degrees. These results are displayed in Fig. 6.6 (see [17] for details) for orientation of the internuclear axis at angles of $0°$, $45°$ and $90°$ relative to the direction of momentum transfer. The electron angular distributions show a characteristic double-lobe structure, with a dominant 'binary peak', corresponding to a clean knock-out of a target electron directed roughly along the direction of momentum transfer $\boldsymbol{q}$ and a recoil peak in the opposite direction, corresponding to electrons that have been backscattered by the ion after undergoing an electron–electron collision. The experimental results are compared to the M3DW scattering calculations [44]. Distinct orientation-dependent effects were observed at practically all electron emission angles, with a parallel orientation of the molecule with respect to the momentum transfer direction favoured over a perpendicular orientation. This trend was quantitatively well replicated in the M3DW calculations, which provided a reasonable description of the shape of the measured angular distributions. The experimental results were also compared to calculations based on the two-centre picture developed by Stia *et al.* [20], which involves multiplying the triple differential cross section for an atomic target by an interference factor. It was found that this model disagreed significantly with the experimental results, suggesting that interference effects are accounted for implicitly in the M3DW model in a more sophisticated manner. Cross sections were also presented in [17] as a function of the emission angle of the protonic fragment relative to the scattering plane for selected kinematics. For

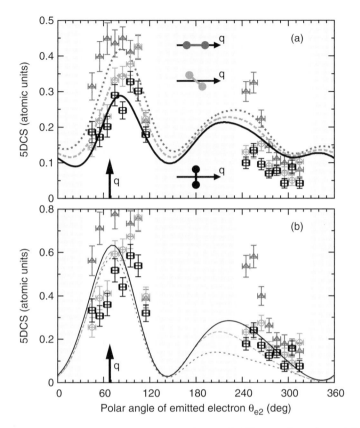

Figure 6.6 Five-fold differential cross sections (5DCS) as a function of the emitted electron's emission angle in the scattering plane. This electron's energy is 3.5 ± 2.0 eV and the projectile scattering angle is $16° \pm 4°$. Points represent experimental results, lines model calculations, which are either (a) molecular three-body distorted wave or (b) the three-Coulomb wave-function approach for a helium target multiplied with the interference factor given by Stia *et al.* (2003). See [17] for details. Reprinted figure with permission from [17]. Copyright (2010) by IOP publishing.

kinematic regions where direct ionization was the dominant channel, proton emission was weakly peaked along q. In contrast, a strong preference for emission along the q axis was observed for dissociation occurring via intermediate autoionizing states of H_2.

In [18], additional results were presented at a 200 eV electron impact energy over an extended kinematic range. The orientation dependence of the ionization cross section on the energy of the emitted electron and on the scattering angle of the projectile electron (momentum transfer) were investigated. It was found that

(i) at higher values of emitted-electron energy the orientation-dependent anisotropy was essentially independent of scattering angle, (ii) the orientation dependence is larger at lower energies of the emitted electron and (iii) for a low energy of electron emission, the anisotropy increases with scattering angle. The behaviour of the 5DCS was also investigated for selected orientations of the molecular axis perpendicular to q but directed out of the scattering plane (geometry defined in Figure 9 of [18]). There the cross section was found to be largely independent of the azimuthal orientation angle ϕ around q.

6.6 Conclusion

This chapter has presented the current status of (e,2e) spectroscopy using fragmentation processes. Effects of molecular orientation on the (e,2e) scattering amplitudes have been unambiguously observed for the H_2 molecule both experimentally and theoretically. Although the currently possible discussion is limited, due mainly to the scarcity of the data, the great potential of the new approach manifests itself in its various forms discussed in the preceding sections. Future efforts should include a better understanding of orientation effects in electron impact ionization not only for diatomic molecules but also for polyatomic molecules that would exhibit interesting electron-induced dynamics. In this respect, the success of the second approach with analysis of the kinetic energy distribution of the fragment ion, described in Section 6.4, is encouraging, because selection of the kinetic energy of the fragment ion is equivalent to highlighting parts of the Franck–Condon transition profiles of particular transitions and hence it would afford the opportunity of measuring the 5DCSs as a function of nuclear-separation length, beyond the framework of the Born–Oppenheimer approximation. For quantitative analysis of such experimental data, further development of theoretical methods is crucial. Strongly aligned or oriented molecular ensembles will also be available for advancing (e,2e) spectroscopy more and more, by using, for instance, a stable 5 kHz femtosecond laser system to generate pump-laser and probe-electron pulses. In this case, fragmentation processes help to identify the head and tail of the aligned heteronuclear molecules. As such, greater information about the target electronic structure and ionization mechanisms will be provided by (e,2e) spectroscopy using fragmentation processes in the future.

Acknowledgements   One of the authors (MT) is grateful to Grant-in-Aids for Scientific Research (S), No. 20225001 and (A), No. 16205006 from the Japan Society for the Promotion of Science (JSPS). JL gratefully acknowledges the support of the Mercator programme of the Deutsche Forschungsgemeinschaft.

References

[1] V. I. Shematovich, R. E. Johnson, M. Michael and J. G. Luhmann, *J. Geophys. Res.*, **108**, 5087 (2003).

[2] J. Liu, F. Sun and H. Yu, *Current Appl. Phys.*, **5**, 625 (2005).

[3] I. Ipolyi *et al.*, Int. J. Mass. Spectrom., **252**, 228 (2006).

[4] I. E. McCarthy and E. Weigold, *Electron-Atom Collisions* (Cambridge: Cambridge University Press, 1995).

[5] X. Ren *et al.*, *Phys. Rev. A*, **82**, 032712 (2010).

[6] C. A. Coulson, *Math. Proc. Cambridge Philos. Soc.*, **37**, 55 (1941).

[7] H. Sakai *et al.*, *J. Chem. Phys.*, **110**, 10235 (1999).

[8] M. Leibscher, I. Sh. Averbukh and H. Rabitz, *Phys. Rev. Lett.*, **90**, 213001 (2003).

[9] J. J. Larsen, K. Hald, N. Bjerre, H. Stapelfeldt and T. Seideman, *Phys. Rev. Lett.*, **85**, 2470 (2000).

[10] H. J. Loesch and A. Remscheid, *J. Chem. Phys.*, **93**, 4779 (1990).

[11] K. H. Kramer and R. B. Bernstein, *J. Chem. Phys.*, **42**, 767 (1965).

[12] R. N. Zare, *Mol. Photochem.*, **4**, 1 (1972).

[13] M. Takahashi *et al.*, *J. Electron. Spectrosc. Relat. Phenom.*, **141**, 83 (2004).

[14] M. Takahashi, N. Watanabe, Y. Khajuria, Y. Udagawa and J. H. D. Eland, *Phys. Rev. Lett.*, **94**, 213202 (2005).

[15] M. Takahashi, *Bull. Chem. Soc. Jpn.*, **82**, 751 (2009).

[16] S. Bellm, J. Lower, E. Weigold and D. W. Mueller, *Phys. Rev. Lett.*, **104**, 023202 (2010).

[17] A. Senftleben *et al.*, *J. Phys. B*, **43**, 081002 (2010).

[18] A. Senftleben *et al.*, *J. Chem. Phys.*, **133**, 044302 (2010).

[19] R. Dörner *et al.*, *Phys. Rep.*, **330**, 95 (2000).

[20] C. R. Stia, O. A. Fojón, P. F. Weck, J. Hanssen and R. D. Rivarola, *J. Phys. B: At. Mol. Opt. Phys.*, **36**, L257 (2003).

[21] J. Gao, J. L. Peacher and D. H. Madison, *J. Chem. Phys.*, **123**, 204302 (2005).

[22] J. Gao, D. H. Madison and J. L. Peacher, *J. Chem. Phys.*, **123**, 204314 (2005).

[23] J. Gao, D. H. Madison and J. L. Peacher, *J. Phys. B*, **39**, 1275 (2006).

[24] O. Al-Hagan, C. Kaiser, D. Madison and A. J. Murray, *Nat. Phys.*, **5**, 59 (2009).

[25] M. S. Pindzola *et al.*, *J. Phys. B*, **40**, R39 (2007).

[26] J. Colgan *et al.*, *Phys. Rev. Lett.*, **101**, 233201 (2008).

[27] J. Colgan and M. S. Pindzola, *J. Phys. Conf. Ser.*, **288**, 012001 (2011).

[28] C. Champion, J. Hanssen and P. A. Hervieux, *Phys. Rev. A*, **63**, 052720 (2001).

[29] C. Champion, J. Hanssen and P. A. Hervieux, *Phys. Rev. A*, **72**, 059906(E) (2005).

[30] C. Champion, D. Oubaziz, H. Aouchiche, Yu. V. Popov and C. Dal Cappello, *Phys. Rev. A*, **81**, 032704 (2010).

[31] C. Champion and R. D. Rivarola, *Phys. Rev. A*, **82**, 042704 (2010).

[32] C. Dal Cappello, I. Kada, A. Mansouri and C. Champion, *J. Phys. Conf. Ser.*, **288**, 012004 (2011).

[33] M. Inokuti, *Rev. Mod. Phys.*, **43**, 297 (1971).

[34] E. Weigold and I. E. McCarthy, *Electron Momentum Spectroscopy* (New York: Kluwer Academic/Plenum Press, 1999).

[35] M. Takahashi and Y. Udagawa, in *Nanoscale Interactions and Their Applications: Essays in Honour of Ian McCarthy*, ed. F. Wang and M. J. Brunger (Kerala, India: Transworld Research Network, 2007), p. 157.

[36] R. J. Le Roy, *Comput. Phys. Commun.*, **52**, 383 (1989).

[37] T. E. Sharp, *Atomic Data*, **2**, 119 (1971).

[38] M. Takahashi, Y. Khajuria and Y. Udagawa, *Phys. Rev. A*, **68**, 042710 (2003).

[39] M. Yamazaki *et al.*, *Meas. Sci. Technol.*, **22**, 075602 (2011).

[40] A. T. J. B. Eppink and D. H. Parker, *Rev. Sci. Instrum.*, **68**, 3477 (1997).

[41] M. Takahashi, J. P. Cave and J. H. D. Eland, *Rev. Sci. Instrum.*, **71**, 1337 (2000).

[42] J. Lower, R. Panajotovic, S. Bellm and E. Weigold, *Rev. Sci. Instrum.*, **78**, 111301 (2007).

[43] G. H. Dunn and L. J. Kieffer, *Phys. Rev.*, **132**, 2109 (1963).

[44] J. Gao, D. H. Madison, J. L. Peacher, A. J. Murray and M. J. H. Hussey, *J. Chem. Phys.*, **124**, 194306 (2006).

7

A coupled pseudostate approach to the calculation of ion–atom fragmentation processes

M. MCGOVERN, H. R. J. WALTERS AND COLM T. WHELAN

7.1 Introduction

In this paper we briefly review a novel approach we and our collaborators have recently developed for the theoretical treatment of ion–atom fragmentation processes [1–6]. We illustrate the power of the method by considering its application to antiproton collisions. Atomic units in which $\hbar = m_e = e = 1$ are used throughout.

7.2 Theory

The method consists of two stages. The first is the extraction of the differential motion of the heavy projectile from the straight-line impact parameter treatment. The second is the extraction of differential electron ejection from the pseudostates.

7.2.1 The impact parameter method and extraction of the differential motion of the projectile

In the impact parameter approximation, the projectile is assumed to move with constant velocity $\boldsymbol{v}_0$ along a straight line at perpendicular distance $\boldsymbol{b}$ (the impact parameter) from the target nucleus, see Fig. 7.1. At time t its position relative to the target nucleus is described by $\boldsymbol{R} = \boldsymbol{v}_0 t + \boldsymbol{b}$. Let Ψ be the electronic wave function at time t. We introduce a set of eigenstates and pseudostates, ψ_n, which together diagonalize the atomic Hamiltonian,

$$\langle \psi_n | H_{\mathrm{A}} | \psi_m \rangle = \epsilon_n \delta_{nm}. \tag{7.1}$$

Fragmentation Processes: Topics in Atomic and Molecular Physics, ed. Colm T. Whelan. Published by Cambridge University Press. © Cambridge University Press 2013.

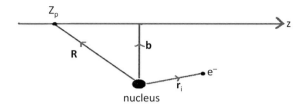

Figure 7.1 It is convenient to work in the relative co-ordinate system in which the target remains at rest and the projectile has a reduced mass μ, $\boldsymbol{R}$ is the position vector of the projectile, $\boldsymbol{r}_i$ the position of the ith electron relative to the target nucleus.

Pseudostates can be thought of as 'clumps' or 'distributions' of eigenstates which give a discretized representation of the continuum [7]. We expand Ψ in the set ψ_n according to

$$\Psi = \sum_n a_n(\boldsymbol{b}, t) e^{-i\epsilon_n t} \psi_n. \tag{7.2}$$

Substitution of (7.2) into the time-dependent Schrödinger equation and projecting with the ψ_n leads to the coupled set of equations:

$$i\frac{da_n}{dt} = \sum_m e^{i(\epsilon_n - \epsilon_m)t} \langle \psi_n | V | \psi_m \rangle a_m, \tag{7.3}$$

where V is the interaction between the projectile and atom. If the projectile is a bare ion of charge Z_P, the interaction takes the form:

$$V = Z_P \left[\frac{Z_T}{R} - \sum_{i=1}^{Z_T} \frac{1}{|\boldsymbol{R} - \boldsymbol{r}_i|} \right], \tag{7.4}$$

where the target nucleus has charge Z_T and $\boldsymbol{r}_i$ is the position of the ith electron relative to it. Note that we retain the interaction $Z_T Z_P/R$ between the projectile and the nucleus. If the atom is initially in the state ψ_0, then we must solve (7.3) subject to the boundary condition

$$a_n(\boldsymbol{b}, -\infty) = \delta_{n0}.$$

In a wave treatment of the collision, the scattering amplitude for exciting the state ψ_n would be (in the relative co-ordinate system of Fig. 7.1):

$$f = \langle e^{i\boldsymbol{k}_n \cdot \boldsymbol{R}} \psi_n | V | \Psi_0^+ \rangle, \tag{7.5}$$

where $\boldsymbol{k}_n$ is the final momentum of the projectile and Ψ_0^+ is the full scattering wave function starting from the state ψ_0. Using first-order perturbation theory in (7.5) we arrive at the first Born approximation (FBA):

$$f^{B1} = \langle e^{ik_n \cdot R} \psi_n | V | e^{ik_0 \cdot R} \psi_0 \rangle, \tag{7.6}$$

where k_0 is the incident momentum of the projectile. Taking the z direction to be along k_0 and $R = b + Z\hat{k}_0$, where $\hat{k}_0$ is a unit vector in the direction of k_0 and perpendicular to b, it follows that:

$$e^{i(k_0 - k_n) \cdot R} = e^{iq \cdot b} e^{iZq \cdot \hat{k}_0}, \tag{7.7}$$

where $q \equiv k_0 - k_n$. Conservation of energy is given by

$$k_n^2 = k_0^2 + 2\mu(\epsilon_0 - \epsilon_n), \tag{7.8}$$

where μ is the reduced mass, hence

$$(k_0 - k_n) \cdot \hat{k}_0 = \frac{(\epsilon_n - \epsilon_0)}{v_0} + O\left(\frac{1}{\mu}\right), \tag{7.9}$$

where $v_0 = \left(\frac{k_0}{\mu}\right)$ is the incident velocity of the projectile. Introducing

$$t = \frac{Z}{v_0}$$

and neglecting terms of $O\left(\frac{1}{\mu}\right)$, it follows that

$$f^{B1} = v_0 \int d^2 b e^{iq \cdot b} \int_{-\infty}^{\infty} e^{i(\epsilon_n - \epsilon_0)t} \langle \psi_n | V(t) | \psi_0 \rangle dt. \tag{7.10}$$

But from (7.3), the first Born approximation in the impact parameter treatment gives:

$$a_n^{B1}(b, \infty) = \delta_{n0} - i \int_{-\infty}^{\infty} e^{i(\epsilon_n - \epsilon_0)t} \langle \psi_n | V(t) | \psi_0 \rangle dt, \tag{7.11}$$

hence

$$f^{B1} = iv_0 \int e^{iq \cdot b} \left[a_n^{B1}(b, \infty) - \delta_{n0} \right] d^2 b. \tag{7.12}$$

In [1] this analysis was extended to higher orders and it was shown that essentially the same result holds for each order. Summing over all Born terms it was deduced that

$$\langle e^{ik_n \cdot R} \psi_n | V | \Psi_0^+ \rangle \rightarrow iv_0 \int e^{iq \cdot b} [a_n(b, \infty) - \delta_{n0}] d^2 b. \tag{7.13}$$

7.2.2 Extracting the differential motion of the ejected electron

The required ionization amplitude is

$$f_{\text{ion}} = \langle e^{i k_{\text{f}} \cdot R} \psi_{\kappa}^{-} | V | \Psi_0^{+} \rangle, \tag{7.14}$$

where ψ_{κ}^{-} is the ionized state of the atom corresponding to an ejected electron with momentum κ and (see Eq. (7.7))

$$k_{\text{f}}^2 = k_0^2 + 2\mu(\epsilon_0 - \kappa^2/2). \tag{7.15}$$

Assuming that the pseudostates form a complete set, Eq. (7.14) may be written:

$$f_{\text{ion}} = \sum_n \langle \psi_{\kappa}^{-} | \psi_n \rangle \langle e^{i k_{\text{f}} \cdot R} \psi_n | V | \Psi_0^{+} \rangle. \tag{7.16}$$

Then, using (7.13) we obtain the core result of our analysis:

$$f_{\text{ion}} = i v_0 \sum_n \langle \psi_{\kappa}^{-} | \psi_n \rangle \int e^{i q \cdot b} [a_n(b, \infty) - \delta_{n0}] d^2 b. \tag{7.17}$$

In [1] it was shown how to perform manageable calculations using (7.17). There is, however, a subtle point that needs to be considered here. Unless deliberately engineered (see Eqs. (7.8) and (7.15)),

$$k_{\text{f}}^2 \neq k_0^2 + 2\mu(\epsilon_0 - \epsilon_n) \tag{7.18}$$

and so the amplitude

$$f_{n0} \equiv \langle e^{i k_{\text{f}} \cdot R} \psi_n | V | \Psi_0^{+} \rangle \tag{7.19}$$

will normally be off energy shell. In [1], the pseudostates were chosen so that one state of each angular symmetry had exactly the right energy to make (7.18) an equality. With this choice it transpired that the overlap integral $\langle \psi_{\kappa}^{-} | \psi_n \rangle$ was negligibly small for all other states, so that, in effect, (7.19) was on the energy shell. This restriction on the choice of pseudostates is of no great matter if one is focused on calculating cross sections at a single ejected electron energy, but is highly inconvenient if a range of ejected energies is required. In [4] a relaxed form of our approximation was introduced that greatly eases the computations. We have used two large basis sets, one of 75 states, the other of 165 states. Details of the calculations and comparison with other theories and experiment are given in [1,4]. We were able to test the method at first Born level for antiproton impact ionization of atomic hydrogen, where an analytic result is available for the triple differential cross section (TDCS). The exact first Born cross section is shown in Fig. 7.2(a). Sitting essentially on top of it is the first Born cross section calculated from (7.17) but with a_n replaced by a_n^{B1} of (7.11). This level of agreement provides very strong support for our method. Also shown in Fig. 7.2(a) is a full coupled

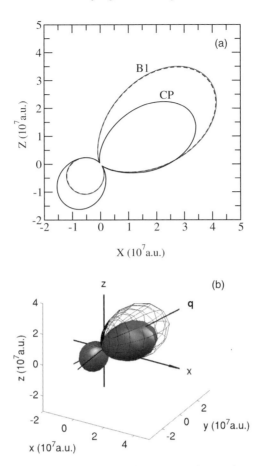

Figure 7.2 Laboratory frame cross sections, a.u., for antiproton ionization of atomic hydrogen at 500 keV and an ejected electron energy of 5 eV. (a) Polar plot of TDCS in scattering plane ($q = 0.25$ a.u.): CP is the coupled pseudostate approximation, solid curve B1 is the exact first Born approximation, dashed curve B1 is the impact parameter pseudostate first Born approximation. (b) 3D plot of TDCS ($q = 0.25$): solid surface is CP, wire surface is exact first Born approximation. Except for exact first Born, all calculations use a 165-pseudostate approximation.

165-state calculation. We were able to establish full three-dimensional TDCSs for a range of kinematics. In Fig. 7.2(b) and Fig. 7.3 we show examples from [1].

There are some implicit assumptions made in our analysis:

- Our approach is impact parameter based.
- Our expansion in (7.2) is clearly single centred based on the target atom and thus makes no explicit allowance for charge exchange to the projectile. This means that as it stands our approximation is reasonable when the projectile is a negatively charged antiproton or a positively charged projectile at an impact energy high enough for charge exchange to be negligible.

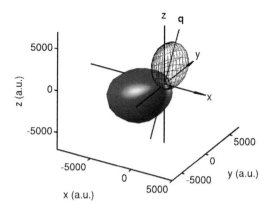

Figure 7.3 TDCS in three dimensions for antiproton impact ionization of atomic hydrogen with an ejected electron energy of 5 eV: impact energy 2 keV, $q = 2.5$ a.u. Wire cage, FBA; solid surface, full coupled pseudostate calculation (CP). The antiproton is incident along the z direction; compare with Fig 7.2(b).

7.3 Antiproton-induced ionization

Our primary motivation for initially focusing on antiproton collisions was as a test ground for developing our theoretical models; it is a 'clean system' in that it does not involve the complication of identical fermions or electron exchange between the target and projectile. As well as these convenient theoretical features, it should be noted that there is increasing interest in measuring antiproton collision processes. Extensive experimental data will shortly become available at the European Centre for Nuclear Research (CERN), where a large amount of focused effort is being devoted to the study of antiproton collisions with atomic and molecular targets [8–10]. In addition, the Facility for Antiproton and Ion Research (FAIR) [11] will soon be coming online and full differential studies of antiproton collisions are of considerable interest to both the FLAIR [12] and SPARC [13] international collaborations. This work is an important complement to the study of antihydrogen, which is relevant to proposed fundamental antimatter tests of the CPT invariance of relativistic quantum mechanics and the Weak Equivalence Principle of General Relativity [14–17], and as an aid to understanding antiproton reactions in the biological context of antiproton radiotherapy [18].

So far, only measurements of total ionization cross sections have been made, but there are plans for kinematically complete differential ionization experiments [8, 19]. Yet, even for total cross section measurements, theoretical information on differential scattering at as detailed a level as possible is valuable in planning the experiments [20]. In Fig. 7.4 we show a comparison between our coupled pseudostate calculations and the latest experiments from CERN [9] for the total

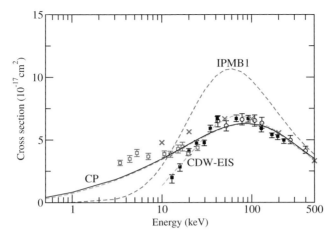

Figure 7.4 Total ionization cross section for antiproton impact on He: solid (dashed) CP curves, coupled pseudostate impact parameter calculation with 165(75)states; IPMB1, first Born cross section in the impact parameter treatment using the 165 set; CDW-EIS, continuum distorted wave approximation from [21]; crosses, calculations of Igarashi *et al.* [22]; open (solid) circles, old experimental data from [23]([24]); open squares, new experimental data from [9] in the energy range 3 to 25 eV.

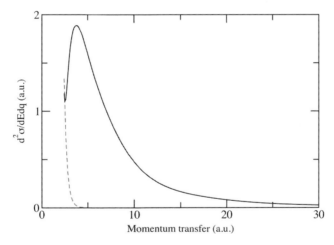

Figure 7.5 Cross section $d^2\sigma/dEdq$ for antiproton ionization of H at an impact energy of 2 keV and for an ejected electron energy of 5 eV: solid curve, CP approximation; dashed curve, exact first Born approximation.

ionization cross section for helium. The new experiments agree significantly better with our calculations than the earlier results from this group [23, 24].

An interesting result is revealed by the double differential cross section $d^2\sigma/dEdq$, which is differential in the energy of the ejected electron (E) and the momentum transfer (q) in the collision. Figure 7.5 shows this cross section for

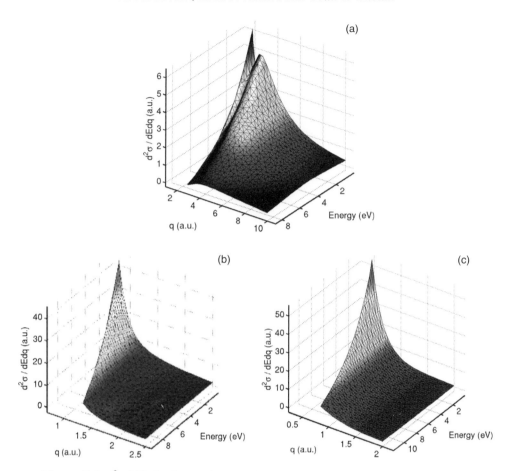

Figure 7.6 $d^2\sigma/dEdq$ for antiproton ionization of H at impact energies of (a) 2 keV, (b) 10 keV and (c) 30 keV, as calculated in the CP approximation.

antiprotons incident upon H at 2 keV and for an ejected energy of 5 eV. In the figure is the full coupled pseudostate approximation (7.17) (CP) and the exact first Born calculation of the same cross section. We see that with increasing momentum transfer q the CP cross section initially falls rapidly but then quickly reaches a sharp minimum followed by a large maximum, and a sustained cross section out to momentum transfers as large as 30 a.u. By contrast, the corresponding first Born cross section just declines rapidly with increasing q. The difference between the two results is attributed to the interaction $Z_P Z_T/R$ between the projectile and the target nucleus in the CP approximation and its absence from the first Born approximation (FBA). In its initial fall, the CP approximation is like the FBA and so in this stage it is presumed that long-range effects, which are taken into account in the FBA, are dominant. However, the quick divergence of the two approximations with increasing q attests to the rapid onset of dominance of the interaction of the projectile with the target nucleus. In Fig. 7.6(a) we show the same CP cross section at 2 keV but now for a range of ejected energies up to 8 eV. In this picture we see a more dramatic representation of the change-over from long-range to nuclear scattering, namely a pronounced crease in the cross section surface. However, as Fig. 7.6(b) shows, by 10 keV the crease has disappeared and it is no longer possible to see a visual separation between long-range and nuclear scattering; they have merged into a continuum. Figure 7.6(c), for an impact energy of 30 keV, is similar to Fig. 7.6(b) but shows a more rapid rate of fall of cross section with both ejected energy and q.

References

[1] M. Mc Govern *et al.*, *Phys. Rev. A*, **79**, 042707 (2009).
[2] M. Mc Govern *et al.*, *J. Phys. Con. Ser.*, **194**, 012042 (2009).
[3] M. Mc Govern *et al.*, *J. Phys. Con. Ser.*, **195**, 012029 (2010).
[4] M. Mc Govern *et al.*, *Phys. Rev. A*, **78**, 032708 (2010).
[5] M. Mc Govern *et al.*, *Phys. Rev. A*, **81**, 042704 (2010).
[6] M. Mc Govern, Colm T. Whelan and H. R. J. Walters, *Phys. Rev. A*, **82**, 032702 (2010).
[7] H. R. J. Walters, *J. Phys. B*, **21**, 1277 (1988).
[8] H. Knudsen, *J. Phys. Con. Ser.*, **194**, 012040 (2009).
[9] H. Knudsen *et al.*, *Phys. Rev. Lett.*, **101**, 043201 (2008).
[10] H. Knudsen *et al.*, *Nucl. Instrum. Meth. B*, **267**, 244 (2009).
[11] http://www.gsi.de/fair/.
[12] Facility for Low-Energy Antiproton and Ion Research: http://www.oeaw.ac.at/smi/flair/.
[13] Stored Particle Atomic Research Collaboration: http://www.gsi.de/fair/experiments/sparc/.
[14] M. Charlton *et al.*, *Phys. Rep.*, **241**, 65 (1994).
[15] L.V. Jørgensen *et al.* (ALPHA collaboration), *Nucl. Instrum. Meth. B*, **266**, 357 (2008).
[16] G. Gabrielse *et al.* (ATRAP collaboration), *Phys. Rev. Lett.*, **100**, 113001 (2008).
[17] A. Kellerbauer *et al.* (AEGIS collaboration), *Nucl. Instrum. Meth. B*, **266**, 351 (2008).
[18] H. V. Knudsen *et al.* (CERN ACE collaboration), *Nucl. Instrum. Meth. B*, **266**, 530 (2008).
[19] C. P. Welch *et al.*, *AIP Conf. Proc.*, **796**, 266 (2005).
[20] Y. Yamazaki, private communication.
[21] P. D. Feinstein *et al.*, *Phys. Rev. A*, **36**, 3639 (1987).
[22] A. Igarashi, S. Nakazaki and A. Ohsaki, *Phys. Rev. A*, **61**, 062712 (2000).
[23] L. H. Andersen *et al.*, *Phys. Rev. A*, **41**, 6536 (1990).
[24] P. Hvenplund *et al.*, *J. Phys. B*, **27**, 924 (1994).

8

Electron impact ionization using (e,2e) coincidence techniques from threshold to intermediate energies

ANDREW JAMES MURRAY

8.1 Introduction

Understanding the collision processes that lead to ionization of atoms and molecules is of great importance, not only in furthering fundamental quantum physics, but also for many applications in technology, industry and science. Applications include lighting, laser development and plasma etching, and these types of collisions occur in stellar and planetary atmospheres, the upper atmosphere of the earth, in a wide range of biological systems, and in the production of greenhouse gases. Indeed, in any area where ionization by electron impact occurs, it is essential to understand the collision mechanisms to fully describe the system under study.

The *probability of ionization* depends on several factors, including the incident energy of the electron projectile, and the nature of the target that is being ionized. Since electrons cannot be created or destroyed during these collisions (the energy is almost always too low for such a process to occur), the incident electron will be scattered from the target through some angle with respect to the incident direction. Ionization then results in one or more electrons being ejected from the target, and these electrons will also emerge at different angles, depending upon the collision dynamics.

In the types of processes discussed here, we confine our studies to *single ionization* (with the subsequent release of only one electron from the target), and only consider the interactions when the incident electron has relatively low energy compared to the ionization potential of the target. We here define this as the region from *threshold* (where the ejected and scattered electrons have almost no energy after the collision) through to *intermediate energies*, where the incident electron has energies typically several times that of the ionization potential (IP). In this energy regime, the electrons spend sufficient time in the region of the target that

Fragmentation Processes: Topics in Atomic and Molecular Physics, ed. Colm T. Whelan. Published by Cambridge University Press. © Cambridge University Press 2013.

impulsive approximations (as successfully used at high incident energies) are no longer applicable. It is hence necessary to carefully consider different mechanisms that lead to ionization.

The collisions considered here are of particular relevance since the probability of ionization is highest in this energy range for all atomic and molecular targets [1]. Hence it is here that most ionization reactions occur in nature.

The physical processes that lead to ionization by electron impact are complex. At high impact energies they are dominated by 'billiard ball' type collisions between the incident electron and ejected electron, the remaining electrons and nucleus of the target largely playing a spectator role. In this case, the electrons are expected to emerge at a mutual angle $\sim \frac{\pi}{2}$ in the forward direction, due to their equal mass. This process can be accurately modeled using plane waves for the incident and emerging electrons, and is used with success in electron momentum studies (EMS). As the energy is lowered, the collision is more accurately modeled by introducing distorted waves for the electrons in both the incident and final channels. Distorted wave Born approximations (DWBA) and other models are then used to produce theoretical data to compare with experiment. As the energy is lowered still further, additional processes must be considered. These include polarization of the target and the resulting ion by the incident and outgoing electrons, exchange, multiple collisions between the electrons and the core, and post-collisional interactions (PCI) between electrons, and between the electrons and ion as they emerge from the interaction. In the intermediate energy range all of these processes need to be considered, and this makes modeling of the interaction very challenging. At threshold, PCI often dominates, and so simpler theories may once again prove quite accurate in describing the interaction.

For atomic targets the interaction can be described using a spherical basis centred on the nucleus of the atom, which introduces symmetries in the collision process that aid in calculation. Molecular targets add considerably more complexity, since their nuclei are distributed throughout the molecule and so provide multiple scattering centres during the interaction. Further, the molecules are often randomly oriented in space with respect to the scattering geometry defined by the experiment, and so theory must *incoherently* sum over all possible orientations of the target to provide accurate comparison to experimental data. The target molecules may also be in different electronic, vibrational and rotational states prior to and after the collision. These additional challenges have only recently been considered, the new molecular ionization models currently being developed having mixed success at the present time.

To fully characterize these interactions, quantum models calculate the cross section σ, which is directly related to the probability of ionization. The most detailed studies establish the cross section as a function of the momentum of the incident electron k_0 with respect to the centre of mass of the target, and the momenta of the outgoing electrons k_1, k_2 and ion k_A that result from the interaction. Since both momentum and energy must be conserved, we may write:

$$k_0 = k_1 + k_2 + k_A \tag{8.1}$$

and also

$$\left. \begin{aligned} \frac{k_0^2}{2m_e} &= \frac{k_1^2}{2m_e} + \frac{k_2^2}{2m_e} + \frac{k_A^2}{2M_A} + IP \\ &\Leftrightarrow E_0 = E_1 + E_2 + E_A + IP \end{aligned} \right\} , \tag{8.2}$$

where m_e, M_A are the masses of the electrons and ion, respectively, and IP is the ionization energy required to ionize the target from its initial state to the final state of the ion. In most cases, both the target and ion are in their ground states prior to and after the collision, respectively; however this is not always the case (see below). Since the mass of the ion is much larger than that of the electrons, the ion recoil energy $E_A = \frac{k_A^2}{2M_A}$ is usually much smaller than that of the electrons or the IP, and so is often neglected.

Whereas Eqs. (8.1) and (8.2) are the basic equations that govern the interaction, they do not reveal the probability of ionization occurring. This is given by a *differential* cross section, and in the usual nomenclature is written as a triple differential cross section (TDCS):

$$\text{TDCS}(k_0, k_1, k_2) = \frac{d^3\sigma(E_0)}{d\Omega_1 d\Omega_2 dE_1}. \tag{8.3}$$

where Ω_1, Ω_2 are the solid angles defining the direction of the outgoing electrons, $\hat{k}_1$, $\hat{k}_2$ with respect to the direction of the incident electron, $\hat{k}_0$, and the energy of the electrons are related by Eq. (8.2).

Since the electrons may emerge in any direction following the collision with different probabilities, it is necessary to define the geometry of any measurements that are being made. The most common is that of Ehrhardt and co-workers [2], as shown in Fig. 8.1. In this configuration, the incident electron and one of the scattered electrons define a scattering plane, with the incident electron projected along the z-axis. The scattered electron then has polar angles $(\theta_1, \phi_1) = (\theta_1, 0)$. The second electron emerges from the interaction region at polar angles (θ_2, ϕ_2). If the second electron emerges in the scattering plane, so that $(\theta_2, \phi_2) = (\theta_2, \pi)$ or $(\theta_2, 0)$, this is referred to as a *coplanar geometry*. For kinematical conditions where $\theta_1 = \theta_2$ and $\phi_2 = \pi$, the geometry is referred to as *coplanar symmetric*.

Although the Ehrhardt geometry is most commonly used, it is not always adopted. In particular, for experiments that study out-of-plane kinematics, an alternative geometry first introduced by the Manchester group [3] is shown in

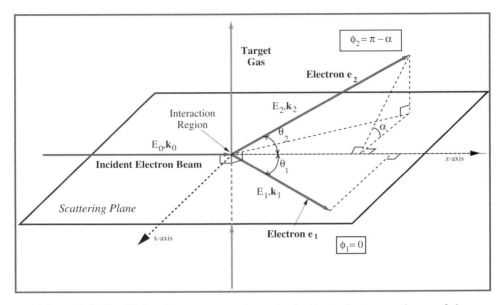

Figure 8.1 The Ehrhardt geometry, where the incident electron and one of the final electrons define the scattering plane.

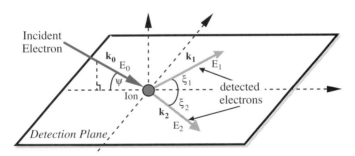

Figure 8.2 The Manchester geometry, where the scattered and ejected electrons define a detection plane. The incident electron then makes an angle ψ with respect to this plane so as to access non-coplanar kinematics.

Fig. 8.2. This geometry defines a detection plane, which is the plane mapped out by k_1 and k_2. The incident electron then can take different angles ψ with respect to this plane. k_1 and k_2 are at angles (ξ_1, ξ_2) with respect to the projection of k_0 onto the detection plane. The relationship between the two geometries is given by [3]:

$$\left. \begin{array}{l} \cos \theta_i = \cos \xi_i \cos \psi \\ \tan \phi_i = cot\xi_i \sin \psi \end{array} \right\}. \qquad (8.4)$$

In the Manchester geometry, coplanar kinematics are defined when $\psi = 0°$, whereas the *perpendicular plane* is defined when $\psi = \frac{\pi}{2}$. A particularly attractive feature of this geometry is that a common normalization point exists for all incident electron angles when $\xi_1 = \xi_2 = \frac{\pi}{2}$ [4]. This allows all measurements from coplanar kinematics through to the perpendicular plane to be inter-normalized, providing a robust set of data over all geometries for comparison to theory.

8.1.1 Description of the experimental coincidence technique

The experimental geometries described in Figs. 8.1 and 8.2 assume that electrons of well-defined momenta k_1 and k_2 are detected to ascertain the corresponding probability distribution, and hence TDCS. In reality, the experiments use a beam of incident electrons of well-defined momentum k_0, which is directed at an interaction region where the targets are located. These targets are usually produced from either an effusive atomic or molecular beam, from a supersonic target beam or from an oven. High-energy experiments may also use a solid or liquid target at the interaction region. Electrons that are scattered and ejected from the interaction may emerge in any direction, and therefore it is important for the experiment to correlate the electrons of interest so as to compare with theory. The electrons must hence be detected with well-defined momenta, and must be time correlated to ensure that each arises from the same scattering event. In all experiments, this requires the use of *coincidence techniques*.

The basic electron impact ionization coincidence experiment is shown in Fig. 8.3 [4]. The incident electron beam is produced from a gun that produces an electron beam of well-defined momentum. Beam currents from pA to μA may be used, depending upon the type of experiment that is being conducted. The atomic or molecular beam produces targets in the interaction region, so as to scatter electrons, resulting in ionization. Two electron analyzers are located around the interaction region to detect electrons of the required momentum. Individual electrons selected by the analyzers are detected using either channel electron multipliers (CEMs) or micro-channel plates (MCPs), so as to produce a fast current pulse when an electron is detected. The output from the electron multipliers then starts and stops a fast clock in order that the time difference between their detection can be ascertained. This clock may be constructed using either analogue components (e.g., using a time to amplitude converter and multi-channel analyzer) or may be constructed of all digital components. In either case, it is necessary to provide sub-ns timing resolution so that the true correlated signal can be distinguished from the random background arising from uncorrelated events. A time delay may be added to one of the signals to ensure that the correlated signal appears centrally in the accumulating signal. In the figure, an additional photomultiplier tube is shown that detects photons emitted

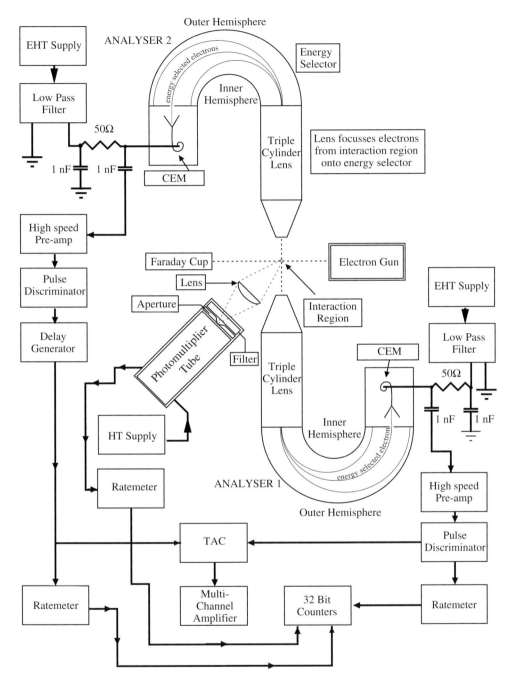

Figure 8.3 The timing and detection electronics required to perform (e,2e) coincidence measurements.

from the interaction region. This is used to aid tuning of the incident electron beam from the gun. A Faraday cup opposite the gun collects electrons that pass directly through the interaction region.

The coincidence technique is used widely in areas ranging from particle physics, through nuclear physics to atomic physics. As such, the control electronics required to ascertain the correlated signal are robust and are available commercially [5]. In practice, the experimental TDCS is determined by accumulating coincidence signals at a given scattering geometry for a given time, then the analyzers are moved to a new geometry and the experiment repeated. By ensuring that the electron and atomic beams are stable throughout data collection, and by ensuring that the electron detector efficiencies remain unchanged, the probability of ionization is accumulated as a function of the different selected scattering geometries. It is this probability map that is directly related to the TDCS calculated from quantum collision theory.

8.1.2 (e,2e) experiments near threshold

Although the physical mechanisms that lead to ionization by electron impact occur at all energies, they have importance to different degrees depending upon the energy regime that is studied. Ionization close to threshold tends to be dominated by post-collisional interactions (PCI), where the Coulomb interaction between the emerging electrons largely defines the final momentum of the detected electrons. This is of particular importance when both electrons have similar energies, since the relative momenta $k_2 - k_1$ of the electrons is then minimized. In this case, the electrons have more time to interact strongly as they leave the interaction region, and their repulsion leads to electrons that tend to emerge at a mutual angle near π. Models that have been used successfully in this energy regime follow the original work of Wannier [6]. There is some contention as to where the simple Wannier model breaks down, and a full quantum treatment becomes necessary [7–9]. This appears to be dependent not only on the momenta of the emerging electrons, but also rather surprisingly on the target that has been ionized.

8.1.3 (e,2e) experiments from threshold to intermediate energies

In the intermediate energy regime all processes described in Section 8.1 must be considered. This makes the problem particularly difficult to study theoreti- cally, and a large effort is continuing to attempt to solve the scattering problem in this regime for a wide range of targets. From an experimental viewpoint, there are considerable challenges dealing with electrons in this energy range, as the electron momenta are strongly influenced by stray magnetic and electric fields that may be present. Accurate control of the electron momenta requires careful design of the spectrometer and the electron optics controlling the incident electron beam and detected electrons. It is imperative to use specialized materials in all parts of the spectrometer to ensure stray fields are minimized. When these conditions are satisfied, it becomes possible to measure electrons with energies of less than 1 eV, and to produce well-collimated electron beams with energies less than ~8 eV.

8.1.4 Summary

In this chapter, the methods used to carry out experimental measurements in the low to intermediate energy regime will be presented, and the different models that have been applied in this regime will be described. Since other chapters specialize in detailed theoretical treatments, selected results from different models will be shown, rather than attempting any detailed derivation. Results from these studies will then be presented both for atomic targets and for molecular targets, so as to highlight successes and show where challenges remain. New experiments carried out from laser-excited atoms will then be presented, where the target is aligned with respect to the scattering geometry. The status of these research studies at the present time will then be summarized, and future directions considered.

8.2 Experimental methods and techniques

The timing electronics that allow coincidence measurements to be conducted are shown in Fig. 8.3. The fast timing electronics are usually obtained commercially. By contrast, the spectrometers in which experiments are carried out are custom designed and built by each research group, and a substantial amount of effort has been injected into the various designs throughout the world so as to ensure results are reproducible and accurate. In particular, in the low to intermediate energy regime it is essential to carefully choose construction materials for individual components and the vacuum chamber, and to carefully design the electron optics used to create the incident electron beam and detect electrons or ions following the collision.

In this section, some of the key methods adopted for the design of an (e,2e) spectrometer operating at low energies are outlined. These techniques are based upon the successful spectrometers developed by the Manchester group, who have for many years specialized in building spectrometers in this energy regime. The techniques presented here are by no means complete, and could be added to by others who work in this area. It does, however, establish essential details that must be considered when designing a spectrometer for this type of work.

8.2.1 Materials

One of the key considerations when building an electron spectrometer is the materials used for construction of different components. The main problems are the reduction to negligible levels of electric and magnetic fields that may influence the trajectories of electrons or ions, the production of a good vacuum that ensures the detected particles are unhindered by background vapour, the use of materials with low secondary electron emission characteristics and low outgassing rates, and a design that ensures the experiment can access as wide an angular range as possible. The method of construction of the electron gun, the electron analyzers, the atomic/molecular beam source and any other components within the spectrometer must also be carefully considered.

The choice of materials is often limited by cost and availability, however, it is essential not to compromise in regions where the electrons pass close to different components. This can be particularly critical for electrons of low energy, since these are easily influenced by the local surface fields of these components.

For the vacuum chamber, the material most commonly used is stainless steel. Different grades are possible, and it is imperative to choose one with a very low residual magnetism and which has low outgassing. Of the different stainless steels adopted in Manchester, 310 grade is found to be the best for chamber construction. However, recent changes to the nickel composition of 310 steel has made some batches more magnetic than others. It is hence recommended that a small quantity of any steel being considered for the spectrometer be obtained and tested before committing to a significant purchase. At Manchester, a high-sensitivity 3-axis Bartington magnetometer [10] with a resolution of 10 V/gauss is used to test the material prior to use (*all* materials should be checked for residual magnetism before use). For low-energy spectrometers the field from any material at its surface should be less than 0.1 mG (i.e., less than 10^{-8} T). Ideally, there should be no response from the magnetometer when the material is tested. Even parts from non-magnetic materials should be tested prior to installation, since the machine tools used in their construction may leave residual quantities of magnetic material on or inside the part.

Other types of stainless steel may also be used. In particular, 316LN, 254SMO and 253MA have proven to have very low residual magnetism, with good surface and structural characteristics that produce low outgassing. These materials are also useful for construction of internal structural components inside the spectrometer, as they machine precisely and are strong. Screws for fitting components together should be carefully assessed for their magnetism. In general, stainless steel screws need to be of at least A4 grade, and each must be individually checked for residual magnetism. For critical components close to the path of any low-energy electrons, the screws should be titanium, or may be custom made. All screw-holes should be accompanied by an outgassing hole to let trapped gas escape once the screw is fitted. Alternatively, the screw should be flattened on one side to eliminate trapping of gases upon assembly. The use of identical materials for the fitting screw and assembled part will result in vacuum welding, which causes significant problems when disassembly is required. This can be avoided by using different materials for the screw and part, or if this is unavoidable, the effects can be reduced by coating the screw with colloidal graphite prior to assembly.

At Manchester, the use of aluminium is avoided for chamber construction and for internal components, even though it is non-magnetic, as we have found that standard stock of this material is porous and produces significant amounts of outgassing, especially when heated (vacuum-compatible aluminium is available, but this is difficult to obtain). Aluminium further has an oxide surface layer that will charge up when struck by stray electrons, producing patch fields that are impossible to control or model. In a similar way, oxygen-free copper has an oxide layer that may produce problems with low-energy electrons, although this material does not outgas significantly. Copper can hence be used inside the vacuum chamber if thoroughly cleaned prior to installation, and then treated with a coating of colloidal graphite to provide a good conducting surface. Titanium is an excellent material in vacuum, is non-magnetic and does not outgas significantly. This material can be used without a graphite coating, as the surface properties are similar to stainless steel and remain conductive. It is, however, more difficult to machine compared to stainless steel, and is more expensive.

For components inside the spectrometer where electrons pass close to surfaces (e.g., in the electron gun and analyzers), molybdenum is the material of choice. Molybdenum is non-magnetic, and is found to have low secondary electron emission characteristics. This is particularly important where electrons strike a surface, as in the electron gun or at defining apertures. Since these secondary electrons tend to have low energies, they can significantly affect the operation of a spectrometer studying collisions in this energy regime. Their production must hence be minimized. Molybdenum is difficult to machine precisely, and so care is needed when producing components from this material. If a part is complex, it is often easier to machine it from 310 stainless steel or titanium, then line the surface exposed to electrons with a thin sheet of molybdenum spot-welded onto the component. Once molybdenum is cleaned, the surface does not need to be coated with graphite (as is required for copper).

For conductive shielding inside the spectrometer, the use of Advance (Constantan) sheet is generally used. This is a non-magnetic resistive material that can be spot-welded, has good secondary electron emission characteristics and can be cut and shaped easily. It is readily available, and can be obtained in sheets of different thickness.

Insulating materials use either ceramic or vacuum-compatible plastics. Machinable ceramics are available, commonly used materials being Macor, Boron Nitride, Shapal M-soft and pyrophyllite. The first three can be machined and used directly, whereas pyrophyllite must be fired after machining (a small amount of shrinkage occurs). Pre-shaped commercially available insulators, rods and spacers constructed from pure Al_2O_3 are also used, as are ruby and sapphire precision-machined balls (e.g., as used in turntable bearings). Vacuum-compatible plastics are limited in availability, the principal ones being PEEK, PTFE, Kapton and Vespel. All are quite expensive, but offer the advantage that precision components can be made easily. PEEK and Vespel provide rigidity, so are the materials of choice (although the latter is very expensive).

For connecting electrical signals to different components, either Kapton coated or PTFE coated wire is generally used. Coaxial cable of 50 ohm impedance can be purchased, which is constructed from either material; however, care needs to be taken to ensure the conductors in these cables are not magnetic. Coaxial cables are necessary for connection of high-speed components inside the spectrometer, including channel electron multipliers, channel plates and photomultiplier tubes. Connector cables to high-voltage and switching components (such as time-of-flight detectors or pulsed electron guns) should also use coaxial feeds to minimize fields. For connections to lens elements, PTFE coated constantan wire [11] is easy to use as it can be spot welded directly (the use of solder inside a spectrometer should be avoided). All insulated wiring must be fully shielded with a good conductor to ensure no plastics are exposed, as these will charge up by stray electrons, producing uncontrolled patch fields. This shielding requirement is essential for all insulating components in the spectrometer that may be hit by electrons, to ensure insulator patch fields are eliminated.

For operation at low energies, it is essential to reduce magnetic fields from external sources to a minimum inside the spectrometer. Magnetic shielding is essential, and this is usually constructed from μ-metal that has been fabricated into the required shape, then annealed in a high-temperature hydrogen atmosphere. The spectrometers in Manchester usually adopt magnetic shielding both inside and outside the vacuum chamber, and may further reduce magnetic fields using large Helmholtz coils outside the chamber perimeter. It is important to ensure that no discontinuous edges of the μ-metal are in proximity with internal elements, as this is where residual fields are concentrated (e.g., holes in the shield where feed-through ports are located). It is also important to ensure the μ-metal is fully overlapping to prevent ingress of fields inside the shielding. By careful construction, the magnetic field inside the spectrometer from all external sources can hence be reduced to much less than 5 mG.

Specialized components may be constructed from other materials. As an example, platinum–iridium may be used for construction of nozzles in atomic or molecular beam sources. Copper–beryllium and phosphor–bronze are non-magnetic materials that can be used to construct springs for use in vacuum components such as translators (almost all commercial springs are magnetic, and so cannot be used). Tantalum is often used in high-temperature ovens inside the chamber as it is machinable and remains ductile after heating to high temperatures. Other materials and compounds can also be used, depending upon the requirements; however, all should be fully assessed for their suitability prior to use.

8.2.2 Design of the electron gun and analyzers

The most critical components inside the spectrometer are the electrostatic optics inside the electron gun that produce the incident electron beam, and the optics that transport energy-selected scattered and ejected electrons to the electron multipliers. Many designs of gun are possible, depending upon the energy resolution that is required and the beam current to be delivered. For (e,2e) experiments that are not studying resonances, the cross sections are usually slowly varying in energy, and so an energy-*unselected* electron gun is often used. This allows maximum current to be delivered to the interaction region, thereby increasing coincidence yield and the accuracy of measurement. Care must be taken with the gun design and operation, as high beam currents at low energies will produce angular divergence of the electron beam due to electron–electron repulsion (space charge effects). This can result in loss of control of the incident momentum vector k_0 and so should be avoided. Further constraints may occur when an unselected electron gun is operated near threshold for ionization of a target, as the cross section may change rapidly over the region spanned by the energy resolution of the gun. In this case, an energy-selected gun may be used, although this will result in significant loss of signal due to lower beam currents.

Figure 8.4 gives an example of an unselected-energy electron gun, as is used in the (e,2e) spectrometers in Manchester. A commercial tungsten hairpin filament cathode produces electrons by passing current from a constant current power supply through the filament. A Pierce grid in combination with the anode shapes the emitted electrons into a quasi-collimated beam that passes through the anode aperture. A 3-element aperture lens with aperture diameter $D = 5$ mm and spacing $A = 2.5$ mm accelerates the electrons from the anode into the field-free region, where two 1 mm defining apertures are located. These apertures define the pencil and beam angles of the electron beam that is emitted from the gun. A second 3-element aperture lens

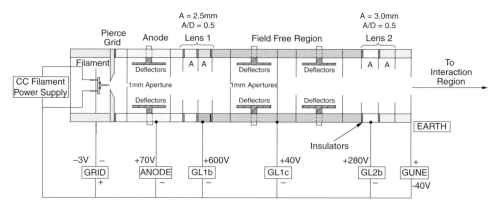

Figure 8.4 The unselected electron gun used in (e,2e) experiments in Manchester.

controls the electron beam that passes through the defining apertures, so as to shape and transport the output beam to the interaction region. The energy of the electron beam at the interaction region is set by the gun energy power supply as shown, this power supply biasing the filament with respect to ground (i.e., with respect to the potential of the spectrometer). Three sets of correction deflectors are placed inside the gun to allow for small deviations of the beam as it traverses through the gun. The final set of deflectors allows the beam to be accurately directed to the interaction region. All elements in the gun are constructed from molybdenum.

Typical voltages used to produce an electron beam of 40 eV are shown in Fig. 8.4. Note that all lens supplies are biased with respect to the gun energy (GUNE), whereas the deflectors are biased with respect to the element in which they are immersed. This electron gun has been used for a wide range of atomic and molecular targets with incident energies from ~8 eV to greater than 100 eV. It can produce beam currents in excess of 6 μA at the interaction region at higher energies, this current reducing to ~100 nA at the lowest energy.

For the electron analyzers, space charge effects do not apply (scattered electron rates are usually many orders of magnitude smaller than the rate of electrons produced by the gun), and so the operation of the electron optics is more controlled than for the gun. In the Manchester spectrometers, a three-element cylindrical electrostatic lens is used to focus the interaction region onto the entrance aperture of the energy selector (see Fig. 8.5). This zoom lens accelerates the electrons, so that electrons of the correct incident energy pass around the energy selector and into the electron multiplier. A set of deflectors is placed at the pass energy to ensure electrons from the interaction region pass into the hemispherical energy selector. Jost correctors are used at the entrance and exit to the energy selector to minimize the effects of field termination inside the hemispheres. Once again, all elements are constructed from molybdenum. The zero of kinetic energy for

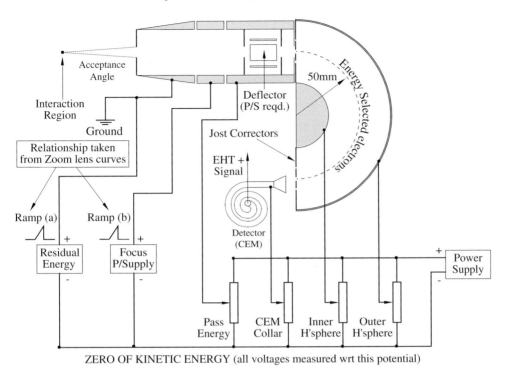

Figure 8.5 The electron analyzer used in (e,2e) experiments in Manchester.

the selected electrons is set by the residual energy power supply, and all lens elements and hemisphere potentials are adjusted with respect to this potential. For scanning of the analyzer residual energy, the focus voltage can be adjusted to follow the zoom lens curves for this lens. A channel electron multiplier (CEM) detects electrons of the required energy, and produces a current pulse for each electron that is detected. This current pulse then passes to the timing electronics, as shown in Fig. 8.3.

8.2.3 Example: the (e,2e) spectrometer in Manchester

Figure 8.6 shows one of the two (e,2e) spectrometers located in Manchester. This spectrometer has been operating since 1990 and has produced some of the benchmark data in this field, as well as being used to make new discoveries in atomic physics [3, 4, 12–15]. It is fully computer controlled and computer optimized, allowing data collection continuously over long periods of time without user intervention. The spectrometer allows measurements to be made from a coplanar geometry to the perpendicular plane, as discussed above. It can be used with gas targets produced from an effusive platinum–iridium needle, or an oven can be installed for

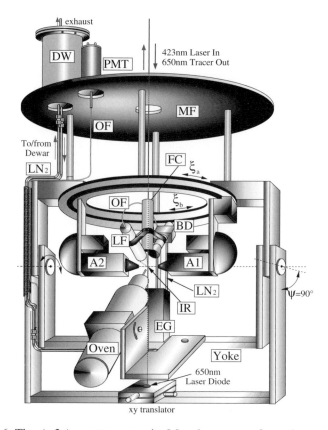

Figure 8.6 The (e,2e) spectrometer in Manchester, configured to study laser-excited calcium targets produced from an atomic beam oven. For details, see text.

metal vapour targets as shown. A window on the top flange allows laser radiation to enter the spectrometer and pass through the interaction region, so that studies of laser-excited targets can also be carried out. This laser beam is directed into the chamber by following a tracer beam from an internal 650 nm laser diode that has been accurately positioned to pass through the interaction region.

The spectrometer operates with incident energies from ~8 eV to >100 eV, and the analyzers can detect electrons with good efficiency down to <1 eV. A photo-multiplier tube is installed to detect photons emitted from the interaction region, allowing tuning of the electron gun optics independently from the analyzers. A magnetic angle changer (MAC) device can also be installed to allow measurements in a coplanar asymmetric geometry over extended angles [16–18]. A Faraday cup opposite the electron gun collects unscattered electrons from the incident electron beam; the Faraday cup, oven, liquid-nitrogen-cooled beam dump and optical fibre delivery system all rotate with the electron gun to access different geometries. This ensures that the interaction region remains unchanged as different experiments are conducted.

This instrument is one of the most sophisticated spectrometers for carrying out this type or research. It can access geometries from coplanar to the perpendicular plane by rotating the gun assembly on the yoke from $\psi = 0°$ to $\psi = 90°$. It is fully computer controlled and the electrostatic lenses are computer optimized, allowing the user to define parameters that ensure the spectrometer remains on-tune throughout data collection. The electron gun can be optimized to maximise output current as detected by the Faraday cup, photon counts from the interaction region, and counts in the analyzers. The input lenses of the analyzers are also optimized onto the interaction region, so as to ensure efficient coincidence yield. The analyzer residual energies and electron gun can also be controlled to ensure the correct overlap of energies so as to satisfy Eq. (8.2), and the computer further controls the angles of the gun and analyzers using stepper motors when data is being collected.

8.2.4 Multi-detection – the COLTRIMS reaction microscope

Spectrometers of the type described above follow a more conventional approach, which moves the analyzers to a selected angle, detects the electrons and establishes the coincidence yield as a function of the scattering geometry. The analyzer lenses define a narrow angular acceptance, and so measurements are carried out in a serial way, angle by angle, to establish the TDCS.

Alternatives exist that collect all electrons from the interaction, and then use correlation techniques to determine the cross section. These multi-detection techniques are powerful, and have been used principally at high incident energies to detect the coincidence yield over a range of energies, or a range of angles. The most successful multi-detection technique is probably that of COLTRIMS (cold target recoil ion momentum spectroscopy) and its derivative, the reaction microscope, which adopt a controlled electric and magnetic field to determine the time of flight of electrons and ions emerging from the interaction region [19 and references therein]. Other authors have detailed these methods, so they are only briefly considered here for completeness. The techniques are mostly used at higher incident energies than considered here, although new measurements are starting to be conducted in the intermediate regime. Like the (e,2e) spectrometer in Manchester, reaction microscopes take measurements over all geometries, and therefore also provide the full TDCS over three dimensions.

Figure 8.7 shows the basic set-up of these techniques. An electric field is superimposed upon a uniform magnetic field inside the spectrometer. The target beam must have well-controlled momentum, and is usually produced from a supersonic beam

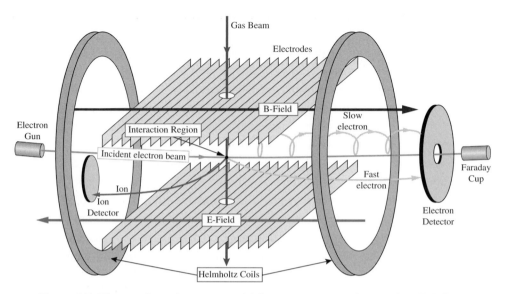

Figure 8.7 The reaction microscope, which uses a target gas beam of well-defined momentum and a pulsed electron or photon beam to initiate ionization. The momentum of the resulting ion is detected on one detector, whereas the momentum of one or more of the electrons is detected on a second. The electric and magnetic fields allow the time of flight of the ions and electrons to be determined so as to yield the TDCS.

that has been skimmed prior to entering the spectrometer. For electron impact ionization experiments, the incident electron beam enters the chamber and is directed to the interaction region. Ionization produces a second electron and a resulting ion. The ion is directed through a curved trajectory by the combined field and hits the channel plate biased for ion detection. The electrons, which are much lighter than the ions, travel through a spiral trajectory and strike a second plate biased to detect electrons. The number of spirals the electrons make depends upon their momentum at the interaction region and the magnitude of the fields. By detecting the position that the electron hits the plate and the time of flight after ionization, the momentum of the detected electron can be calculated. By time correlating the detected ion and detected electron, the TDCS can be determined.

A number of challenges arise with these techniques, making this a complex experimental method. For accurate determination of the final momentum of the ion, the initial momentum of the target must be accurately controlled, and this usually requires the use of supersonic expansion and skimmers prior to the interaction region. Even with such techniques, this momentum is difficult to define accurately. A second challenge is with the incident electron beam, which is directed into the chamber as shown. This beam will follow a curved path to reach the interaction region, thus alignment can prove difficult (especially for low-energy electrons).

The unscattered electron beam must be efficiently dumped in a Faraday cup so that there is a low background count in the detectors. The incident electron beam must be pulsed with high temporal resolution (often sub-ns) to ensure accurate measurement of the time of flight of the detected particles.

Despite these challenges, these methods have proven highly successful and continue to produce new data over all scattering geometries. The difficulties with the initial target beam momentum can be resolved by detecting the two electrons in coincidence (as in a conventional (e,2e) spectrometer); however, this requires measurements to be performed with unequal scattered and ejected electron energies in order to determine the time of flight of each electron. A new method using laser-cooled and trapped atoms has also been demonstrated to resolve the momentum of the targets (although this has not yet been applied to (e,2e) studies).

Experimental studies using conventional spectrometers and reaction microscope techniques as described here have been conducted over a wide range of energies and geometries for over 30 years. A large body of data has now been accumulated for comparison to different models that have been developed in parallel with the experiments. The methods continue to evolve as new demands are placed on the required data. Examples where this evolution is taking place are the studies of molecular ionization, studies which leave the ion in an excited state, and studies from laser-prepared and laser-cooled targets. Such experiments become possible as the technology they use advances and the speed of detection electronics increases. This ensures that this research area remains active and produces new discoveries in atomic and molecular physics.

8.3 Theoretical models

8.3.1 Near threshold

For ionizing collisions where the incident electron energy is close to the IP, the outgoing electrons only carry a small amount of energy away from the interaction region. In this regime, post-collisional interactions (PCI) are the dominant mechanism that produce the final (asymptotic) momenta of the electrons that are subsequently detected. This *threshold region* was first studied by Wannier [6], who considered the interaction to be purely Coulombic between the electrons, while assuming the mass of the resulting ion was infinite. Wannier showed that the probability of ionization depends critically on the energy shared between the electrons, and that the maximum probability occurs when the outgoing electrons share the excess energy equally. He further calculated the probability of ionization as a function of the mutual angle between the electrons, leading to predictions for the relative TDCS as a function of the momenta of the incident and outgoing electrons.

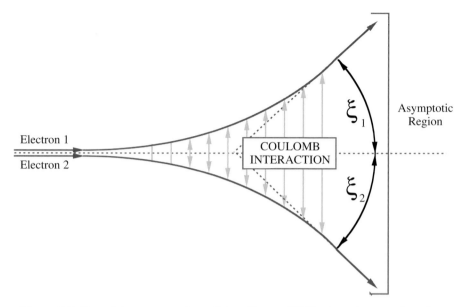

Figure 8.8 Pictorial representation of the effects of PCI when the outgoing electrons initially move in the same direction.

Since these first calculations were made, several authors have treated this region, both semi-classically and fully quantum mechanically [7–9]. These theories extend the original work and can predict the TDCS over a much wider range of incident energies. At the same time, experimental measurements in this energy region commenced, principally in France, Germany and the UK. These measurements studied the interaction at incident energies as low as 0.5 eV above the IP, as well as at higher energies in order to determine where other factors become important. The experiments were carried out for a range of atomic targets, and studies were made over different geometries from the coplanar to the perpendicular plane [20, 21].

The effects of PCI can be described in a simple way, as shown in Fig. 8.8. In this figure, the two electrons are depicted to leave the target region in the same initial direction ($\Delta\xi_{\text{initial}} = \xi_1 + \xi_2 = 0°$), their mutual repulsion driving the electrons apart as they move into the asymptotic region. If one electron leaves with a significantly greater relative velocity, their mutual repulsion will accelerate the fast electron and retard the slower electron. In this case, the slow electron may then not have sufficient energy to escape the field of the ion core, so that ionization is prevented. The highest probability for ionization hence occurs when both electrons have equal energy, and when they maintain this equality into the asymptotic region. The effect of the initial mutual angle between the electrons will also play a role, since for $\Delta\xi_{\text{initial}} \neq 0°$ they will move apart more quickly,

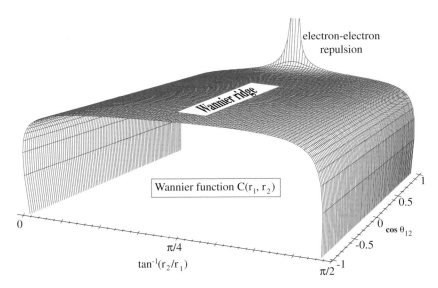

Figure 8.9 The Wannier ridge, showing the region of stability for ionization at low energies. The axes depict the Wannier function with respect to the relative distance of the two electrons from the nucleus ($\tan^{-1}(r_2/r_1)$) and the relative angle between them ($\cos\theta_{12}$). When $\cos\theta_{12} = 1$, $\theta_{12} = 0$ and the electrons move together, so strong repulsion occurs. When $\cos\theta_{12} = -1$, $\theta_{12} = \pi$ and electron repulsion is minimized. The ridge of stability occurs when $\tan^{-1}(r_2/r_1) = \pi/4$; $\rightarrow r_2 = r_1$.

resulting in less interaction. The extreme case occurs when the electrons leave the interaction region at a mutual angle, $\Delta\xi_{\text{initial}} = 180°$, since the effect of PCI is then minimized. Even in this case, an energy imbalance between the electrons may lead to retardation of one electron in the field of the ion, preventing ionization. Again, the maximum probability of ionization will occur when the electrons have equal energy. A delicate balance clearly exists between the momenta of the escaping electrons for ionization to occur.

Wannier described these interactions using hyper-spherical co-ordinates that depicts the stability condition for ionization as a ridge along which the electrons traverse, as shown in Fig. 8.9 [22]. For electrons whose co-ordinates remain on this ridge as they move outwards, ionization will occur. For all other conditions, Coulomb repulsion will lead to one electron gaining energy and escaping, the other remaining bound to the ion core. Modern theories consider this problem quantum mechanically and include the effects of angular momentum during the interaction. The agreement between theory and experiment for atomic targets that have been studied is generally excellent, so ionization in this energy regime is now considered as largely solved.

As the outgoing electron energies are reduced (Fig. 8.8), the effects of PCI increase the probability that the ionized and scattered electrons emerge at a mutual angle $\Delta \xi = 180°$. Wannier theory indicates that the highest probability then occurs for detected electrons of equal energy. One uncertainty in these studies is the energy extent above threshold where the effects of PCI dominate (the Wannier region). New measurements for molecular targets indicate that this region depends on the target under study. In these studies, it was found that the Wannier energy region for helium extends to $\sim$2 eV above threshold [23], compared to $\sim$3.3 eV for molecular hydrogen [24]. This difference was explained by noting that the molecular wave function extends further in space than for atomic helium, thereby increasing the sensitivity of the interaction to the momentum balance along the Wannier ridge.

The success of threshold theories for electron impact ionization matches that of high-energy theories, which model the interaction using impulse approximations. In both cases, the extremes of outgoing electron energies mean that simpler physical processes dominate, and thus can be modelled with a high degree of accuracy (i.e., PCI dominates at low energies, whereas simple binary knockout collisions dominate at high energies). Unfortunately, the threshold and high-energy regions are where the probability of ionization is smallest. The ionization probability is highest between these extremes, and therefore it is in this intermediate energy regime that most ionizing collisions take place in nature. It is hence essential to formulate quantum models that can accurately predict the ionization process in this intermediate regime, and it is here that much of the current experimental and theoretical research is now concentrated.

8.3.2 The intermediate energy regime

The intermediate regime can be characterized by the energy region where both threshold and high-energy theories fail to predict experimental data. This is clearly somewhat subjective, and depends upon the degree of accuracy of both the experimental data and the model. It is therefore imperative that data from experiments has a high degree of precision, and that data can be compared to theory over a wide range of kinematical conditions.

A number of different theories have been developed for ionization in this energy regime. Distorted wave Born approximations (DWBA) are a variant on high-energy impulse approximations that describe the electrons using plane waves. By contrast, DWBA theories use *distorted* waves to describe the incident and outgoing electrons [25]. These distortions arise from Coulombic and other interactions between participating electrons and the nucleus, which then modifies the momenta and hence wave functions describing the particles. Convergent close coupling (CCC) theories describe the target using a discrete basis of states both for the bound electrons ('real' states), and for the ionization continuum ('pseudo' states) [26]. CCC theories have proven very successful in describing excitation of atoms and these models have now been extended to describe the ionization region with good results for a number of atomic targets. Time dependent close coupling (TDCC) theories solve the time-dependent Schrödinger equation describing the interaction, so are computationally very intensive [27]. These calculations have had some spectacular successes for both light atomic and molecular targets, although some notable discrepancies are still found even for these simple targets.

The key elements described by these theories are complex. The models must consider the effects of distortions in the wave functions used to describe the incident electrons, the electrons in the target, the ion core that remains after ionization and the electrons that emerge after the collision. Exchange between electrons may occur, and the target may be polarized by the field of the electrons both prior to and after the collision occurs. Post-collisional interactions will further influence the detected outcome as described above, and multiple collisions between electrons and the nucleus must be considered. All of these processes have to be carefully treated for the model to provide an accurate comparison to experiment.

Further complexity arises for molecular targets and for targets that are oriented in space, since this reduces the symmetry of the interaction. Molecular targets have nuclei that are distributed within the target region, so they provide multiple scattering centres that can interact with the electrons during scattering and ionization. The wave functions describing the electrons in the target also do not possess spherical symmetry and therefore a Cartesian or cylindrical co-ordinate system is required, making the calculations computationally intensive. Higher order cross sections are hence needed to fully describe the ionization process; these cross sections depend on the orientation of the target with respect to the scattering geometry, as well as the electron momenta k_0, k_1 and k_2.

Although the challenges facing theory are considerable, the models outlined above (and their variants) have gradually improved in accuracy as they are refined and more data becomes available. No single model has yet found success in predicting all data that has been accumulated by experiment. The DWBA model has recently been applied to molecular targets, including polyatomic systems, as to calculate the (e,2e) cross section for these targets [25 and references therein]. CCC proves accurate for a range of atoms, but has not yet been applied to molecules. The TDCC model has been applied to atomic targets and to the simplest molecule H_2, and is proven to be accurate over a range of scattering kinematics. The intense computational requirements for the TDCC model precludes it from being applied to more complex molecular targets at the present time.

In the remaining sections, selected experimental results will be compared to theory to highlight successes and differences that are found. The data are principally taken from experiments at Manchester, which have been performed for a range of atomic and molecular targets over a period of 20 years. The depicted

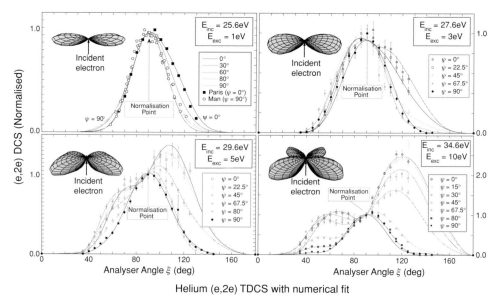

Helium (e,2e) TDCS with numerical fit

Figure 8.10 TDCS measurements from helium taken at energies from 1 eV to 10 eV above the *IP*.

results are by no means complete; there is a large body of experimental data from other research groups that also contrast the successes and limitations of different models. Results from other research groups are presented in different parts of this book.

8.4 Atomic targets

8.4.1 Near-threshold measurements on helium

Figure 8.10 shows the results of experimental studies carried out from a helium target at energies close to threshold, from 1 eV excess energy through to 10 eV excess energy [20, 21]. The data were taken from a coplanar geometry ($\psi = 0°$) to the perpendicular plane ($\psi = 90°$), with both outgoing electrons carrying equal energy away from the interaction. The electrons were detected with equal angles so that $\xi_1 = \xi_2 = \xi$. At the lowest energy, the results in a coplanar geometry were taken by the Paris group [21], and the results in the perpendicular plane by the Manchester group [20]. All other data were taken by the group in Manchester. A numerical fit to the data was obtained using the parameterization of the TDCS due to Klar and Fehr [28] as detailed in [29], and by using Wannier theory at the lowest energy [21]. These numerical fits were then used to obtain the three-dimensional representation of the cross section shown in the accompanying figure, adopting the symmetries that exist in the experiment to provide a complete cross section over all angles.

The results clearly show how the threshold region evolves into the intermediate regime. At the lowest energy a single structure is seen, with the electrons emerging from the interaction region with the highest probability at a mutual angle $2\xi = 180°$. This is due to their mutual repulsion dominating all other processes, and is represented in the three-dimensional TDCS as two lobes, one on either side of the incident beam direction (due to the symmetry of the experiment). As the excess energy increases, the effects of other processes become increasingly important. Four lobes can be seen to emerge, two in the forward direction, and two in the backward direction. From the data, it can be seen that even at 10 eV excess energy, there is significant probability that the electrons will emerge when $2\xi = 180°$. The forward scattering lobes are due to a binary collision between the incident electron and the ionized electron, so that their mutual angle is $\sim 2\xi = 90°$. The lobes in the backward direction principally arise from the incident electron initially scattering elastically from the target, followed by a binary collision. In this case, their mutual angle will be $\sim 2\xi = 270°$. In the data for helium it has been found that the backscattering lobes are larger than the forward lobes up to ~ 30 eV excess energy, after which the forward lobes dominate (see below) [3]. Since the electrons carry equal energy away from the interaction, the TDCS at a mutual angle of $0°$ and $180°$ must be zero due to post-collisional interactions. The numerical fit is hence constrained to zero at these extremes.

A number of different theoretical studies have been conducted in this region, including DWBA [24] and CCC [30] models, although no results have been formally presented over the full range of scattering angles and energies that have been measured. Examples of the predictions of the different models for an excess energy of 2 eV are shown in Fig. 8.11, the results being compared to data from the Ehrhardt group in a coplanar geometry [31] and the Manchester group in the perpendicular plane [20].

From these results it is clear that both models closely emulate the experimental data in both shape and magnitude. The DWBA theory appears to more closely resemble the data in the perpendicular plane at this energy. The results agree closely with the Wannier model, both showing a dominance of PCI as expected. It is clear that both models include the important physical processes that describe the interaction.

8.4.2 *Measurements on helium at intermediate energies*

As the energy increases beyond the threshold region, the effects of PCI start to diminish in comparison to other mechanisms, although PCI must always be

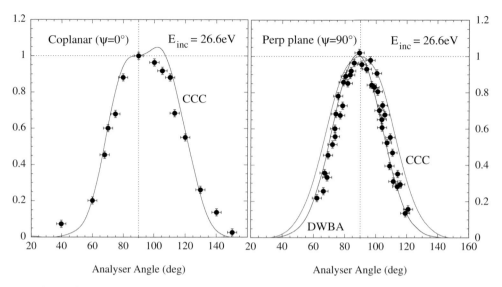

Figure 8.11 TDCS measurements from helium for an excess energy 2 eV above the *IP*. The results are shown compared to the CCC model in both coplanar and perpendicular plane geometries, and to the DWBA model in the perpendicular plane.

included to correctly predict the cross section for all scattering angles. This is particularly important when the selected electron energies are equal in magnitude, as has been detailed above. Figure 8.12 shows how the cross section varies from the coplanar geometry to the perpendicular plane for a helium target, where the excess energy is 20 eV to 50 eV above the *I P*. As the energy increases, the lobes in the forward direction become dominant, indicating that a simple binary collision between the incident and ejected electrons has the highest probability. At 30 eV excess energy, the probability of ionization in the forward and backward scattering directions is approximately equal.

Whereas the overall structure in the data is predicted by different theories, there are subtleties in the measurements that appear when the cross section is studied in more detail. In particular, a deep minimum was discovered in the cross section at an energy 40 eV above the *IP* for an incident electron angle $\psi \sim 67.5°$ [13], which has lead to new models being developed. This deep minimum is depicted in Fig. 8.13, the most remarkable consequence being that the TDCS is found to reduce to zero when the experimental resolution is considered. This minimum has been predicted using different quantum calculations, an example being shown in Fig. 8.13, where comparison is made with the TDCC calculations of the Los Alamos group [32]. The calculations show close agreement with the data, indicating the accuracy of the model for this target.

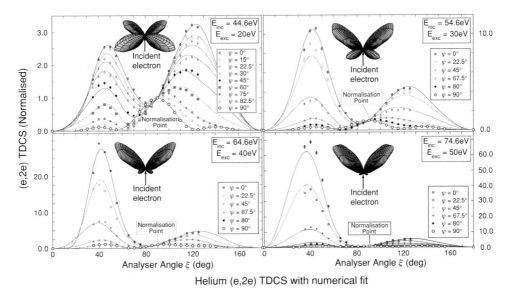

Helium (e,2e) TDCS with numerical fit

Figure 8.12 TDCS measurements from helium for excess energies from 20 eV to 50 eV above the *I P*, where the detected electrons carry equal energy and have the same scattering angle $\xi_1 = \xi_2 = \xi$ in the detection plane. The results show that a simple binary collision dominates as the energy increases towards the high-energy regime. All data are normalized to unity at the common point where $\xi_1 = \xi_2 = 90°$, which is independent of the incident electron angle ψ.

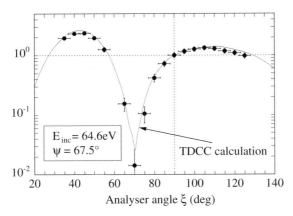

Figure 8.13 TDCS from helium for an excess energy 40 eV above the *I P*, where the detected electrons carry equal energy and have the same angle $\xi_1 = \xi_2 = \xi$ in the detection plane. The results show the deep minimum discovered in the cross section at an incident electron angle $\psi = 67.5°$. The data are normalized to unity at $\xi_1 = \xi_2 = 90°$. The calculated cross section from the TDCC calculation is also shown.

Although the most sophisticated quantum models are successful at predicting the TDCS for helium when compared to theory, as shown in Fig. 8.13, the physical reason why the dip occurs is masked by the complexity of the calculations. For such a deep and sharp minimum to occur, both real and imaginary parts of the scattering amplitudes that produce the cross section must independently sum to zero at this specific point. Given the complexity of the interactions expected in this energy regime for these non-coplanar kinematics, such a result is completely unexpected. There must be a simpler scattering mechanism being revealed by the data and producing such a remarkable result.

Macek, Briggs, Feagin and others [33, 34] have recently considered this problem and have postulated that quantum vortices in the scattering wave functions are responsible for the zeros observed in the ionization cross section. They show that these vortices are related to the angular momentum transferred between the electrons that are described by a correlated two-electron wave function. Their model hence predicts an entirely new quantum process leading to ionization, which would probably not have been considered without the experimental data presented above.

The many experimental and theoretical studies of the ionization of helium indicate that the physical mechanisms for this target are well understood. Indeed, the precision with which the calculations compare to data have led to claims that our understanding of this process is now complete [30]. Whereas it is true that the comparisons are remarkably accurate, there are some minor differences that remain, particularly for conditions where the electrons carry equal energy away from the interaction [35]. Such discrepancies may be due to experimental conditions, or may be due to minor differences in the calculations.

Of particular note is how the calculations have steadily converged onto the experimental data as more accurate descriptions of the scattering mechanisms are included. It is also notable that different models that provide quite different ways of calculating the scattering problem have largely converged. Close coupling theories (such as CCC and TDCC models), as well as distorted wave (DWBA) theories, are all now in close agreement over a wide range of energies and kinematics, at least in predicting the shape of the cross sections that are measured. It would certainly be interesting to see a full comparison of theory over all the kinematical conditions and energies that have been measured by experiment to ascertain where any discrepancies remain.

In summary, the key mechanisms responsible for ionization by electron impact include:

- Post-collision interactions, which must always be included when the outgoing electrons have equal energies, and are particularly important in the threshold region.

- Single binary scattering of the incident electron from the ionized electron, leading to forward scattering, with the electrons leaving at $\sim 90°$ to each other.
- Multiple scattering, where the electron scatters from the nucleus, then an ionizing collision occurs (this may be followed by further scattering from the core).
- Polarization of the target both prior to and after the collision has occurred.
- Wavefront distortions and interference effects in the quantum mechanical description of the interaction.
- Exchange between the incident electron and target electrons.
- Quantum vortices leading to zeros in the cross section for specific kinematical conditions.

It is expected that these mechanisms will also play a role when describing ionization of all other atomic targets, and that under high and low energies the principal mechanisms (PCI, binary collision processes) should still apply. For heavier targets, relativistic effects will become important, since the speed of the electrons close to the nucleus becomes significant compared to the speed of light. In the next section, selected results from different noble gas targets are detailed to establish whether the results can be described by these mechanisms, or whether a more complex analysis is needed. These measurements were taken in the perpendicular plane, since this provides a sensitive probe of the different mechanisms that contribute.

8.4.3 *Measurements on the noble gases in the perpendicular plane*

Figure 8.14 shows selected results for the ionization of helium in the perpendicular plane, here used as a reference to describe the key mechanisms leading to ionization in this geometry. The data span from threshold through to relatively high incident energies, and illustrate the processes that lead to ionization under these kinematic conditions. All data are normalized to unity at the peak so that a comparison of the TDCS structure can be made.

Note that for correlated electrons to be detected in the perpendicular plane, it is necessary for *multiple* scattering to occur. This may take on different forms, however, it is generally accepted that the highest probability requires elastic scattering of the incident electron from the target in order to be deflected into the perpendicular plane, followed by subsequent collisions with the bound electrons and the target nucleus. Studies in the perpendicular plane are therefore particularly sensitive to these multiple collision processes.

At threshold, the perpendicular plane data show the characteristic Wannier shape, with the cross section being a single broad peak located at $\phi = 180°$. At low energies this peak arises mainly due to PCI as detailed above, however, recent calculations show that other mechanisms also contribute to this structure. In particular, it has

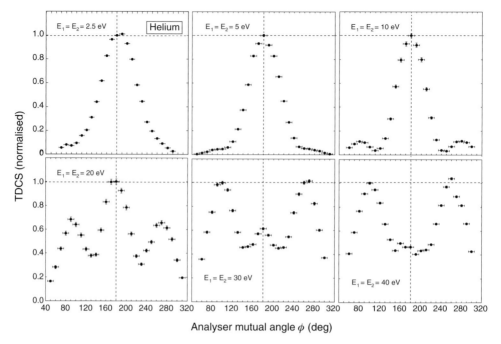

Figure 8.14 TDCS from helium in the perpendicular plane ($\psi = 90°$), where the detected electrons carry equal energy away from the interaction. In this plane only the mutual angle $\phi = \xi_1 + \xi_2$ has meaning. The results are shown from 5 eV excess energy to 80 eV excess energy, all data being normalized to unity at the peak of the cross section.

been shown that a *triple* scattering mechanism has a high probability of producing a peak centred around $\phi = 180°$ in this plane. In this mechanism, the electron initially scatters elastically from the target to enter the detection plane. A binary collision then occurs between the incident electron and a target electron. One of the electrons then undergoes further elastic scattering from the core of the target so as to leave the interaction in the opposite direction to the other electron. Both PCI and triple scattering mechanisms then contribute to produce a peak around $\phi = 180°$.

As the incident electron energy increases, the effects of PCI reduce in magnitude, leaving the triple scattering process as the dominant mechanism that produces the peak at $\phi = 180°$. The data show that two other peaks emerge at $\phi \sim 90°$ and $\phi \sim 270°$ when the excess energy increases above 5 eV. These are found to result from a single binary collision process (after initial scattering of the incident electron into the perpendicular plane), which occurs between target and ionized electrons. Since the electrons have equal mass, their mutual angle will be $\xi_1 + \xi_2 \sim 90°$, as observed. The peak at $\phi \sim 270°$ is then equivalent to that at $\phi \sim 90°$ due to rotation

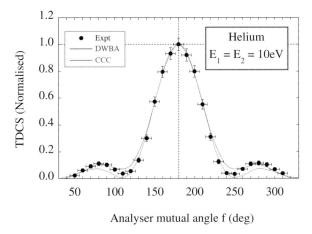

Figure 8.15 TDCS from helium in the perpendicular plane ($\psi = 90°$) for an excess energy of 20 eV, compared to calculations using CCC and DWBA theories.

symmetry around the detection plane. As the incident energy increases further, the relative probability of triple scattering (producing the peak at $\phi = 180°$) compared to the simple binary collision process (producing peaks at $\phi \sim 90°$, $270°$) is found to diminish. At an excess energy of 80 eV, the data show almost complete domination by the binary collision mechanism.

The data from helium in the perpendicular plane as presented here has been predicted well by CCC, TDCC and DWBA models. Figure 8.15 shows an example where the data at 20 eV excess energy is compared to CCC [30] and DWBA [37] calculations (all are normalized to unity at the peak). In this case, both models closely emulate the structure around $\phi = 180°$, whereas the DWBA calculation appears to be more closely aligned to the data for the side peaks.

The results for helium in the perpendicular plane illustrate the important scattering mechanisms required to describe the data. Interestingly, these descriptions do not require a detailed structure of the helium target to be known (although this clearly plays a role in determining the relative magnitudes of the contributing processes). It might hence be expected that ionization of other atomic targets in the perpendicular plane will follow a similar pattern, with low-energy data being dominated by a peak at $\phi = 180°$, whose relative magnitude decreases as the energy increases, together with binary peaks at $\phi \sim 90°$, $270°$, which evolve as the energy increases.

Such a study has recently been carried out for all the stable noble gas targets [15] over a similar energy range as carried out for helium. Examples of these results are presented below for argon and for xenon to contrast differences that are observed between a light target, a target with medium mass and a heavy target. Figure 8.16

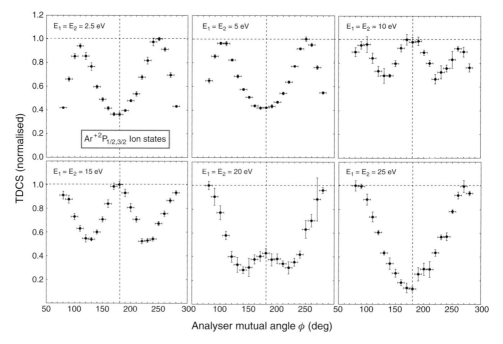

Figure 8.16 TDCS from argon in the perpendicular plane ($\psi = 90°$), where the detected electrons carry equal energy away from the interaction. In this plane only the mutual angle $\phi = \xi_1 + \xi_2$ has meaning. The results are shown from 5 eV excess energy to 50 eV excess energy, all data being normalized to unity at the peak of the cross section.

shows the results for argon and Fig. 8.17 the results for xenon. Note that for xenon it was possible to resolve the fine structure splitting of the ion core, whereas this was not possible for argon.

The data presented for argon are very different to helium. At 5 eV excess energy there is already a dip in the cross section at $\phi = 180°$, although this does start to fill in at 2 eV excess energy [15] (not shown here). Since PCI must dominate at threshold, these results show the extent of the threshold region is quite different for argon compared to helium. It appears that binary collisions between the scattered and ejected electrons strongly dominate at this energy, and that the effects of the triple scattering process seen in helium is weak. At 10 eV excess energy a peak does emerge at $\phi = 180°$, which remains visible until the incident electron has an energy 50 eV above the ionization threshold. At this energy the binary scattering peaks around $\phi \sim 90°$, $270°$ once more dominate, as is found for helium at higher energies.

Whereas the argon results are significantly different to those for helium, the mechanisms that produce ionization are probably similar. It may be that the triple

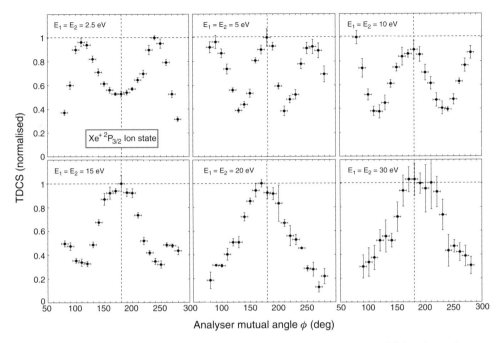

Figure 8.17 TDCS from xenon in the perpendicular plane ($\psi = 90°$), where the detected electrons carry equal energy away from the interaction. In this plane only the mutual angle $\phi = \xi_1 + \xi_2$ has meaning. The results are shown from 5 eV excess energy to 60 eV excess energy, all data being normalized to unity at the peak of the cross section.

scattering process which produces such a dominant peak at $\phi = 180°$ in helium at low energies is weakened by the greater number of electrons in the core for argon; however, since no theoretical calculations have been carried out for this target this can only be speculative at the present time. Calculations are clearly needed to test which mechanisms are playing a role in these interactions. Whereas the argon results can probably be explained using similar mechanisms as for helium, the data from ionization of xenon cannot. Figure 8.17 shows a selected range of results for this target taken from [15], again from low energy to relatively high energy. At the lowest energy the data are similar to argon, showing a dominance of binary scattering compared to triple scattering or PCI. The relative onset of the peak at $\phi = 180°$ is however at the much lower excess energy of 3.5 eV [15] (not shown), and the relative strength of this peak then grows with respect to the binary peaks, again as for argon. However, unlike all other targets that have been studied, the relative strength of the central peak compared to the binary peaks does not then decrease as the energy increases. Indeed, at the highest incident energy this becomes the dominant feature in the TDCS, in contrast to what is expected.

Clearly, additional mechanisms must be playing a role for this target, which require explanation from theory.

The sample of results presented here show that much has yet to be learned about the ionization process. Whereas ionization of helium is well understood, data from heavier targets present subtleties and unusual features that are unexpected. This is particularly noted for the heaviest target, xenon, and it will be informative to see how well future calculations succeed in explaining these unusual features.

Many other experiments have been performed from atomic targets in this energy regime that are not described here, using both conventional spectrometers and COLTRIMS techniques. These include data using asymmetric kinematics, and data from metal vapour targets. For the lighter elements, theory often succeeds in emulating the data well, however, for heavier targets some features remain unexplained. It is a goal of current research to provide accurate experimental data for these targets so that the models can be extended to produce as successful a comparison with experiment as is now found for helium.

8.5 Molecular targets

New experiments have now commenced studying the ionization of diatomic and polyatomic molecules in this energy regime. The process of ionization for molecules is considerably more complex, as the nucleus is no longer central with respect to the target charge cloud, making it impossible to invoke spherical symmetry in the collision when fully characterizing the interaction. Each nucleus will act as a scattering centre, the electron distribution will no longer be spherical and the molecule and associated ion may have rotational and vibrational energy before and after the collision. To fully characterize these types of interaction, it is necessary to adopt a cross section of higher order than the TDCS used for atoms to account for orientation of the target with respect to the interaction geometry. This is a challenging problem to solve theoretically.

Further complexities arise since almost all experiments so far conducted cannot resolve the orientation of the target prior to the interaction; a full comparison of theory to measurement requires the model to integrate the results over all possible target orientations. This has been achieved for only a limited number of targets so far. The most amenable molecule for this study is hydrogen, which has been considered in detail using the TDCC theory, the results being compared to recent experiments carried out from the coplanar to the perpendicular plane [37]. This theory calculates the cross section for different orientations of the target, then incoherently sums the results over all orientations to compare with experiment. The TDCC model therefore provides a direct comparison to the data.

An alternative approach has been developed which uses distorted waves to calculate the cross section. In this theory (called the 3DW model) [36], the target wave function is initially averaged over all space to provide a new spherically averaged wave function for input to the model. The calculation then proceeds as for an atomic target, distorted waves being used to represent the ingoing and outgoing particles. The 3DW model has the attraction that the scattering problem is then once more spherically symmetric and can be used for both diatomic and polyatomic targets.

Both approaches have their limitations. The TDCC theory is capable of determining the fully differential cross section that is required, but is presently limited to due to the scale of the numerical computations that are needed. By contrast, the 3DW model can be applied to a wide range of targets; however, it is as yet unclear how much the spherical pre-averaging approximation affects the final outcome. For H_2, the 3DW model provides reasonable comparison with the data as shown below, however, this agreement is poorer for more complex targets.

8.5.1 Measurements from H_2

In this section, examples are presented for ionization of H_2 from near threshold to the intermediate energy regime, over a range of scattering geometries. Comparisons with both TDCC and 3DW models are then given. Examples of ionization of the polyatomic molecule, CH_4 are then presented, with a comparison to results from the 3DW model.

Figure 8.18 shows a comparison of experimental results carried out from the coplanar to the perpendicular plane geometry ionizing H_2 with the incident electron 20 eV above the IP for this target. The outgoing electrons were detected at equal scattering angles in the detection plane so that $\xi_1 = \xi_2 = \xi$ and carried equal energy away from the interaction. A range of incident electron angles ψ was chosen. All data were inter-normalized to the common point when $\xi_1 = \xi_2 = 90°$, as discussed in Section 8.4.2. The calculations show the TDCC theory averaged over all orientations of the target, and the 3DW theory both with and without a correlation–polarization potential (CP) included. Since the data are not on an absolute scale, the best visual fit to theory is made in the perpendicular plane. The common point at $\xi_1 = \xi_2 = 90°$ then sets the magnitude of the data for all incident angles ψ.

The most immediate observation from this figure is how well the TDCC theory fits the data in the perpendicular plane, and continues to do so as the incident angle decreases to $\psi = 45°$. However, as this angle decreases further towards coplanar geometry, the TDCC theory diverges from the experimental data in magnitude, although the shape of the cross section is similar. The 3DW theory including the

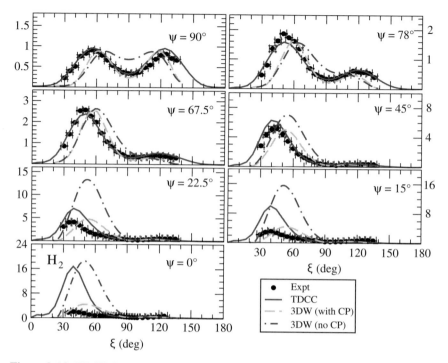

Figure 8.18 TDCS from the perpendicular plane ($\psi = 90°$) to a coplanar geometry ($\psi = 0°$), where the detected electrons carry equal energy of 10 eV away from the interaction. The results are compared to TDCC and 3DW models. For details, see text.

correlation–polarization potential produces a much closer comparison with respect to magnitude, but does not predict the position of the peaks well. When CP is excluded, the 3DW model does not fit the data.

These results are surprising, as the TDCC model is one of the most sophisticated calculations and includes orientation of the target prior to integration. It clearly predicts the data well in the perpendicular plane where multiple scattering occurs, yet does not obtain the correct magnitude in a coplanar geometry (the alternative viewpoint that the model fits the data in a coplanar geometry but not in the perpendicular plane is also valid, however, the overall fit is best for $\psi = 90°$). Indeed, when the TDCC is compared to other data for unequal electron energies (not shown here), it once again produces results in close agreement with the data for $\psi > 45°$, but not for lower angles [37]. At present it is unknown why these differences occur, however, it is clear the model is not emulating the data as well as for atomic targets.

8.5.2 Measurements from polyatomic molecules

The only theory currently available that attempts to describe ionization of poly-atomic molecules in this energy regime is the 3DW model. As noted, this theory

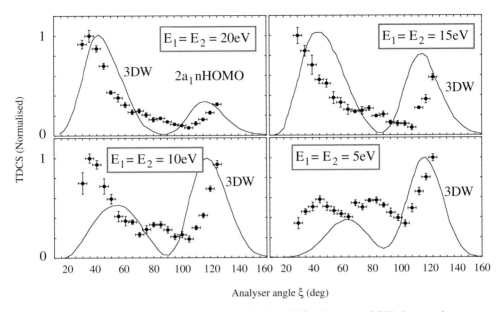

Figure 8.19 TDCS for ionization from the $2a_1$ nHOMO state of CH_4 in a coplanar ($\psi = 0°$) symmetric geometry, with the detected electrons carrying equal energy from the interaction. The results are compared to the 3DW model. For details, see text.

spherically averages the molecular wave function prior to modeling the collision, after which the same methods as apply for atomic systems are adopted. The nuclear charges are described as thin spheres of charge and the electron charge density becomes radial in profile [36].

The spherical averaging technique clearly provides a severe approximation to reality; however, it allows the calculations to become tractable using current computational facilities. The method produces reasonable results for simpler diatomic targets, as shown above. It is less accurate for larger polyatomic targets, however, as can be seen in the example shown in Fig. 8.19. In this case the target was CH_4, the state from which ionization occurred being the $2a_1$ nHOMO state. A coplanar symmetric geometry was adopted with the outgoing electrons carrying equal energy. The $2a_1$ state was chosen as it has a high degree of inherent radial symmetry. This was then expected to minimize the effects of the spherical averaging approximation that is used in the 3DW model.

It is clear from the figure that the 3DW model is not emulating the data well. The TDCS is found to develop three distinct peaks as the energy is lowered, which are not reproduced by theory.

The relative peak heights at the highest energy are reproduced, but the peak positions are incorrect in all cases. Theory predicts a *minimum* in the cross section around $\xi = 90°$, whereas the data finds a maximum. There must be other factors playing a role in the ionization process that are not being included in the 3DW model at present.

The relatively poor comparison between theory and experiment is not entirely surprising given the approximations that are being made and the complexity of the target under study. Such discrepancies were seen in early calculations for atomic targets, however, steady improvements in understanding the different mechanisms involved, as well as increased computational power, have led to models for atomic ionization that closely emulate data. Additional experiments from molecular targets are clearly required to provide a robust test of developing models. New ideas about the scattering mechanisms are also required, and more powerful computational facilities are needed. By employing such methods it is expected that theory and experiment will converge for these complex molecular targets in the near future.

8.6 Experiments from laser-aligned atoms

One of the main constraints when studying molecular ionization by electron impact is the orientation of the target prior to the collision, as discussed above. Experimentally, the molecular beam is produced from either an effusive beam or supersonic expansion, both of which produce a target ensemble with no preferred direction in space. Some experimental groups are investigating methods that detect the alignment direction of the molecule post-ionization (e.g., using Coulomb break-up of a repulsive ion state); however, this requires a triple coincidence measurement (ion, scattered and ejected electrons), a process which has low yield. Other groups are investigating whether the molecule can be pre-aligned prior to the collision (e.g., by a high-intensity laser field); however, once again these are very difficult techniques when applied to (e,2e) measurements due to the small amount of time the laser field is on (typically 50 ns/second).

To try to understand the differences between theory and experiment, recent measurements have been carried out not from aligned *molecules*, but from aligned *atoms* [38]. In this case, the atomic system prior to the collision is prepared in an excited state using linearly polarized laser radiation. Ionization then occurs from the aligned and excited target and (e,2e) coincidence measurements determined. Since the target electron is now pre-aligned with respect to the scattering geometry, the cross section must consider both the scattering geometry of the ionization process, as well as the polar angle of the charge cloud with respect to this geometry. This requires a quadruple cross section:

$$QDCS(\boldsymbol{k}_0, \boldsymbol{k}_1, \boldsymbol{k}_2, \boldsymbol{k}_T) = QDCS(\Omega_T, \Omega_1, \Omega_2, E_0, E_1, E_2)$$
$$= \frac{d^4\sigma(E_0)}{d\Omega_T, d\Omega_1, d\Omega_2, dE_2}, \tag{8.5}$$

where $\Omega_T(\theta_T, \phi_T)$ defines the target alignment angle $\hat{k}_T$, and the energies of the electrons are measured with respect to the ionization potential of the laser-excited state. Note that in this case the nucleus remains at the centre of the interaction, so spherical co-ordinates can be used. Such measurements therefore allow existing atomic ionization models to be applied, but they must carefully consider the bound electron charge cloud alignment in detail. As such, these new experiments test developing theories for aligned targets with respect to the electron charge cloud distribution, without needing to consider the effects of the distributed nuclear charges found in molecules.

8.6.1 The laser excitation process

For (e,2e) measurements to proceed from laser-aligned atoms, it is essential that the laser interaction produces a high density of excited targets in the interaction region. This density must remain unchanged over the lifetime of data collection, which is typically several days for any given measurement. Stabilization of the laser wavelength and intensity is therefore essential for long periods of time, and this is a very demanding task. Modern lasers now can be operated which satisfy these conditions, with feedback control from the experiment ensuring the laser remains on resonance for long periods of time. These advances in laser technology now allow such difficult experiments to be conducted.

To determine the relative population of atoms created during the laser excitation process, a density matrix formalism is adopted to describe the excited atomic state. For atoms in a ground state that has no angular momentum, excitation with single-mode continuous wave polarized (CW) laser radiation produces an excited P-state, whose alignment direction $\hat{k}_T$ is along the axis of the laser polarization vector ϵ. The state can then be described by a wave function in J representation (assuming no hyperfine structure) given by:

$$|\psi\rangle^{P\text{-state}} = |J = 1, m_J = 0\rangle \Leftrightarrow \rho^{P\text{-state}}(t) = \begin{bmatrix} 0 & 0 & 0 \\ 0 & \rho_T^{\text{Las}}(t) & 0 \\ 0 & 0 & 0 \end{bmatrix}, \quad (8.6)$$

where $\rho_T^{\text{Las}}(t)$ is the relative population of excited targets produced by the laser prior to the collision. This population is time dependent due to Rabi oscillations that occur as stimulated emission and absorption take place. The relative population can be determined from calculations of the laser interaction, as described in [39, 40].

Equation (8.6) is written in a framework where the quantization axis is along the polarization vector of the laser, which may be different to that of the collision process. The density matrix $\rho^{P\text{-state}}(t)$ can hence be rewritten for a different quantization axis by noting that:

$$\rho_{kl}^{z_1} = \sum_{m,n} D_{km}^J(\alpha, \beta, \gamma) D_{ln}^{J*}(\alpha, \beta, \gamma) \rho_{mn}^{z_2}, \tag{8.7}$$

where $D_{km}^J(\alpha, \beta, \gamma)$ are rotation matrices, the Euler angles (α, β, γ) rotating the quantization axis z_2 to z_1. The associated density matrix in the collision frame can then be directly input to the scattering model.

8.6.2 Ionization from laser-excited magnesium

The first (e,2e) measurements from aligned atoms were recently published using a magnesium target [38]. This atom was chosen as it can efficiently be excited by laser radiation at 285.296458 nm, which was available from a frequency-doubled CW dye laser. The population of excited targets was found to be $26\% \pm 4\%$ of the total number of atoms in the interaction region, and the laser was directed into the experiment along the incident electron beam. Since the direction of the incident electron beam is the 'natural' quantization axis for the collision geometry, Eq. (8.7) can then be used to describe the excited state density matrix in this frame. The result of applying Eq. (8.7) for the 3^1P_1 excited state of magnesium is then:

$$\rho_{3^1P_1}^{\text{Las}}(t) = \frac{\rho_T^{\text{Las}}(t)}{2} \begin{bmatrix} 1 & 0 & e^{+2i\epsilon} \\ 0 & 0 & 0 \\ e^{-2i\epsilon} & 0 & 1 \end{bmatrix}, \tag{8.8}$$

where ϵ is the angle of the laser polarization vector with respect to the scattering plane. When $\epsilon = 0°$, the polarization vector and the aligned state are in the scattering plane, whereas when $\epsilon = 90°$ they are orthogonal to this plane.

The magnesium experiments were carried out using three laser polarization angles, $\epsilon = 0°, 45°$ and $90°$. Since the relative population of excited targets was known, it was possible to calculate the magnitude of the QDCS relative to the TDCS for the 3^1S_0 ground state. A direct comparison between ionization from both ground and excited targets could hence be made, all results being placed on a common scale. The experiments were carried out in a coplanar asymmetric geometry with one analyzer fixed at a scattering angle of $30°$, the other moving around the plane. Both analyzers detected electrons of equal energy, the excess energy being 40 eV above the ionization potential of the 3^1P_1 state when studying excited targets, and 40 eV above the IP of the 3^1S_0 state for ionization from the ground state.

Figure 8.20 shows the result of these studies. The TDCS measurements are normalized to unity at the peak of the cross section, and the QDCS data are

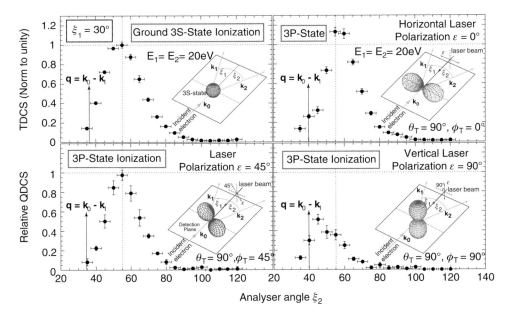

Figure 8.20 TDCS and QDCS measurements from magnesium, normalized to unity at the peak of the TDCS, with the relative magnitude of the QDCS being determined from the population of laser-excited targets. The QDCS is shown for alignment angles $\epsilon = 0°, 45°$ and $90°$ of the $3P$-state. The uncertainty in the relative normalization compared to the ground state is $\pm 9\%$ for $\epsilon = 0°$, $\pm 7\%$ for $\epsilon = 45°$ and $\pm 22\%$ for $\epsilon = 90°$.

then calculated from the relative population of laser-excited targets. No theoretical calculations exist as yet to compare with these results. The data show that the QDCS is ~13% larger than for the ground state when the target is pre-aligned in the scattering plane. The QDCS changes in both shape and magnitude as the target alignment direction varies, and is smallest when the target is aligned orthogonal to the plane. For target alignment orthogonal to the scattering plane, multiple scattering must occur for the electron to be redirected back to a coplanar geometry, so this mechanism must be included in any future calculations. It will be interesting to see how well current models of the interaction compare to these results.

8.7 Future work and conclusions

This chapter has described the ionization processes that occur when atomic and molecular targets are ionized by an incident electron. The discussions have been limited to the threshold and intermediate energy regimes, where the ionization cross section is highest and the interaction occurs through different and often complex mechanisms. This makes theoretical calculations of these processes challenging.

Several examples have been chosen that illustrate the success of the models when compared to experimental data, particularly for lighter atomic targets. Examples have also been given where even the most sophisticated models do not agree with experiment, illustrating that more work is needed.

There are clearly a number of challenges that face researchers in this area. Ionization from heavy atomic targets has yet to be fully understood, the results for the heaviest stable noble gas being very different from lighter targets where models have achieved excellent results. Theoretical calculations for molecular targets are still in their infancy in this energy regime and it will be interesting to see how the models develop for even more complex polyatomic targets. It is likely that the limitations facing existing theories will be eased when more powerful computers are commissioned, and it is hoped that the calculations will converge towards the experimental data at that time.

Experiments in this area also face new challenges. In particular, the requirement to resolve the alignment and orientation of molecular targets is very difficult. This is clearly needed to fully compare with theory and several groups are actively involved in attempting to solve this problem. The study of aligned atomic targets provides new data with which to compare to developing theories and these results should help to solve some of the challenges with theory. Such experiments are difficult, but have provided new data of high precision. It is expected that theory will guide such experiments in the future so that different scattering mechanisms can be studied in detail.

In summary, the study of electron impact ionization is a field with a proven history of success. It has allowed quantum scattering theories to be developed for a wide range of complex targets that have direct relevance in many different areas. New scattering mechanisms have been discovered and new experimental techniques developed. The field continues to flourish and is expected to provide researchers with interesting new challenges well into the future.

Acknowledgements   The experiments in Manchester could not be accomplished without a team of dedicated and talented researchers. These include Professor Frank Read FRS, Professor Nick Bowring, Dr Martyn Hussey, Dr Christian Kaiser, Dr Danica Cvejanovic and Dr Kate Nixon. It would be impossible to carry out these experiments without the skills and expertise of the mechanical and electronic technicians who have worked on and maintained these instruments. These include Mr Alan Venables, Mr Dave Coleman and Mr Mike Needham. We are most fortunate to collaborate with theoreticians throughout the world, who have aided our understanding of the collision processes we study. We would therefore also like to thank Professor Don Madison, Dr James Colgan, Professor John Briggs and Professor Igor Bray for many useful and fruitful discussions. Finally, we would like to thank the EPSRC in the UK for providing funding for these projects and the University of Manchester for allowing access to the laser system used in the magnesium studies.

References

[1] B. H. Bransden and C. J. Joachain, *Physics of Atoms and Molecules* (Harlow: Prentice Hall, 2002).
[2] H. Ehrhardt, K. Jung, G. Knoth and P., Schlemmer, *Z. Phys. D*, **1**, 3 (1986).
[3] A. J. Murray and F. H. Read, *Phys. Rev. A*, **47**, 3724 (1993).
[4] A. J. Murray, B. C. H. Turton and F. H. Read, *Rev. Sci. Inst.*, **63**, 3346 (1992).
[5] See, for example, www.ortec-online.com
[6] G. H. Wannier, *Phys. Rev.*, **90**, 817 (1953).
[7] A. R. P. Rau, *Phys. Rev. A*, **4**, 207 (1971).
[8] R. J. Peterkop, *Phys. B*, **4**, 513 (1971).
[9] D. S. Condren, J. F. McCann and D. S. F. Crothers, *J. Phys. B*, **39**, 3639 (2006).
[10] www.Bartington.com
[11] See, for example, www.polux.co.uk
[12] A. J. Murray and F. H. Read, *Phys. Rev. Lett.*, **69**, 2912 (1992).
[13] A. J. Murray and F. H. Read, *J. Phys. B*, **26**, L359 (1993).
[14] A. J. Murray, *Phys. Rev. A*, **72**, 062711 (2005).
[15] K. L. Nixon, A. J. Murray and C. Kaiser, *J. Phys. B*, **43**, 085202 (2010).
[16] F. H. Read and J. M. Channing, *Rev. Sci. Inst.*, **67**, 2372 (1996).
[17] A. J. Murray, M. H. Hussey, J. Gao and D. H. Madison, *J. Phys. B*, **39**, 3945 (2006).
[18] M. A. Stevenson and B. Lohmann, *Phys Rev A*, **73**, 020701 (2006).
[19] J. Ullrich *et al.*, *Rep. Prog. Phys.*, **66**, 1463 (2003).
[20] T. J. Hawley-Jones, F. H. Read, S. Cvejanovic, P. Hammond and G. C. King, *J. Phys. B*, **25**, 2393 (1992).
[21] P. Selles, A. Huetz and J. Mazeau, *J. Phys. B*, **20**, 5195 (1987).
[22] F. H. Read, in *Electron Impact Ionization*, ed. T. D. Märk and G. H. Dunn (New York: Springer, 1985), p. 42.
[23] O. Al-Hagan, A. J. Murray, C. Kaiser, J. Colgan and D. H. Madison, *Phys. Rev. A*, **81**, 030701 (2010).
[24] J. M. Martinez, H. R. J. Walters and C. T. Whelan, *J. Phys. B*, **41**, 065202 (2008).
[25] D. H. Madison and O. Al-Hagan, *J. At Mol. Opt. Phys.*, **2010**, 367180 (2010).
[26] I. Bray, D. V. Fursa, A. S. Kheifets and A. T. Stelbovics, *J. Phys. B*, **35**, R117 (2002).
[27] M. S. Pindzola *et al.*, *J. Phys. B*, **40**, R39 (2007).
[28] H. Klar and M. Fehr, *Z. Phys. D*, **23**, 295 (1992).
[29] A. J. Murray, F. H. Read and N. J. Bowring, *Phys. Rev. A*, **49**, R3162 (1994).
[30] I. Bray, D. V. Fursa, A. S. Kadyrov and A. T. Stebovics, *Phys. Rev. A*, **81**, 062704 (2010).
[31] T. Rösel, J. Röder, L. Frost, K. Jung and H. Ehrhardt, *Phys. Rev. A*, **46**, 2539 (1992).
[32] J. Colgan, O. Al-Hagan, D. H. Madison, A. J. Murray and M. S. Pindzola, *J. Phys. B*, **42**, 171001 (2009).
[33] J. H. Macek, J. B. Sternberg, S. Y. Ovchinnikov and J. S. Briggs, *Phys. Rev. Lett.*, **104**, 033201 (2010).
[34] J. M. Feagin, *J. Phys. B*, **44**, 011001 (2011).
[35] J. Colgan, I. Bray, K. L. Nixon and A. J. Murray, *J. Phys. B: At Mol. Opt. Phys.*, in preparation.

[36] O. Al-Hagan, C. Kaiser, D. H. Madison and A. J. Murray, *Nature Phys.*, **5**, 59 (2008).

[37] J. Colgan *et al.*, *Phys. Rev. A*, **79**, 052704 (2009).

[38] K. L. Nixon and A. J. Murray, *Phys. Rev. Lett.*, **106**, 123201 (2011).

[39] P. M. Farrell, W. R. MacGillivray and M. C. Standage, *Phys. Rev. A*, **44**, 1828 (1991).

[40] A. J. Murray, M. J. Hussey and M. Needham, *Meas. Sci. Tech.*, **17**, 3094 (2006).

9

(e,2e) processes on atomic inner shells

COLM T. WHELAN

9.1 (e,2e) processes – an overview

An (e,2e) process is one where an electron, of well-defined energy and momentum, is fired at a target, ionizes it and the two exiting electrons are detected in coincidence. The energies and positions in space of these electrons are determined by the experiment, so in effect all but the spin quantum numbers are then known. We can, therefore, describe it as a kinematically complete experiment; if we could also measure all the spins we would have all the information from a scattering experiment that quantum mechanics will allow. The technique offers both the possibility of a direct determination of the target wave function and a profound insights into the nature of few-body interactions. What information you extract from such an experiment really depends on the kinematics you chose and the target you use. Integrated cross sections can be crude things and you need the full power of a highly differential measurement to tease out the delicacies of the interactions. Indeed, often the most intriguing effects turn up in peculiar geometries where the cross sections are small and where a number of relatively subtle few-body interactions are present.

In recent years, attempts to give a complete numerical treatment of electron impact ionization have made considerable progress. In particular, one should mention the pioneering close coupling work of Curran and Walters [1–3], the convergent close coupling approach, [4], the complex exterior scaling calculations, [5], and the propagating exterior complex scaling method, [6]. For an up to date review, see [7]. This intensive computational work is of enormous value but has been restricted almost entirely to one and two electron targets at very low energies. In this paper we are concerned with the description of the inner-shell ionization of multi-electron atoms, where a full *ab initio* calculation is entirely impractical. Our

Fragmentation Processes: Topics in Atomic and Molecular Physics, ed. Colm T. Whelan. Published by Cambridge University Press. © Cambridge University Press 2013.

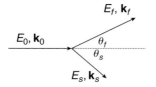

Figure 9.1 Schematic diagram of a coplanar (e, 2e) experiment. Showing the energies momenta and angles of the incoming and two outgoing electrons, respectively. The subscript 0 applies to the incoming, f to the fast and s to the slow outgoing electron.

approach, by necessity, will be mechanistic. Our ambition is to find the simplest theoretical model consistent with the physics. Even then, there are non-trivial computational challenges to be met. This is especially true when considering collisions at relativistic energies on heavy metal targets. Atomic units are used throughout ($m_e = e = \hbar = 1$), so the numerical value of the speed of light in the vacuum is approximately 137.

9.2 Non-relativistic theory

In this section will review the non-relativistic theory. Let us begin by considering the electron impact ionization of atomic hydrogen. Our approach here follows the finite-range formalism of [1, 8–10], in other words, we assume that the Coulomb potentials are cut off to zero at some very large but finite distance. Suppose we have an electron with momentum k_0, energy E_0, which collides with the atom that is initially in the state ψ_0, and two electrons, (k_f, E_f) and (k_s, E_s), are detected in coincidence. The Hamiltonian for the system can be written:

$$H = T_s + T_f + V_s + V_f + V_{sf}, \tag{9.1}$$

where

$$T_s = -\frac{1}{2}\nabla_s^2, \quad T_f = -\frac{1}{2}\nabla_f^2$$

$$V_s = -\frac{1}{r_s}, \quad V_f = -\frac{1}{r_f}$$

$$V_{sf} = +\frac{1}{\|r_f - r_s\|}$$

Let Ψ_i^+ be the exact solution of the Schrödinger equation corresponding to outgoing boundary conditions appropriate to the initial state,

$$\Phi_i = \frac{1}{(2\pi)^{3/2}}\psi_0(r_s)e^{ik_0 \cdot r_f}. \tag{9.2}$$

Then following [1, 8–10] we may write the direct amplitude as

$$f(\mathbf{k}_f, \mathbf{k}_s) = -\langle \Phi_j | V_j | \Psi_i^+ \rangle, \tag{9.3}$$

where

$$\Phi_j = \frac{1}{(2\pi)^3} e^{i\mathbf{k}_f \cdot \mathbf{r}_f} e^{i\mathbf{k}_s \cdot \mathbf{r}_s},$$

$$V_j = V_s + V_f + V_{sf},$$

and the exchange amplitude is given by $g(\mathbf{k}_s, \mathbf{k}_f) = f(\mathbf{k}_f, \mathbf{k}_s)$. The spin-averaged TDCS is defined as

$$\frac{d^3\sigma}{d\Omega_s d\Omega_f dE_s} = (2\pi)^4 \frac{k_s k_f}{4k_0} \left(|f + g|^2 + 3|f - g|^2 \right), \tag{9.4}$$

then introducing some interaction W_j into the left-hand side it is found that [1,8,10]

$$f(\mathbf{k}_s, \mathbf{k}_f) = -\left\langle \xi_j^- \left| V_j - W_j \right| \Psi_i^+ \right\rangle - \left\langle \xi_j^- \left| V_i - V_j + W_j \right| \Phi_i \right\rangle, \tag{9.5}$$

where $V_i = V_f + V_{sf}$. $\xi_j^-(\mathbf{r}_s, \mathbf{r}_f)$ is the wave function for the scattering of initially free electrons with momenta $\mathbf{k}_s, \mathbf{k}_f$ by W_j, that is,

$$\xi_j^- = \left[1 + \left(E - T_s - T_f - W_j - i\eta \right)^{-1} W_j \right] \Phi_j$$

$$= \left[1 + \left(E + \frac{\nabla_s^2}{2} + \frac{\nabla_f^2}{2} - W_j - i\eta \right)^{-1} W_j \right] \Phi_j. \tag{9.6}$$

Clearly, if we take $W_j = V_j$, then the direct ionization amplitude can be written,

$$f(\mathbf{k}_s, \mathbf{k}_f) = -\langle \Psi_j^- | V_i | \Phi_i \rangle. \tag{9.7}$$

Starting from Eq. (9.7), we could also go backwards and incorporate some interaction $W_i(\mathbf{r}_s, \mathbf{r}_f)$ into the initial state, see [9,10] for details. It is clear from the analysis that once we treat the formalism with due respect we can move the interactions as we please from one side of the matrix element to the other. The equivalence between Eqs. (9.3) and (9.7) is the well-known 'post-prior' equivalence. Let us emphasize once again that Ψ_i^+, Ψ_j^- are exact and this freedom to move the interaction and the wave functions will be lost once we approximate either.

If, further, W_j is separable, i.e., if we can write

$$W_j(\mathbf{r}_s, \mathbf{r}_f) = V_a(\mathbf{r}_f) + V_b(\mathbf{r}_s), \tag{9.8}$$

where the potentials $V_a(\mathbf{r}_f), V_b(\mathbf{r}_s)$ act on the fast and slow electrons, respectively, then for the choice (9.8), it follows that

$$\xi_j^-(\mathbf{r}_f, \mathbf{r}_s) = \chi_a^-(\mathbf{k}_f, \mathbf{r}_f) \chi_b^-(\mathbf{k}_s, \mathbf{r}_s), \tag{9.9}$$

where

$$\left(T_f + V_a(\boldsymbol{r}_f) - \frac{1}{2}k_f^2\right)\chi_a^-(\boldsymbol{k}_f, \boldsymbol{r}_f) = 0, \tag{9.10}$$

$\chi_a^-(\boldsymbol{k}_f, \boldsymbol{r}_f)$ having ingoing scattered waves, originating from the state

$$(2\pi)^{-3/2}e^{i\boldsymbol{k}_f\cdot\boldsymbol{r}_f}$$

of Φ_j, and similarly for $\chi_b^-(\boldsymbol{k}_s, \boldsymbol{r}_s)$. Then it is shown in [8,9] that (9.5) becomes

$$f(\boldsymbol{k}_s, \boldsymbol{k}_f) = -\left\langle \chi_a^-(\boldsymbol{k}_f, \boldsymbol{r}_f)\chi_b^-(\boldsymbol{k}_s, \boldsymbol{r}_s)\,\middle|\,V_j - V_a - V_b\,\middle|\,\Psi_i^+\right\rangle. \tag{9.11}$$

We could, for example, take

$$V_a(\boldsymbol{r}_f) = \frac{z_f}{r_f} \tag{9.12}$$

$$V_b(\boldsymbol{r}_s) = \frac{z_s}{r_s}$$

then [8],

$$f(\boldsymbol{k}_s, \boldsymbol{k}_f) = -\int \psi_C^{-*}(z_s, \boldsymbol{k}_s, \boldsymbol{r}_s)\psi_C^{-*}(z_f, \boldsymbol{k}_f, \boldsymbol{r}_f)$$

$$\times \left[V_{sf} - \frac{1 - z_s}{r_s} - \frac{1 - z_f}{r_f}\right]\Psi^+(\boldsymbol{r}_f, \boldsymbol{r}_s)d^3r_s d^3r_f, \tag{9.13}$$

where $\psi_C^-(z, \boldsymbol{k}, \boldsymbol{r})$ defines a continuum Coulomb function with ingoing waves and normalization:

$$\langle\psi_C^\pm(z, \boldsymbol{k}, \boldsymbol{r})|\psi_C^\pm(z, \boldsymbol{k}', \boldsymbol{r})\rangle = \delta(\boldsymbol{k} - \boldsymbol{k}')$$

the effective charges z_f, z_s are not functions of $\boldsymbol{r}_f, \boldsymbol{r}_s$ but can be functions of $\boldsymbol{k}_f, \boldsymbol{k}_s$. The importance of the effective charges and the difference between the finite-range formalism and taking the Coulomb potentials strictly to infinity as in [7,11,12] is discussed in detail elsewhere [8,10]; here we will only make a few remarks. The wave functions χ_a^-, χ_b^- are really quite general and once we have the exact wave function Ψ_i^+ on the left-hand side of Eq. (9.11) we will get the exact scattering amplitude. In order to perform a calculation we will need to make an approximation. The simplest possible is to take Ψ_i^+ to be the unperturbed initial state Φ_i, then (9.13) becomes [8]:

$$f(\boldsymbol{k}_s, \boldsymbol{k}_f) = \frac{-1}{(2\pi)^{3/2}}\int \psi_C^{-*}(z_s, \boldsymbol{k}_s, \boldsymbol{r}_s)\psi_C^{-*}(z_f, \boldsymbol{k}_f, \boldsymbol{r}_f)$$

$$\times \left[V_{sf} - \frac{1 - z_s}{r_s} - \frac{1 - z_f}{r_f}\right]\psi_0(\boldsymbol{r}_s)e^{i\boldsymbol{k}_0\cdot\boldsymbol{r}_f}d^3r_s d^3r_f. \tag{9.14}$$

If we take $z_s = z_f = 0$, then we have the plane wave Born approximation, PWBA:

$$f^{\text{PWBA}}(\boldsymbol{k}_s, \boldsymbol{k}_f) = \frac{1}{(2\pi)^{9/2}} \int e^{-i(\boldsymbol{k}_s \cdot \boldsymbol{r}_s)} e^{-i(\boldsymbol{k}_f \cdot \boldsymbol{r}_f)}$$

$$\times \left[V_{\text{sf}} - \frac{1}{r_s} - \frac{1}{r_f} \right] \psi_0(\boldsymbol{r}_s) e^{i\boldsymbol{k}_0 \cdot \boldsymbol{r}_f} d^3 r_s d^3 r_f. \tag{9.15}$$

If we take $z_s = 1$, $z_f = 0$, we have the first Born approximation, FBA:

$$f^{\text{FBA}}(\boldsymbol{k}_s, \boldsymbol{k}_f) = -\frac{1}{(2\pi)^3} \int \psi^{-*}(1, \boldsymbol{k}_s, \boldsymbol{r}_s) e^{-i\boldsymbol{k}_f \cdot \boldsymbol{r}_f} [V_{\text{sf}}] \psi_0(\boldsymbol{r}_s) e^{i\boldsymbol{k}_0 \cdot \boldsymbol{r}_f} d^3 r_s d^3 r_f. \tag{9.16}$$

Notice that the nuclear potential, $1/r_f$, does not contribute because of the orthogonality between $\psi_0(\boldsymbol{r}_s)$ and $\psi^-(1, \boldsymbol{k}_s, \boldsymbol{r}_s)$. If one argues that electron–electron interaction, V_{sf}, will be dominant and neglects the nuclear potentials in (9.15), one arrives at the plane wave impulse approximation, PWIA:

$$f^{\text{PWIA}}(\boldsymbol{k}_s, \boldsymbol{k}_f) = -\frac{1}{(2\pi)^{9/2}} \int e^{-i(\boldsymbol{k}_s \cdot \boldsymbol{r}_s)} e^{-i(\boldsymbol{k}_f \cdot \boldsymbol{r}_f)} [V_{\text{sf}}] \psi_0(\boldsymbol{r}_s) e^{i\boldsymbol{k}_0 \cdot \boldsymbol{r}_f} d^3 r_s d^3 r_f. \tag{9.17}$$

A common characteristic of the three approximations (9.15–9.17) is that they involve the integral:

$$\int d^3 r_f e^{i(\boldsymbol{k}_0 - \boldsymbol{k}_f) \cdot \boldsymbol{r}_f} \frac{1}{\| \boldsymbol{r}_f - \boldsymbol{r}_s \|}. \tag{9.18}$$

This is the well-known Bethe integral [13] and it can be evaluated,

$$\int d^3 r_f e^{i(\boldsymbol{k}_0 - \boldsymbol{k}_f) \cdot \boldsymbol{r}_f} \frac{1}{\| \boldsymbol{r}_f - \boldsymbol{r}_s \|} = \frac{4\pi}{q^2} e^{i(\boldsymbol{q} \cdot \boldsymbol{r}_s)}, \tag{9.19}$$

where we have introduced the momentum transfer vector:

$$\boldsymbol{q} = \boldsymbol{k}_0 - \boldsymbol{k}_f. \tag{9.20}$$

It follows, applying (9.19) to (9.17), that

$$f^{\text{PWIA}} = -\frac{2}{(2\pi q)^2} \left(\frac{1}{(2\pi)^{3/2}} \int d^3 r e^{i(\boldsymbol{k}_0 - \boldsymbol{k}_f - \boldsymbol{k}_s) \cdot \boldsymbol{r}} \psi_0(\boldsymbol{r}) \right)$$

$$= -\frac{2}{(2\pi q)^2} \hat{\psi}_0(\boldsymbol{p}),$$

where $\hat{\psi}_0(\boldsymbol{p})$ is the momentum space wave function, see for example [14], of the target evaluated at the ion recoil momentum,

$$\boldsymbol{p} = \boldsymbol{k}_0 - \boldsymbol{k}_f - \boldsymbol{k}_s, \tag{9.21}$$

and it follows that the triple differential cross section is proportional to the square of this momentum space wave function, i.e.,

$$\frac{d^3\sigma^{\text{PWIA}}}{d\Omega_f d\Omega_s dE} \propto |\hat{\psi}_0(\boldsymbol{p})|^2. \tag{9.22}$$

Thus, if the electron is initially in a state n, l, m we have

$$\frac{d^3\sigma^{\text{PWIA}}}{d\Omega_f d\Omega_s dE} \propto |\hat{\psi}_{nlm}(\boldsymbol{p})|^2. \tag{9.23}$$

A hydrogen atom wave function in momentum space admits an analytic solution [14],

$$\hat{\psi}_{nlm}(\boldsymbol{p}) = F_{nl}(p)Y_{lm}(\hat{\boldsymbol{p}}), \tag{9.24}$$

where Y_{lm} is the usual spherical harmonic and F_{nl} is a function of p, the magnitude of $\boldsymbol{p}$. It is instructive to consider the form of the function F_{nl} for an s state, $1s$ say, and a p state, $2p$ say, i.e.,

$$F_{1s}(p) = \frac{2^{5/2}}{\sqrt{\pi}} \frac{1}{(1+p^2)^2},$$

$$F_{2p}(p) = \frac{128}{\sqrt{3\pi}} \frac{p}{(4p^2+1)^3}. \tag{9.25}$$

Clearly $F_{1s}(p)$ has a maximum at $p = 0$ and then decreases, while $F_{2p}(p)$ has a minimum at $p = 0$ and rises to a local maximum at $p_{\max} = \sqrt{.05}$ and then decreases monotonically. The condition,

$$\boldsymbol{p} = \boldsymbol{k}_0 - \boldsymbol{k}_f - \boldsymbol{k}_s = 0, \tag{9.26}$$

defines the Bethe ridge and corresponds to zero recoil of the ion. The TDCS is a maximum for the $1s$ state at this point; by contrast it is zero at this point for the $2p$ state and increases away from it as we vary $p = \|\boldsymbol{p}\|$ until we reach the local maximum. This dependence on the target wave function in momentum space has long been observed. In Fig. 9.2 we show a comparison between experiment [15] and the PWIA. The TDCS was measured in non-coplanar symmetric geometry with all electrons detected at the same polar angle but different azimuthal angles. Shown are results for the ionization of ground states of helium and neon $2p$.

9.3 The distorted wave Born approximation

The distorted wave Born approximation has been applied to electron atom scattering for quite some time. The first detailed account was given in [17]. For a full

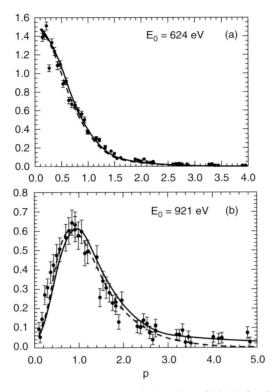

Figure 9.2 TDCS for the electron impact ionization of (a) the $2s$ electron of atomic helium at an impact energy of $E_0 = 624$ eV. The residual ion is in the $2s^2 S_{\frac{1}{2}}$ ground state. (b) the $2p$ electron of neon at an impact energy of $E_0 = 921$ eV. The TDCS is plotted as a function of the magnitude of the recoil momentum $\boldsymbol{p} = \boldsymbol{k}_0 - \boldsymbol{k}_f - \boldsymbol{k}_s$. The experiments are from [15], shown are theoretical calculations in the plane wave impulse approximation, dashed lines, and the solid lines are the Distorted wave Born approximation. In both cases, the target orbitals are taken from the Hartree Fock wave functions of [16].

discussion of the method, its strengths, weaknesses and computational implementation, see [18]. For a one-electron target, the DWBA can be understood in terms of the formalism of Section 9.2, as follows: Ψ_i^+ in (9.11) is approximated by

$$\Psi_i^+(\boldsymbol{r}_s, \boldsymbol{r}_f) \approx \zeta_0^+(\boldsymbol{k}_0, \boldsymbol{r}_f)\psi_0(\boldsymbol{r}_s) \pm \zeta_0^+(\boldsymbol{k}_0, \boldsymbol{r}_s)\psi_0(\boldsymbol{r}_f), \qquad (9.27)$$

where $\pm$ denotes spin singlet$^+$/triplet$^-$ static-exchange wave function for electron scattering by the H atom in the state ψ_0. Whereas (9.11) is quite general once we have the exact wave function Ψ_i^+ on the left-hand side when we make the approximation (9.27), we have, in effect, produced a first-order approximation in the $1/r_{\mathrm{sf}}$ potential. As pointed out by Whelan *et al.* [19], in this case we have recovered our post-prior equivalence that was characteristic of the full problem and one can interpret $\zeta_0^+(k_0, r_f)\psi(r_s)$ as being associated with the incident channel and χ_a^-, χ_b^- as being associated with the final channel. The picture one has is of the incoming electron being scattered by a distorting potential, in this case the static-exchange potential of the hydrogen atom, the ionizing electron–electron interaction occurring once and the two exiting electrons elastically scattering in the potentials V_a and V_b on their way to the detectors. So after making the approximation, the character of χ_a^-, χ_b^- changes. They are no longer 'projectors' that act on Ψ_i^+ to pick out the exact scattering amplitude they are now an intrinsic part of the approximation and it is important to include as much physics as possible in the choice of V_a, V_b. In the DWBA, one takes χ_a^-, χ_b^- to be the wave functions, ζ_a^-, ζ_b^-, calculated in the static-exchange potential of the atom or ion; for hydrogen, the ion potential is simply the Coulombic potential of the proton.

We remark that a feature of this approximation is that ζ_0^+ will be different for singlet and triplet scattering; this will result in different direct and exchange amplitudes, which we will denote by f^t, g^t, f^s, g^s, with t and s denoting triplet and singlet. This is discussed in detail in [2] and [18], and the triple differential cross section in the DWBA for hydrogen is given by [18]:

$$\frac{d^3\sigma^{\mathrm{DWBA}}}{d\Omega_f d\Omega_s dE} = (2\pi)^4 \frac{k_s k_f}{4k_0}\left(|f^s + g^s|^2 + 3|f^t - g^t|^2\right). \tag{9.28}$$

where we have summed over all final and averaged over all initial spin states, and

$$f^{s,t} = \langle \zeta_a^-(r_f)\zeta_b^-(r_s)|V_{\mathrm{sf}}|\zeta^{+s,t}(r_f)\psi_0(r_s)\rangle,$$
$$g^{s,t} = \langle \zeta_a^-(r_f)\zeta_b^-(r_s)|V_{\mathrm{sf}}|\zeta^{+s,t}(r_s)\psi_0(r_f)\rangle. \tag{9.29}$$

We note that had we used the exact wave antisymmetric function, $|\Psi_i^+\rangle$, there would have been no distinction between the singlet and triplet terms. The potential ambiguity for a hydrogen atom is rather neatly avoided by calculating the TDCS using (9.28). However, for closed shell targets this ambiguity is more severe in the final channels and indeed is inherent in the DWBA. Typically, one simply makes a choice. Further, in actual calculations, it is not usual to work with the full non-local exchange potential; rather, one employs a localized version [18, 20, 21].

9.3.1 Geometries

Let us emphasize, once again, that by carefully choosing the geometrical arrangement and the energies of the incoming and exiting electrons, one can predicate the physics that will dominate the shape and magnitude of the triple differential cross section observed. The geometry of the experiment and the physics it reveals are inexorably mixed. To illustrate this, let us consider three examples of geometries that have been used in studying low-energy (e,2e) processes.

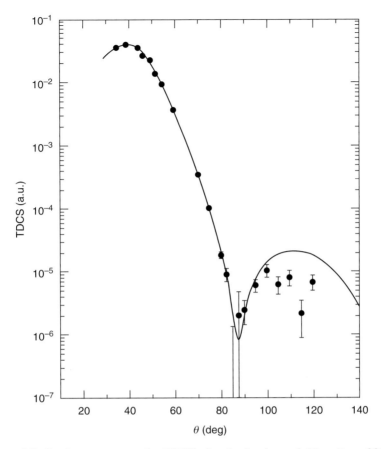

Figure 9.3 Coplanar symmetric TDCS for ionization of He, $E_0 = 200\,\text{eV}$. Experiment from [27], theory DWBA [26].

- Coplanar asymmetric geometry:
  In coplanar asymmetric geometry [22], often called Ehrhardt geometry, the fast outgoing electron is detected with a small angle with respect to the primary beam and the TDCS is measured as a function of the slow ejected angle, θ_s. Naively, one would expect this to be an ideal geometry for the study of perturbative effects, and it has proved to be so at least for outer-shell ionization at intermediate impact energies [2, 23].
- Coplanar symmetric geometry:
  In this geometry, known as Pochat geometry [24], the two outgoing electrons have equal energies and their momenta lie in the same plane as the incident momentum. They are detected with the same angle, θ, left and right of the beam direction. This is a very 'hard' collision, with the incident electron losing over half its initial momentum. If this were a collision between two free particles,

conservation of energy and momentum would mean that the two electrons would be detected at $90°$ to each other, i.e., at $\theta = 45°$. In the experiment of Pochat *et al.* [24], a helium target was used and thus the target electron is not free but in a bound state of the helium atom and thus has a momentum distribution before the collision, and the nucleus may take an active part in the process. Let us assume that ionization occurs as the result of a single electron–electron interaction, then, intuitively, one would expect the cross section to be greatest when the angle between the two particles is approximately $90°$. Now in this geometry this can happen when $\theta = 45°$ and $\theta = 135°$. In the original experiments, the TDCSs were measured at impact energies of 100, 150 and 200 eV. The cross sections were found to be peaked around the $45°$ point and there was a clear indication of a rise at the larger angle. This prompted Whelan and Walters [25], to speculate that a multiple scattering mechanism might be at work. They argued that the large-angle structure could be interpreted in terms of the incident electron first elastically back-scattering from the atom, essentially the nucleus, and then colliding with the target electron. Motivated by this simple intuitive model, a distorted wave Born approximation [26], was applied and produced excellent agreement with experiment.

- Coplanar constant Θ_{fs} geometry:
  In this geometry, both electrons are detected with equal energies and the angle between them is held fixed, and the TDCS given as a function of one of their scattering angles as both are rotated around the beam direction. This geometry was proposed [28, 29] for studying Coulomb three-body effects at low energy. Its principal significance at these energies is that one may, as a first approximation, treat the final-state electron–electron interaction as separable and thus one may concentrate on initial channel effects, such as target polarization.

9.3.2 The ionization of the $2p$ state of argon

There have been a number of interesting experiments on the inner shells of argon. Here we will be particularly interested in the reaction:

$$e^- + \text{Ar}(1s^2, 2s^2, 2p^6, 3s^2, 3p^6) \rightarrow 2e^- + \text{Ar}(1s^2, 2s^2, 2p^5, 3s^2, 3p^6). \quad (9.30)$$

The first such measurements were performed at an impact energy of 8 kev [30, 31], and then by Hink and collaborators at lower energies, in the range of 1.5–3 keV [32–34]. All of these experiments were performed in coplanar asymmetric geometry and the TDCS was given as a function of θ_s of the slow electron for a given fixed angle, θ_f, of the fast. Both angles are given with respect to the beam direction, with θ_f being measured in a clockwise, and θ_s being measured in anticlockwise

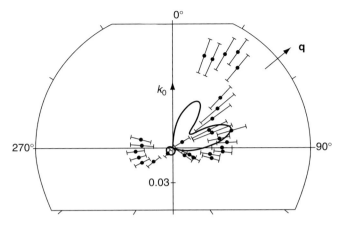

Figure 9.4 Coplanar TDCS for ionization of Ar(2p) at $E_0 = 1949\,\text{eV}$, $E_s = 500\,\text{eV}$, $\theta_f = 30°$ Experiment from [32]; theory FBA-CW.

sense. As we noted above, in a PWIA treatment we would expect a minimum in the TDCS when

$$p = \|\boldsymbol{k}_0 - \boldsymbol{k}_f - \boldsymbol{k}_s\| = 0.$$

Thus, if the kinematics is such that this point is accessible then we would expect a double peak structure around this point.

It is the lower energy, Hink,experiments, which will concern us here. These have been compared with the plane wave impulse approximation, and with various forms of the first Born [32, 35–37]. None of these approximations were satisfactory. All were especially poor in predicting the ratio of recoil to binary. In Fig. 9.4 we show a comparison between the experimental data and the FBA-CW calculations [35]. This approximation differs from the conventional first Born in that the outgoing slow electron sees a pure Coulomb field corresponding to an effective charge Z_{eff}. Z_{eff} was treated as an adjustable parameter.

Since the measurements are relative, they have been normalized to theory at the split binary peak. It is clear that agreement is very poor. In the FBA-CW, the wave functions representing the incident and fast scattered electrons are plane waves, while the slower electron is represented by a Coulomb wave with an adjustable effective charge. Thus this approximation does not allow, in any way, for elastic scattering of the incident electron by the atom prior to ionization, nor the effect of the ion on the fast outgoing electron. To some extent, it does include the influence of the ion on the slower outgoing electron through the effective Coulomb potential. It does get the split binary peak in the momentum transfer direction but it gives very poor agreement in the recoil direction. Zhang *et al.* [34] argued that one could not neglect the influence of the atomic potential on any of the electrons until much higher energies. This they claimed was especially true for inner-shell ionization since one would expect that the ionization process to occur relatively close to the nucleus, i.e., in a region where the static potential is at its strongest. Again, it was argued that the simplest viable approximation was the DWBA. The TDCS is given by:

$$\frac{d^3\sigma^{\text{DWBA}}}{d\Omega_f d\Omega_s dE} = 2(2\pi)^4 \frac{k_s k_f}{k_0} \sum_m (|f_{nlm}|^2 + |g_{nlm}|^2 - \text{Re}(f_{nlm}^* g_{nlm})), \qquad (9.31)$$

where the direct and exchange ionization amplitudes are given by:

$$f_{nlm} = \langle \zeta^-(\mathbf{k}_f, \mathbf{r}_f) \zeta^-(\mathbf{k}_s, \mathbf{r}_s) | \frac{1}{r_{\text{sf}}} | \zeta^+(\mathbf{k}_0, \mathbf{r}_f) \psi_{nlm}(\mathbf{r}_s) \rangle,$$

$$g_{nlm} = \langle \chi^-(\mathbf{k}_f, \mathbf{r}_s) \zeta^-(\mathbf{k}_s, \mathbf{r}_f)) | \frac{1}{r_{sf}} | \zeta^+(\mathbf{k}_0, \mathbf{r}_f) \psi_{nlm}(\mathbf{r}_s) \rangle. \qquad (9.32)$$

The incoming electron distorted wave was calculated in the static-exchange potential of the atom, as was the distorted wave for the outgoing fast electron, while the slow electron wave function was calculated in the spherically averaged singlet static-exchange potential of the ion.

In Fig. 9.5 we show a selection of these results as compared with experiment. Theory and experiment are now in much better agreement.

Zhang *et al.* [34] explored the effect of elastic scattering by 'switching off' the interaction between the incident electron and the atom, i.e., essentially replacing the distorted wave ζ^+ by a plane wave. The result of this calculation is shown in Fig. 9.6, where it is compared with the full DWBA calculation. We see at once that this model calculation has, now, acquired something of the character of the FBA-CW, and that it is the *binary peak* which is most strongly influenced by the distortion effects. Recall that these experiments were relative and all that theory had to compare with was the relative behaviour of the different maxima and minima. Let us emphasize once again that the first Born approximation will predict a recoil peak, even though the incident and fast electrons do not interact with the nucleus, i.e., are represented by plane waves. This can be understood in terms of the relative motion of the target electron and the ion about their centre of mass prior to the collision. For example, Whelan *et al.* [8] considered the ionization of H in coplanar asymmetric geometry for a range of impact energies and scattering angles, θ_f. They pointed out that in the hydrogen atom the average value of the momentum of the electron/proton with respect to the centre of mass was one atomic unit, and they remarked that when $|\mathbf{k}_{\text{recoil}}| < 1$ there was a significant recoil peak

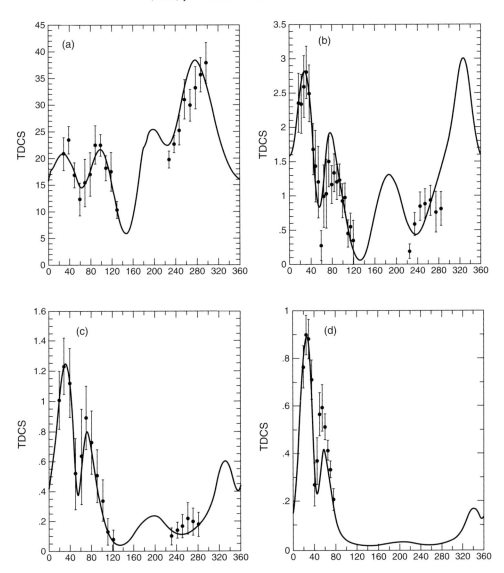

Figure 9.5 Coplanar TDCS for ionization of $Ar(2p)$ (in units of 10^{-3}a.u.) as a function of θ_s at (a) $E - 0 = 1949\,eV$, $E_f = 1550\,eV$, $\theta_f = 15.6°$; (b) $E - 0 = 1949\,eV$, $E_f = 1200\,eV$, $\theta_f = 30°$;   (c) $E - 0 = 2549\,eV$, $E_f = 1500\,eV$ $\theta_f = 33.8°$; (d) $E - 0 = 3249\,eV$, $E_f = 1500\,eV$, $\theta_f = 42.6°$.

in the experimental data and when $|\mathbf{k}_{recoil}| > 1$ the peak became negligibly small. In heuristic terms, an electron coming out in the recoil direction in the second case would correspond to a change in the average momentum, which would be characteristic of the proton experiencing a force. Thus the presence of a significant recoil peak in asymmetric geometry does, not in itself, mean that one should expect

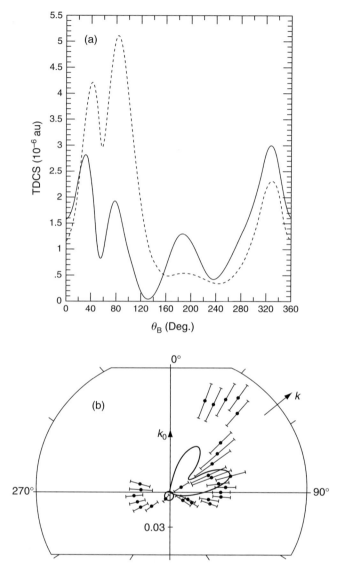

Figure 9.6 Coplanar TDCS for ionization of Ar($2p$) at $E_0 = 1949\,\mathrm{eV}$, $E_\mathrm{s} = 500\,\mathrm{eV}$, $\theta_\mathrm{f} = 30°$. (a) Full curve, full DWBA calculation; dashed curve, model calculation in which incoming distorted wave ζ^+ in the incident channel is replaced by a plane wave. (b) Data from [32] comparing in polar form (in units of 10^{-3} a.u.) their experimental data with the FBA-CW approximation.

strong distortion effects. A pronounced enhancement above the first Born level is a better indication of a distortion effect, but even that may have other origins for the hydrogen case; we have already mentioned there is a noticeable enhancement due primarily to post-collisional electron–electron interaction, [1–3].

9.4 Inner-shell ionization of heavy metal targets at relativistic impact energies

In the previous sections we have considered electron impact ionization at low energies where relativistic effects will be entirely negligible. There have been, however, studies at much higher energies on the deep inner shells of heavy targets; these began in 1982 with (e,2e) experiments by Schüle and Nakel [38, 39] at an incident energy of 500 keV on the K shell of silver. The description of such processes opens up a whole new range of problems. Relativistic effects are globally important: the target electron is in a deep inner shell, the incident and final electrons have velocities that are a significant proportion of the speed of light; retardation and spin-dependent interactions need to be considered. We have seen above that at low energies the DWBA is the simplest possible approximation that we can use to include multiple scattering effects in both the incident and final channels. It works well for a range of geometries where higher order effects, for example, polarization of the target in the incident channel or electron–electron repulsion terms in the final are weak; it gives excellent agreement with inner-shell ionization experiments on argon and it has also been applied with good effect to magnesium ionization [40]. Clearly, therefore, it is a prime candidate as a theoretical model for the relativistic inner-shell problem. Indeed, there are some simplifications: at the energies we are working, exchange in the elastic channels is likely to be negligible and final-state $e^- - e^-$ repulsion will certainly play no role. However, we are now dealing with a fully relativistic problem. This means that we will have to solve for Dirac spinors rather than Schrödinger wave functions and we will need to include the full QED photon propagator. In the earliest attempts to model these processes a number of assumptions were made, which subsequently turned out to be false, and it is perhaps valuable to list these here. The simplest approximation one could use is the plane wave impulse approximation, PWIA, in which all the electron wave functions are represented by plane waves. At non-relativistic energies, this approximation has been used extensively in impulsive experimental arrangements and, when it is valid, it allows the use of the (e,2e) method as a means of mapping the momentum distribution of the target electron. As we have shown above, a nice feature of the low-energy PWIA approximation is that it factorizes into two terms, one which is independent of the target wave function and the other is in essence the wave function of the target in momentum space. In Bell [41], a relativistic version was proposed in which the TDCS was taken to be the product of the free first-order electron–electron (Møller) cross section and the momentum profile of the bound state. Keller and Whelan [42], analyzed the relativistic plane wave impulse approximation and concluded that in fact it did not factorize, and that for any given kinematical arrangement and bound state energy, the 'cross section function' part of the TDCS depended on the spinor structure of the bound state. The results obtained in the relativistic plane wave impulse approximation are substantially different from those found in the impulsive treatment of Bell [41], though both theories share the same, factorized, form in the non-relativistic limit. A number of semi-relativistic variants on the Born approximation were tried. Das and Konar [43] employed a semi-relativistic Sommerfeld Maue function [44] for one of the outgoing electrons, and Jakubaßa-Amundsen, in a series of calculations, studied the influence of different approximate semi-relativistic scattering and bound-state wave functions on the TDCS [45–47]. We will not discuss the validity of using semi-relativistic wave functions here and only remark that they can easily lead one to error; see [48]. Another common feature of all these approximations was that they only included those spin channels that would contribute at low energies; the other relativistic 'spin–flip' channels were not included. In Walters *et al.* [49], it was shown that this was an invalid assumption and that, especially in the symmetric case, they made a very large contribution to the TDCS.

It was clear that only a fully relativistic approximation would stand any chance of describing the TDCS. Further, it was also evident that distortion effects needed to be included: the strong nuclear field could not be neglected. In [50,51] a relativistic version of the DWBA was developed (rDWBA). In the rest of this chapter we will continue to use atomic units and take the metric tensor to be:

$$\text{diag}(g_{\mu\nu}) = (1, -1, -1, -1). \tag{9.33}$$

Contravariant four vectors are written $x^{\mu} = (ct, \boldsymbol{x})$ and the summation convention is understood.

9.4.1 Relativistic, distorted wave Born approximation

The TDCS for the relativistic (e,2e) process, where the spins are not resolved, can be written quite generally as:

$$\frac{d\sigma}{d\Omega_f d\Omega_s d E_s} = \frac{2\pi^2}{c^6} \frac{k_f k_s}{k_0} E_0 E_f E_s \frac{N_\kappa}{2N_m} \sum_{\epsilon_f \epsilon_0 \epsilon_s \epsilon_b} |< k_f \epsilon_f, k_s \epsilon_s \mid \hat{S} \mid k_0 \epsilon_0, \kappa \epsilon_b >|^2,$$

$$\tag{9.34}$$

where $\hat{S}$ is the S-matrix operator; 0, f, s and b refer to the incoming, the two outgoing and the initially bound electron, respectively, E_0, E_f, E_s and k_0, k_f, k_s are the on shell total energies and momenta of the unbound particles where $E^2 = k^2 c^2 + c^4$. We are using κ to denote the quantum numbers of the atomic bound states and ϵ for the spin projection operators with respect to the quantization axis, which we take in the beam direction. In the form (9.34), the TDCS is insensitive to spin polarization: we have averaged over the initial spins, ϵ_b, ϵ_0, and summed over the final ϵ_f, ϵ_s. (Hence the factor $\frac{N_\kappa}{N_m}$; N_κ is the occupation number of state κ and N_m the number of degenerate states with this set of quantum numbers.) In [52] a discussion of explicitly spin-dependent measurements will be presented; here we will assume that all experiments are in no way spin resolved. In [50, 51], the rDWBA approximation was devised which assumed that the ionizing interaction between the two electrons could be treated in first-order perturbation theory, but the strong field of the atom was retained to all orders. It was considered absolutely vital that no unwarranted non-relativistic assumption be made. In [51], the rDWBA is defined with great care; here we will only give a brief summary of the assumptions made and outline some of the main features of the calculations. It was assumed that the electromagnetic field of the nucleus and the atomic electrons could be incorporated in the form of a classical field. In all calculations, to date, this field was assumed to be purely electrostatic and radially symmetric in the reference frame of the calculations (which was taken to be the rest frame of the nucleus). Consequently, the Dirac equation, which had to be solved in the elastic channels, could be separated in spherical co-ordinates. In most of the calculations reported, a self-consistent relativistic Kohn–Sham local density approximation potential was used [53,54] to represent the atomic potential. The full free QED photon propagator, in the Feynman gauge, was included. In relativistic physics, the interaction between the two electrons is represented by photon exchange between the four currents of the two electrons; in contrast, the non-relativistic Coulomb interaction can be thought of as an instantaneous interaction between point charges. By using the full propagator, one automatically includes the magnetic interactions between the currents of the particles and retardation effects. The S matrix in (9.34) can then be written:

$$< k_f\epsilon_f, k_s\epsilon_s \mid \hat{S} \mid k_0\epsilon_0, \kappa\epsilon_b >^{\text{DWBA}} = S^{\text{dir}} - S^{\text{ex}}, \tag{9.35}$$

with

$$S^{\text{dir}} = -i \int d^4x\, \bar{\psi}_f(x)\gamma_\mu A^\mu(x)\psi_0(x), \tag{9.36}$$

where the four-potential $A^\mu(x)$ is given by

$$A^\mu(x) = \int d^4y\, D^0_{\mu\nu}(x - y)J^\nu(y), \tag{9.37}$$

and $D^0_{\mu\nu}(x - y)$ is our free photon propagator, which we can write as

$$D^0_{\mu\nu}(x - y) = \frac{-4\pi}{c}ig_{\mu\nu} \int \frac{d^4k}{(2\pi)^4}\frac{e^{ik(x-y)}}{k^2 + i\epsilon} \tag{9.38}$$

and $J^\nu(y)$ is the fermion current

$$J^\nu = \bar{\psi}_s(y)\gamma^\nu\psi_b(y) \tag{9.39}$$

(we are using the notation of [55]). In other words,

$$S^{\text{dir}} = i\int d^4x \int d^4y \; \psi^\dagger_{k_f,\epsilon_f}(x)\gamma^0\gamma^\mu\psi_{k_0,\epsilon_0}(x)D^0_{\mu\nu}(x-y)\psi^\dagger_{k_s,\epsilon_s}(y)\gamma^0\gamma^\nu\psi_{k,\epsilon_b}(y), \tag{9.40}$$

where γ are the usual Dirac matrices and $\psi_{k,\epsilon}$ is an exact stationary solution of the first quantized Dirac equation, with the spherically symmetric Kohn–Sham potential representing the interaction with the atom; in essence, $\psi_{k,\epsilon}$ are the equivalent of our distorted waves $\chi^\pm$ in (9.32). For this choice of potential, the Dirac equation is separable and solutions can be written:

$$\psi(x) = e^{iEt/c}\begin{pmatrix} g_\kappa(r_x)\Xi_{\kappa,\mu}(\hat{\Omega}_x) \\ if_\kappa(r_x)\Xi_{-\kappa,\mu}(\hat{\Omega}_x) \end{pmatrix}, \tag{9.41}$$

where now the quantum number $\kappa = -(2j+1)(j-l)$ represents good total angular momentum j and parity $(-1)^l$.

The photon propagator can be represented in polar co-ordinates and then expanded in multipoles. Keller *et al.* [51] used the program of Salvat and Mayol [56] to evaluate numerically the radial functions $\psi_{k,\epsilon}$, and the Clebsch–Gordan coefficients were extracted from Burgess and Whelan [57]. The most serious numerical challenge faced in developing the rDWBA code was that one ended up needing to evaluate to a high accuracy a large number of integrals, typically of the order of 100 000, of the form

$$R = \int_0^\infty dr_x\, r_x^2 \int_0^\infty dr_y\, r_y^2\; j_n(\rho r_<)$$
$$\times\left[-y_n(\rho r_>) + ij_n(\rho r_>)\right]\phi_f(r_x)\phi_0(r_x)\phi_s(r_y)\phi_b(r_y), \tag{9.42}$$

where j, y are regular and irregular Bessel functions, $\rho = (E_0 - E_f)/c$, $r_> = \max(r_x, r_y)$, $r_< = \min(r_x, r_y)$ and each ϕ represents a large or small component of the radial Dirac solution. There were some significant numerical challenges to be dealt with in evaluating these integrals:

- The unbound wave function components are only known analytically in the asymptotic region. It was therefore impossible to tailor the integration algorithm explicitly to the analytic structure of the integrand in the vicinity of the nucleus where one would expect the dominant contributions.

- Only the term ϕ_b is bounded, the ϕ's corresponding to the free particles are highly oscillatory, as are the spherical Bessel functions arising from the multipole expansion of the potential. These oscillations are intensified as the angular momentum and polarity are increased.

The major computational problem faced was being sure that the integrals evaluated with sufficient accuracy. Due to the large number of radial matrix elements and the relative signs entering through the vector coupling coefficients, cancellation effects could be very severe. To illustrate, let us switch off the distorting potentials in the incident channel; this gives us back our old friend, the plane wave Born approximation, and the TDCS can be worked out in closed form [59]. Then (9.42) can be written, in essence,

$$R = A \int_0^\infty dr_x r_x^2 \int_0^\infty dr_y r_y^2 j_n(\rho r_>)$$
$$\times (-y_n(\rho r_<) + i j_n(\rho r_<)) j_{l_1}(k_f r_x) j_{l_0}(k_0 r_x) j_{l_2}(k_2 r_y) \phi_b(r_y), \quad (9.43)$$

where A is constant. Then for the larger l values, which after all are the main problem, one can extend the analysis of [58] and apply a relativistic analog of the Bethe approximation, i.e., assume $r_x > r_y$ everywhere, which in our case is a very good approximation, and then

$$R = ABC,$$

where C contains a bounded part and

$$B = \int_0^\infty dr_x r_x^2 (-y_n(\rho r_x) + i \, j_n(\rho r_x)) j_{l_1}(k_1 r_x) j_{l_0}(k_0 r_x). \quad (9.44)$$

Consider the term

$$\int_0^\infty dr_x \, r_x^2 j_n(\rho r_x) j_{l_f}(k_f r_x) j_{l_0}(k_0 r_x). \quad (9.45)$$

This integral is characteristic of the entire problem. In Fig. 9.7, we plot its integrand for a typical set of parameters and the reader will appreciate just how oscillatory these functions are. Whelan [59] succeeded in finding an analytic solution to (9.45). This allowed a crucial benchmark for the numerical methods. The reader is referred to [51] for full details of the computational approach and the extensive testing of the program that was undertaken.

It should be noted that in [51] and subsequent papers it was assumed that all distorted waves could be generated in the field of a neutral atom. The complication of adding Coulomb boundary conditions for one or both electrons has been discussed elsewhere [48]; as one might expect, the effect of the fast-moving outgoing electrons asymptotically seeing a residual unit charge is entirely negligible; what

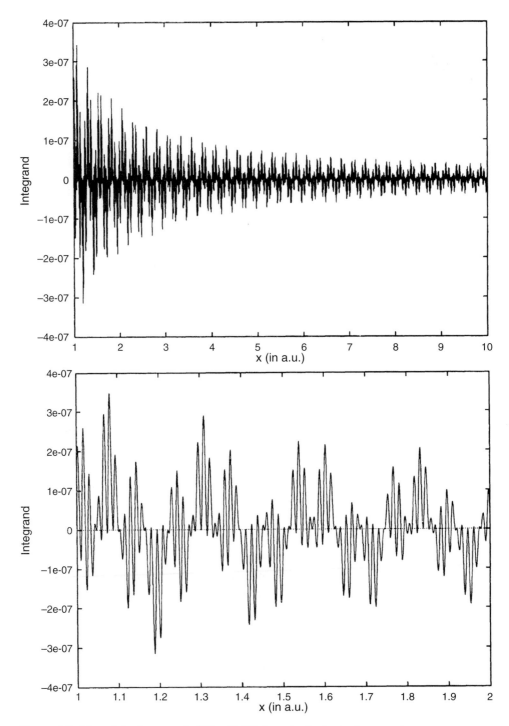

Figure 9.7 Plot of the integrand of Eq. (9.45) for a typical set of parameters.

matters is the influence of the highly charge-nucleus when the electrons are in its vicinity.

In summary, the success of the DWBA at low energies provoked an attempt to develop a relativistic equivalent. The rDWBA is, however, a *fully* relativistic approximation where the same philosophy was used as in the non-relativistic case, but no attempt was made to retain the mathematical formalism that was suitable at low energies: no semi-relativistic wave functions were employed, spin–flip terms were always included, and the full QED photon propagator was used. There is another significant difference. In the low-energy DWBA work, the incoming electron and exiting electrons were scattered in the *static-exchange* potential of the atom/ion, while at high energies we used a pure *static* atomic potential in all channels. It is important to be clear: in both the low- and high-energy work we allow for the indistinguishability of the two final-state electrons, but in the low-energy case we also allow for exchange effects in the elastic scattering channels. For example, at low energies, the approximation allows for the slow-exiting electron to exchange with the remaining atomic electrons on its way to the detector; the rDWBA does not. This simplification is justified when one considers the speed of the ejected electron and the binding energies of the bound electrons in the relativistic case, however, this difference between the two approximations has to be borne in mind when one interprets low- and high-energy spin asymmetry experiments; see [52].

One of the key features of the relativistic experiments was that many of them were absolute [39]. This meant that, unlike the low-energy inner-shell experiments, discussed in Section 9.3.2 above, the theory had both to predict the size as well as the shape of the TDCS. There were a number of attempts to do just this; all got the general shape of the cross section, more or less, right but there were enormous differences in absolute size. The rDWBA does an adequate job and the other simpler approximations generally do less well; this yields information both as to the nature of the dominant interactions and how one should go about the description of a relativistic atomic collision. We will argue that the relativistic distorted wave Born approximation is the simplest possible means at our disposal to describe deep inner-shell electron impact ionization. As we have continually stressed, a measurement of the TDCS gives a very refined picture of the ionization process and by carefully exploiting the geometrical and kinematical set-up of these experiments, one can focus on fine details of these interactions, which can be very revealing. Spin is a characteristic of relativistic phenomena and the possibility of doing experiments that are spin resolved stretches the theoretical description to its limit. A discussion of measurements with polarized beams and the advances that have been made in the theoretical understanding of such a process is given in [52]. Here we will restrict

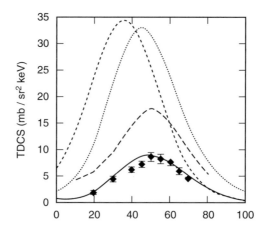

Figure 9.8 Absolute TDCS as a function of θ_s for gold ($Z = 79$) in coplanar asymmetric geometry, $E_0 = 500$ keV, $E_s = 100$ keV, $\theta_f = -15°$. Symbols experimental data: [60]; shorted dashed curve: semi-relativistic first Born approximation of [43]; dotted: semi-relativistic first Born approximation of [49]; long dashed: semi-relativistic Coulomb–Born approximation; [45, 46] solid line: rDWBA [42].

our discussion to the study of the ionization of unpolarized targets by unpolarized beams.

9.5   General features of the cross section

For K-shell ionization, the TDCS as measured in either coplanar asymmetric or symmetric geometry exhibits only a little structure. One expects to see in Ehrhardt geometry one peak in the direction of momentum transfer and possibly a second in the recoil direction, and in the Pochat geometry a binary and a large angle peak associated with elastic scattering from the nucleus. The target wave function has a high degree of symmetry and this symmetry tends to be reflected in the simplicity of the TDCS shape. Thus, the major way that relativistic, distortion or other effects manifest themselves is in the *size* of the cross section and hence the significance of *absolute* experiments [39].

9.5.1   *Coplanar asymmetric-Ehrhardt-geometry*

In Fig. 9.8 we compare the experimental results for a gold target ($E_0 = 500$ keV, $E_1 = 100$ keV, $\theta_1 = -15°$) with theoretical results by Das and Konar [43], who use a semi-relativistic Sommerfeld–Maue Coulomb wave function to represent the slow outgoing electron and the Ochkur approximation for the exchange term, and the first Born results of Walters *et al.* [49], where a Darwin–Coulomb wave is used for this electron. Both theories employ plane waves to describe the incoming and fast outgoing electron, and they both overestimate the experiment by about a factor of roughly four. (It is not clear why the calculation of Das and Konar mis-predicts the location of the maximum.) Agreement between experiment and the rDWBA is satisfactory. Figure 9.8 also shows the result of the semi-relativistic Coulomb–Born calculation of Jakubaßa-Amundsen [45, 46]. The results of this theory, which may be understood as an approximation to the full DWBA, is much closer to the experimental data, but still over-predicts them by a factor of approximately two. Finally, we show the rDWBA calculation. A self-consistent relativistic Kohn–Sham LDA potential of a neutral gold atom has been used in all channels. The bound state was described by a relativistic hydrogenic $1s_{\frac{1}{2}}$ wave function. The effects of Slater screening and of orthogonalizing the outgoing distorted waves to the bound state were found to be negligible because the overlap integrals between the bound state and the scattering partial waves are extremely small.

9.5.2 *Coplanar symmetric-Pochat-geometry*

In the relativistic regime, energy-sharing collisions have been measured on an absolute scale [39]. From the theoretical side, TDCS have been calculated in plane wave impulse approximation (PWIA) [41, 42, 63]; semi-classical impulse approximation [62]; in first Born approximation calculations using a semi-relativistic Coulomb function for one outgoing electron [43, 49]; and in a semi-relativistic Coulomb–Born approximation [46]. None of these theoretical approaches could satisfactorily describe the experimentally determined TDCS for gold, either in shape or in magnitude; see Fig. 9.9. Walters *et al.* [49] showed that it is important to include all possible spin–flip channels in the calculation and their inclusion widens the gap between experimental and theoretical results.

The agreement with the relativistic DWBA is almost perfect; it is the only calculation available that is able to reproduce the decrease of the experimental data for detection angles smaller than 30°. This effect is not due to no post-collisional interaction, PCI, which is neglected in the rDWBA. The observed increase of the theoretical data for detection angles smaller than 15° is not in conflict with the experimental observations. We believe that this increase is not an artefact of the lack of PCI. In the region between 40° and 70°, all theories predict a decrease of the TDCS, which is confirmed by the experimental observations. For larger detection angles, triply differential cross sections calculated in PWBA and first Born approximation decrease steadily, whereas the rDWBA predicts a large angle structure.

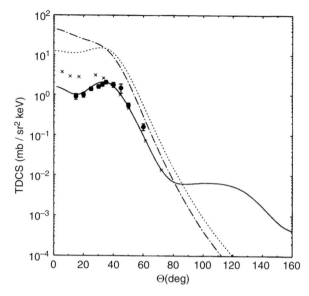

Figure 9.9 Absolute TDCS for $(e,2e)$ on gold in coplanar symmetric geometry, plotted against the detection angle of one of the electrons. Experiment: [49, 60]. Theory: plane wave calculation [42] dashed-dotted; semi-relativistic first Born [49], dotted; Coulomb–Born approximation [46], +++; rDWBA, solid line.

9.6 Special features

As well as their absolute size, the TDCS exhibits some very interesting features that are characteristic of relativistic and distortion effects. In the non-relativistic first Born approximation, the TDCS for a $1s$ target is symmetric about the direction of momentum transfer $q = k_0 - k_f$. For a K shell, this results in the observation of two peaks, one at q, the binary, and another about the recoil direction, $-q$. In the measurements of Bonfert *et al.* [60] on gold at 500 keV, the binary peak is observed to be shifted away from the direction of momentum transfer. Such a shift has been observed at much lower energies. For example, the experiments [22] on H, at energies of 250 eV and below, show the symmetry about q being broken, with the maxima shifted, the binary peak reduced and the recoil enhanced. The hydrogen experiments have been analyzed by a number of authors [2, 8, 23, 64], and it is now generally accepted that the observed shifts in the binary and recoil peaks are due to post-collisional interactions between the slow ejected and the fast scattered electrons; this interaction is implicitly contained in the second Born term and entirely absent from the standard DWBA calculations. In the case of the Bonfert it et al. data, it is clear that the shift has another origin, outside of the fact that PCI effects will be entirely negligible at these energies, there is already a clear indication of the shift in both semi-relativistic and fully relativistic first Born calculations [45, 49, 61]. In order to analyze the shift, Ast *et al.* [65] considered

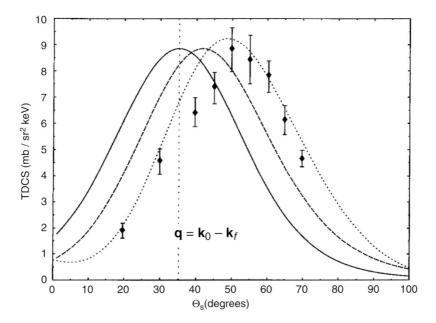

Figure 9.10 Absolute TDCS for the electron impact K-shell ionization of gold, $E_0 = 500\,\text{keV}$, $E_s = 100\,\text{keV}$, $\theta_f = -15°$. Experiment: [60]; first Born approximation [49] normalized to experimental data, dashed line; model calculation normalized to data, solid line; rDWBA, dotted line.

the fully relativistic first Born approximation, which is essentially the same as the rDWBA, where the distorting potentials for the fast incident and scattered electrons are set to zero and the bound and the ejected electrons see a potential which is Coulombic with an effective charge $Z_{\text{eff}} = Z - 0.3$ [66]. Crucially, it contains the full photon propagator. In order to understand the role of magnetic and retardation effects, Ast *et al.* [65] considered the propagator corresponding to the non-relativistic Coulomb potential,

$$D_{00}^{\text{Coul}}(x - y) = -4\pi\,\delta(x_0 - y_0) \lim_{\eta \to 0} \int \frac{d^3k}{k^2 + \eta^2} e^{i\mathbf{k}\cdot(\mathbf{x}-\mathbf{y})}, \qquad (9.46)$$

and calculated the TDCS for the gold, 500 keV case, in asymmetric geometry. We reproduce these results here; Fig. 9.10. Clearly the difference,

$$\tilde{D}_{\mu\nu}(x - y) = D_{\mu\nu}^0 - D_{00}^{\text{Coul}}(x - y), \qquad (9.47)$$

represents the relativistic contributions. The spatial components of this propagator $\tilde{D}_{jk}(x - y)$ describe the magnetic interaction between the currents of the two moving charges, the component $\tilde{D}_{00}(x - y)$ incorporates retardation effects. One notes that the position of the peak in the model calculation agrees with the direction of momentum transfer ($\theta_2 = 35.2°$). Clearly the shift obtained in the first Born

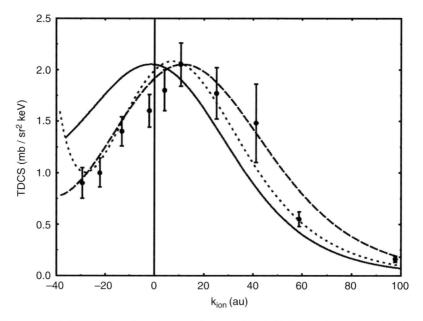

Figure 9.11 TDCS for the electron-impact K-shell ionization of gold, $E_0 =$ 500 keV, $E_f = E_s = 209.65$ keV symmetric geometry. TDCS plotted as a function of the recoil momentum, k_{ion}. Experiment: [49]. Theory: fully relativistic first Born approximation, dashed line (normalized to the experimental data) [61]; model calculation, solid line (normalized to experimental data) [65]; rDWBA, dotted line.

$(\theta_{2,\max} = 42°)$ is caused by magnetic and retardation effects. The position of the binary maxima is insensitive to the use of a semi-relativistic wave function [45]. It is only by doing a full rDWBA calculation can one achieve the full shift to the experimentally observed value, $\theta_{s,\max}^{\mathrm{expt}} = 50°$. Indeed, only then can we achieve agreement with absolute size of the cross section. Ast *et al.* [65] performed a similar analysis for equal energy-sharing collisions. They chose to plot the results in terms of $p = k_{\mathrm{recoil}} = k_0 - k_f - k_s$. Clearly the primary maximum in this geometry, in the non-relativistic first Born approximation, corresponds to $k_{\mathrm{recoil}} = \mathbf{0}$. For the fully relativistic first Born case, there is a significant shift away from zero recoil; see Fig. 9.11. The semi-relativistic approximations and fully relativistic calculations do not predict the same peak position. For a fuller discussion, see [65].

9.6.1 Spin-dependent effects using unpolarized beams on unpolarized targets

In relativistic physics, the spin projection quantum number is no longer a good quantum number. In a collision process, a flip of the spin of one or both involved particles is always possible.

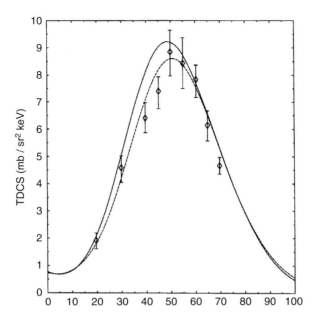

Figure 9.12 Absolute TDCS for the K-shell ionization of gold, $E_0 = 500\,\text{keV}$, $E_s = 100\,\text{keV}$, $\theta_f = -15°$. Experiment: [60]. Theory: full rDWBA calculation, solid line; rDWBA calculation but non-spin–flip channels only, dotted line.

In the relativistic regime, experiments have been performed under either highly asymmetric conditions or fully symmetric. In Fig. 9.12 an asymmetric collision is considered. The TDCS obtained in rDWBA are plotted against the detection angle of the slow outgoing electron for a K-shell ionization of gold at an impact energy of 500 keV. The detection angle of the fast outgoing electron ($E_1 = 319.3$ keV) is fixed at $\theta_f = -15°$; the slow electron carries an energy of 100 keV. The solid curve represents the complete rDWBA TDCS for all channels and the dashed one the partial rDWBA TDCS, including only the non-spin–flip contributions. Here, at these impact energies, spin-flip processes contribute only a little to the TDCS.

However, the picture changes completely for a fully symmetric collision. If we consider Eq. (9.34) there are 16 possible spin channels, only 6 of which would exist in a non-relativistic theory, i.e.,

$$\text{(i)} \quad s_f = s_0 = s_s = s_b,$$
$$\text{(ii)} \quad s_f = s_0 = -s_s = -s_b,$$
$$\text{(iii)} \quad s_f = -s_0 = -s_s = s_b,$$

where s_f is either $+1/2$ or $-1/2$. For channel (i) above, both the direct and exchange S matrices are non-spin–flip while for (ii) S^{ex} and (iii) S^{dir} are spin–flip channels.

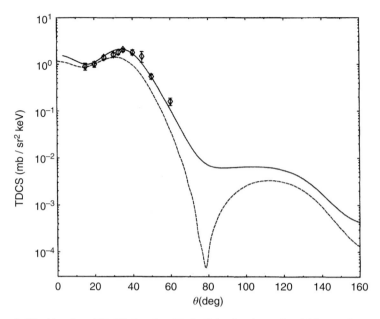

Figure 9.13 Absolute TDCS for the K-shell ionization of gold in coplanar symmetric geometry, $E_0 = 500\,\text{keV}$. Experiment: [60]. Theory: full rDWBA calculation, solid line; rDWBA calculation but non-spin–flip channels only, dotted line.

In coplanar symmetric geometry it is clear that channel (i) is zero, i.e., the non-spin–flip channels with equal spins for both electrons are blocked. Therefore, the relative importance of the spin–flip channels is enhanced compared to asymmetric collisions. This was first noted by Walters *et al.* [49] while investigating the semi-relativistic first Born approximation. In Fig. 9.13, the TDCS for gold, in coplanar symmetric geometry, at an impact energy of 500 keV, is plotted against the detection angle θ of one of the outgoing electrons that share energy ($E_s = E_f = 209.65\,\text{keV}$). The solid line represents the TDCS obtained in rDWBA, including all spin–flip contributions, and the dashed curve represents the partial TDCS for the non-spin–flip channels only. The spin–flip channels contribute up to 40% to the TDCS in the binary region and several orders of magnitude at the minimum point. It is instructive to compare these results with those of Fig. 9.3. We note that just as in the non-relativistic case, the non-spin–flip calculation produces a very distinct minimum between the two peaks, but the addition of the spin–flip terms act to significantly fill up the minimum. Both calculated cross sections are compared with the absolute experimental results. Whereas the calculated rDWBA TDCS (solid line) agrees very well with the experimental results, the partial non-spin–flip TDCS underestimates the experimental results significantly.

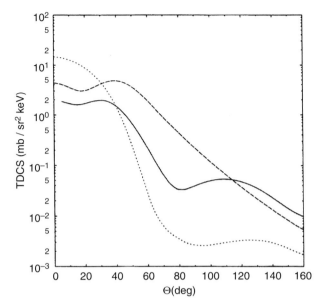

Figure 9.14 Absolute TDCS for the K-shell ionization of gold in coplanar symmetric geometry, $E_0 = 500$ keV. Experiment: [60]. Theory: full rDWBA calculation, solid line; rDWBA with a plane wave in the incident channel but distorted waves in the final, dashed lines; rDWBA with a distorted wave for the incoming but plane waves for the final outgoing electrons, dotted line.

9.6.2 Distortion effects

To investigate the physics of the distortion potentials in the incident and final channels, Whelan *et al.* [67] have looked closer into the collision with gold at an impact energy of 300 keV and performed two additional calculations: in the first, the distorted wave in the incident channel was replaced by a plane wave and the distorted waves for the outgoing electrons retained, and in the second, the distorted wave in the incident channel was retained but two plane waves were used in the outgoing channel, each orthogonalized to the bound state; see Fig. 9.14. These model calculations were used as a means of exploring the influence of the elastic scattering of the electrons in the atomic field separately for the incoming and outgoing electrons and the interference between these effects. The large angle peak was only found if there was a distortion in the incoming channel, i.e., if we have no elastic scattering in the incident channel then there is no large angle peak, exactly as is observed in the non-relativistic case. However, in these relativistic calculations evidence for strong interference effects was observed: the position of the large angle peak was shifted significantly when distortion was included in all the unbound channels compared to the model situation with only a distortion in the incident channel. In the latter, the large angle peak is maximal at about $130°$.

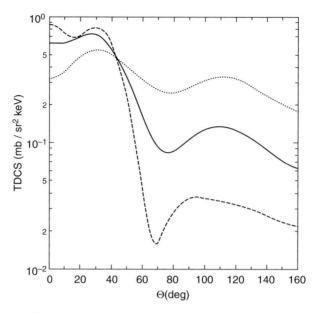

Figure 9.15 TDCS for the K-shell ionization of uranium in coplanar symmetric geometry calculated in the rDWBA approximation $E_0 = 500\,\text{keV}$, dashed lines; $300\,\text{keV}$, solid line; $220\,\text{keV}$, dotted line.

This is more or less the position predicted by the intuitive model discussed above. Having distorted waves in all channels shifted the maximum of the peak to $110°$ ($70°$ seen with respect to the backward direction). This additional shift goes beyond the simple model; interference effects and the influence of the heavy target atom are clearly significant. Whelan *et al.* [67] suggested that since the large angle peak is clearly driven by the interaction of the atom with the incoming and outgoing electrons, a set-up with maximal distortion could lead to a larger TDCS for the large angle region. With this in mind they performed calculations on uranium, which is the heaviest element suitable for an (e,2e) experiment with the present experimental set-ups and timescales. Shown in Fig. 9.15 is the coplanar symmetric TDCS for the K-shell ionization of uranium, calculated with the rDWBA for impact energies of $500\,\text{keV}$ (dashed curve), $300\,\text{keV}$ (full curve) and $200\,\text{keV}$ (dotted curve). The large angle peak is indeed more pronounced and gains in height having a cross section of more than $0.3\,\text{mb/sr}^2\text{keV}$, which is six times bigger than in the collision with gold at $300\,\text{keV}$. Another interesting observation could be made here. Compared with gold, the large angle peak is shifted to even smaller scattering angles. This is another strong indication of the importance of the influence of the strong atomic potential on the process. Clearly, it would also be of interest to make measurements at the minimum point between the binary and large angle peaks. The theoretical

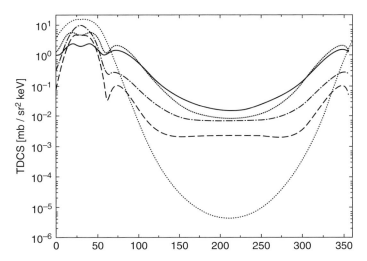

Figure 9.16 TDCS plotted against the detection of one of the electrons for gold at 300 keV the outgoing electrons share energy and their mutual angle is fixed at $\Theta_{fs} = 60°$: rDWBA, solid line; plane waves in all channels, dotted lines; distorted wave in incident channel and plane waves in all final, long dashed line; plane waves in the incident, distorted in both final channels, shorter dashed line; relativistic first Born, dashed-dotted line.

calculations predict that spin–flip channels will give a maximal contribution at this point.

It is of interest to look for geometrical arrangements where distorting effects will be particularly strong. With this in mind it is valuable to consider coplanar experiments where the angle between the exiting electron is held fixed and both are rotated around the beam direction. This geometry was earlier considered by [28, 29] while studying Coulomb three-body effects at low energy. Its principal significance at these energies is that one may, as a first approximation, treat the final-state electron–electron interaction as separable and thus one may concentrate on initial channel effects. In the relativistic case, neither electron–electron repulsion, post-collisionally, nor polarization in the incident channel are expected to be of any significance. The reasons for the use of this geometry in a relativistic situation are completely different. The principal advantage of the new geometry is that it allows one to study the role of strong distortion in the incident and final channels and the effect of their interference. We remark that for existing experiments the TDCS exhibits only a small amount of structure, and the difference between the various theoretical approximations manifests itself mainly through differences in absolute size. In contrast, for the new geometry, much structure is observed and this is crucially dependent on the degree of distortion in the individual channels. Following Whelan *et al.* [68], we show in Fig. 9.16

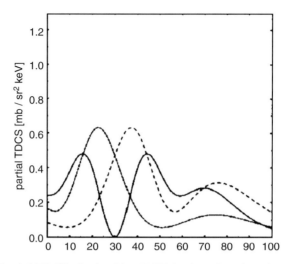

Figure 9.17 Partial TDCS obtained in rDWBA plotted against detection angle of one of the electrons for gold at 300 keV, $\Theta_{fs,} = 60°$: partial cross section channel (i) only, solid line; partial cross section channel (ii) only, dotted line; partial cross section channel (iii) only, dashed line.

the TDCS for the K-shell ionization of gold obtained in the relativistic distorted wave Born approximation (rDWBA), plotted against the detection angle of the second electron. Also shown are various model calculations where some or all the distorted waves are replaced by plane waves. The relative angle $\Theta_{f,s}$ between the two outgoing electrons is fixed to 60°, the impact energy is 300 keV and the two outgoing electrons share energy ($E_1 = E_2 = 109.65$ keV). One recognizes a double-peaked primary structure and a secondary maximum left and right of the primary. Each point accessed in the coplanar constant $\Theta_{f,s}$ geometry has one point in common with a specific Ehrhardt geometry. By comparison with the TDCS obtained in Ehrhardt geometry, one recognizes that the primary structure samples over binary and the secondary maxima over recoil peaks in different asymmetric geometries.

We note further that the saddle point in the primary structure has the co-ordinates $-\theta_s = \theta_f = \frac{\Theta_{s,f}}{2}$ and is therefore also the point in common with the 'Pochat' geometry. Because of the symmetry of the problem, the point $-\theta_f = \theta_f = \frac{\Theta_{1,2}}{2}$ must be either an extremum or the TDCS must be a constant in the neighbourhood of this point. It is very instructive to see why a minimum is realized here. In the calculation of the TDCS one sums over asymptotic spin projection channels for the outgoing electrons and averages over the initial ones. Now, as before, the six channels labelled (i), (ii), (iii), $s_f = \pm\frac{1}{2}$ are dominant. The remaining channels that contain only spin–flip amplitudes (which would not exist in a non-relativistic

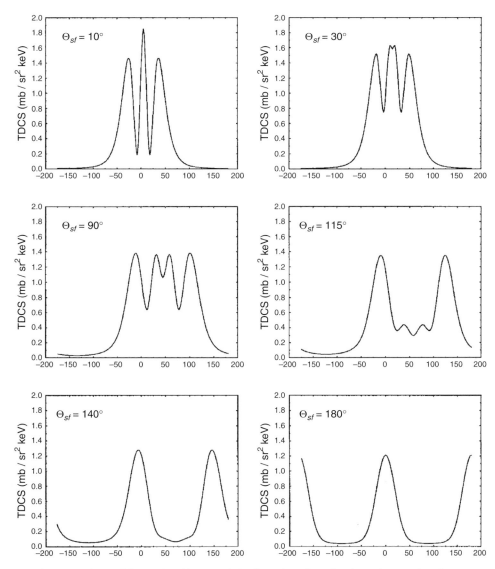

Figure 9.18 TDCS obtained in rDWBA plotted against the detection angle of one of the electrons, the relative angle Θ_{fs} being fixed at different values as shown in the figure.

theory) make a much smaller contribution. For channel (i), where all spins are equal, it is clear that at the symmetric point the S matrix is exactly zero. However, for the two other channels, the S matrix elements are finite at this point. Indeed for (ii) $|S^D|^2 \gg |S^{ex}|^2$ and for (iii) $|S^D|^2 \ll |S^{ex}|^2$ over the entire angular range. Spin–flip amplitudes are smaller compared to non-spin–flip amplitudes

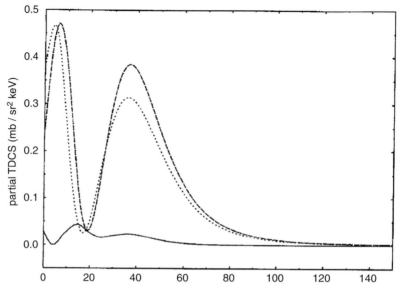

Figure 9.19 Same as Fig. 9.17 but $\Theta_{fs} = 10°$.

at the impact energies under consideration here. In Fig. 9.17, we plot the partial TDCS in rDWBA corresponding to the spin channels (i), (ii) and (iii) with $s_f = \frac{1}{2}$ and $\Theta_{f,s} = 60°$ for gold at 300 keV impact energy. We see that the dip observed in the TDCS arises because of the relative importance of the individual channel contributions as we move away from the point of symmetry.

It is now instructive to vary the relative angle between the two outgoing electrons. In Fig. 9.18, the TDCS is plotted as a function of the detection angle θ_s for gold at 300 keV impact energy for (a) $\Theta_{f,s} = 10°$, (b) 30°, (c) 90°, (d) 115°, (e) 140° and (f) 180°. Significant changes in the shape of the curve are observed if the relative angle $\Theta_{f,s}$ is enlarged or decreased with respect to our earlier example $\Theta_{f,s} = 60°$. If $\Theta_{f,s}$ is enlarged, the primary structure decreases with respect to the secondary maxima, which remains more or less unchanged, and for $\Theta_{f,s} = 140°$ the primary structure has completely vanished. The case (f), $\Theta_{f,s} = 180°$, is somewhat special because for an energy, sharing collision, the direction of the momentum transfer **q** with respect to one electron coincides with minus the direction of momentum transfer $-\mathbf{q}$ with respect to the other electron. In these circumstances the terms binary and recoil become essentially meaningless.

However, the physically more interesting case is the decrease of the fixed relative angle $\Theta_{f,s}$. Here, the primary structure remains strong, but the dip vanishes and the point of symmetry converts from a local minimum to a maximum. To investigate this in greater detail consider once more the contribution of the different spin channels to the TDCS for the case $\Theta_{f,s} = 10°$. From Fig. 9.19 we learn that the contribution of spin channel (i) is dramatically decreased in absolute size and with respect to the channels (ii) and (iii), not only in the neighbourhood of the point of symmetry $-\theta_f = \theta_s = \frac{\Theta_{fs}}{2}$, but over the entire angular range. This is a direct result of the anti-symmetry of the two fermion wave functions. Despite the fact that we have neglected electrostatic repulsion, the possibility of finding two *continuum* electrons in the same spin state decreases as $\Theta_{f,s} \to 0$. This effect is an example of Pauli blocking, familiar from atomic structure physics.

References

[1] E. P. Curran, unpublished PhD thesis, The Queen's University, Belfast (1986).
[2] E. P. Curran and H. R. J. Walters, *J. Phys. B*, **20**, 333 (1989).
[3] E. P. Curran, Colm T. Whelan and H. R. J. Walters, *J. Phys. B*, **24**, L19 (1991).
[4] I. Bray and A. T. Stelbovics, *Comput. Phys. Commun.*, **85**, 1 (1995).
[5] T. N. Resigno *et al.*, *Science*, **286**, 2474 (1999).
[6] P. L. Bartlett, *J. Phys. B*, **39**, R379 (2006).
[7] P. L. Bartlett and A. T. Stelbovics, this volume.
[8] Colm T., Whelan *et al.* in *(e,2e) and Related Processes*, ed. Colm T. Whelan, H. R. J. Walters, A. Lahman-Bannani and H. Ehrhardt (Dordrecht: Kluwer, 1993), p. 1.
[9] H. R. J. Walters, Colm T. Whelan and X. Zhang, in *(e,2e) and Related Processes*, ed. Colm T. Whelan, H. R. J. Walters, A. Lahham-Bennani and H. Ehrhardt (Dordrecht: Kluwer, 1993).
[10] S. P. Lucey, J. Rasch and Colm T. Whelan, *Proc. Royal Soc. (London) A*, **455**, 349 (1999).
[11] R. K. Peterkop, Opt. Spectros., **13**, 87, 1962
[12] M. R. H. Rudge and M. J. Seaton, *Proc. Royal Soc. (London) A*, **283**, 262 (1965).
[13] H. Bethe, *Ann. Physik*, **5**, 325 (1930).
[14] B. H. Bransden and C. J. Joachain, *Physics of Atoms and Molecules* (Harlow: Prentice Hall, 2003).
[15] A. A. Pinkás *et al.* in *New Directions in Atomic Physics*, ed. Colm T. Whelan *et al.*, (New York: Kluwer/ Plenum, 1999).
[16] E. Clementi and C. Roetti, *At. Data and Nucl. Data Tables*, **14**, 177 (1974).
[17] R. V. Calhoun, D. H. Madison and W. N. Shelton, *J. Phys. B*, **10**, 3523 (1977).
[18] Jens Rasch, unpublished PhD thesis, University of Cambridge (1996).
[19] Colm T. Whelan *et al.*, *Aust. J. Phys.*, **44**, 39 (1991).
[20] J. B. Furness and I. E. Mc Carthy, *J. Phys. B*, **6**, 2280 (1973).
[21] M. E. Riley and D. G. Truhlar, *J. Chem. Phys.*, **63**, 2182 (1975).
[22] H. Ehrhardt, K. Jung, G. Knoth and P. Schlemmer, *Z. Phys. D*, **1**, 3 (1986).
[23] F. W. Byron Jr., C. J. Joachain and B. Piraux, *J. Phys. B*, **18**, 3203 (1985).
[24] A. Pochat *et al.*, *J. Phys. B*, **16**, L775 (1983).
[25] Colm T. Whelan and H. R. J. Walters, *J. Phys. B*, **23**, 2989 (1990).
[26] X. Zhang, Colm T. Whelan and H. R. J. Walters, *J. Phys. B*, **23**, L509 (1990).
[27] L. Frost, P. Freienstein and M. Wagner, *J. Phys. B*, **23**, L715 (1990).
[28] C. T. Whelan *et al.*, *Phys. Rev. A*, **50**, 4394 (1994).
[29] J. Röder *et al.*, *Phys. Rev. A*, **53**, 225 (1996).
[30] A. Lahmam-Bennani *et al.*, *Phys. Rev. A*, **30**, 1511 (1984).
[31] G. Stefani *et al.*, *J. Phys. B*, **19**, 3787 (1986).
[32] P. Bickert *et al.*, *J. Phys. B*, **24**, 4603 (1991).

[33] P. Bickert, W. Hink and S. Schönberger, in *Proc. 17th ICPEAC, Brisbane*, ed. I. E. McCarthy, W. R. Mac Gillivray and M. C. Standage, (Brisbane: Griffith University, 1991). Abstracts, 180.

[34] X. Zhang *et al.*, *J. Phys. B*, **25**, 4235 (1992).

[35] C. Dal Capello *et al.*, *J. Phys. B*, **17**, 4557 (1984).

[36] A. N. Grum-Grzhimiliao, *J. Phys. B*, **18**, L695 (1985).

[37] M. J. Brothers and R. A. Bonham, *J. Phys. B*, **19**, 3809 (1986).

[38] E. Schüle and W. Nakel, *J. Phys. B*, **15**, L639 (1982).

[39] W. Nakel and Colm T. Whelan, *Phys. Reports*, **315**, 499 (1999).

[40] P. Bolognesi *et al.*, *J. Phys. B*, **41**, 015201 (2008).

[41] F. Bell, *J. Phys. B*, **22**, 287 (1989).

[42] S. Keller and Colm T. Whelan, *J. Phys. B*, **27**, L771 (1994).

[43] J. N. Das and A. N. Konar, *J. Phys. B*, **7**, 247 (1974).

[44] M. E. Rose, *Relativistic Electron Theory*, 2nd edn (New York: Wiley, 1961).

[45] D. H. Jakubaßa-Amundsen, *Z. Phys. D*, **11**, 305 (1989).

[46] D. H. Jakubaßa-Amundsen, *J. Phys. B*, **25**, 1297 (1992).

[47] D. H. Jakubaßa-Amundsen, *J. Phys. B*, **28**, 259 (1995).

[48] L. U. Ancarani *et al.* in *Coincidence Studies of Electron and Photon Impact Ionization*, ed. Colm T. Whelan and H. R. J. Walters (New York: Plenum, 1997), p. 215.

[49] H. R. J. Walters *et al.*, *Z. Phys. D*, **23**, 353 (1992).

[50] H. Ast *et al.*, *Phys. Rev. A*, **50**, R1 (1994).

[51] S. Keller *et al.*, *Phys. Rev. A*, **50**, 3865 (1994).

[52] J. Lower and Colm T. Whelan, this volume.

[53] R. M. Dreizler and E. K. U. Gross, *Density Functional Theory* (Berlin: Springer, 1990).

[54] M. Kampp *et al.*, in *Many-Particle Spectroscopy of Atoms, Molecules and Surfaces*, ed. J. Berakdar and J. Kirschner (New York: Kluwer/Plenum, 2001), p. 91.

[55] J. D. Bjorken and S. D. Drell, *Realtivistic Quantum Mechanics* (New York: McGraw-Hill, 1964).

[56] F. Salvat and R. Mayol, *Comput. Phys. Commun.*, **62**, 65 (1991).

[57] A. Burgess and Colm T. Whelan, *Comput. Phys. Commun.*, **47**, 295 (1987).

[58] Colm T. Whelan, *J. Phys. B*, **19**, 2343 (1986).

[59] Colm T. Whelan, *J. Phys. B*, **26**, L823 (1993).

[60] J. Bonfert, H. Graf and W. Nakel, *J. Phys. B*, **24**, 1423 (1991).

[61] S. Keller *et al.*, *Z. Phys. D*, **37**, 191 (1996).

[62] A. Cavalli and L. Avaldi, *Nuovo Cimento Soc. Ital. Fis. D*, **16**, 1 (1994).

[63] I. Fuss, J. Mitroy and B. M. Spicer, *J. Phys. B*, **15**, 3321 (1992).

[64] H. Klar *et al.*, *J. Phys. B*, **20**, 821 (1987).

[65] H. Ast *et al.*, *J. Phys. B*, **29**, L585 (1996).

[66] J. C. Slater, *Phys. Rev.*, **36**, 57 (1930).

[67] Colm T. Whelan *et al.*, *J. Phys. B* **28**, L33 (1995).

[68] Colm T. Whelan *et al.*, *Phys. Rev. A*, **53**, 3262 (1996).

10

Spin-resolved atomic (e,2e) processes

JULIAN LOWER AND COLM T. WHELAN

10.1 Introduction

The (e,2e) process for an atom describes an electron-impact-induced ionization event in which the momentum states of the incident and two outgoing electrons are defined, i.e., the reaction kinematics is fully specified. Due to its highly differential nature, the cross section describing this process provides a stringent test of electron-scattering theory. However, a quantum mechanically complete description of the (e,2e) process requires additional variables to be specified, namely the spin projection states of the continuum electrons, as well as angular momentum, and its projection state for the target atom before and the residual ion after the collision, respectively. While the goal of performing such a complete measurement is presently beyond experimental capabilities, (e,2e) experiments for which a subset of the quantum mechanical variables were determined have been performed. All employed beams of polarized electrons, enabling cross sections to be determined individually for the two spin states of the projectile (namely $m_s = \pm\frac{1}{2}$); others additionally resolved the angular momentum state of the target atom prior to the collision. In this chapter we will illustrate how the resolution of angular momentum states can powerfully highlight and provide new insight into specific aspects of the (e,2e) collision dynamics.

Electron spin emerges naturally from the relativistic treatment of quantum mechanics and, as a consequence, spin-resolved experiments are ideally suited to probe aspects of relativity in electron–atom scattering. Less obvious is that in the non-relativistic limit, spin-resolved measurements provide a sensitive probe to the nature of electron exchange processes in the (e,2e) ionization dynamics. This arises from the fact that, in the absence of explicit spin-dependent forces, the projection state of the incident electron remains unchanged throughout the collision,

Fragmentation Processes: Topics in Atomic and Molecular Physics, ed. Colm T. Whelan. Published by Cambridge University Press. © Cambridge University Press 2013.

i.e., its identity can be 'tagged' by its spin state and can thereby be distinguished from that of the electron it ejects if of opposite spin projection. However, in relativistic quantum mechanics, spin is not a good quantum number outside of the rest frame of the particle and care is needed in formulating the theoretical framework in which one can properly describe spin and angular-momentum sensitive (e,2e) experiments over the whole kinematic range from low to high impact energies and for low- and high-Z atomic targets alike.

The chapter is organized in the following manner. In Section 10.2 the concept of a spin-polarized ensemble of electrons is presented. Parameters which enable spin-dependent aspects of scattering to be highlighted are defined and their relation to experimentally determined quantities explained. The method of generating beams of polarized electrons using circularly polarized light and a semiconductor photocathode is discussed. Section 10.3 reviews (e,2e) experiments involving low-Z, hydrogen-like, spin-polarized target atoms and low electron impact energies. Such collisions can be described in the non-relativistic limit in which explicit spin-dependent forces acting on the continuum electrons are neglected and the projections of spin and orbital angular momentum are separately conserved. Under these conditions, the measured spin asymmetries provide information on the nature of electron exchange between the incident and the ejected target electron and on angular-momentum transfer from the target to the continuum-electron pair.

Section 10.4 describes low impact energy experiments on heavy noble-gas atoms. In this case, target relativistic effects play an important role in determining the form of the orbital wave functions, and for valence shell ionization, the fine-structure splitting in the residual ion is sufficiently large for an (e,2e) experiment to resolve. For these total-spin-zero targets, strong non-zero values of spin asymmetry are observed due to a complicated interplay between electron exchange, angular-momentum coupling and collision-induced orientation of the residual ion. Comparison of the experimental results with semi-relativistic scattering theories indicates that while bound-state relativistic effects play an important role, the experimentally derived spin asymmetries remain largely uninfluenced by continuum relativistic effects over the kinematic range of existing measurements. In this case, the derived spin asymmetries provide information on the degree of orientation acquired by the ion and on the many-electron exchange processes contributing to the (e,2e) cross section. The development of distorted wave Born approximation (DWBA) calculations in various degrees of sophistication is discussed and its ability to describe the experimental data is assessed.

In Section 10.5, the case of high impact energy collisions on heavy target atoms is considered. To describe high-energy kinematics, a fully relativistic scattering theory must be employed. To this end, the relativistic distorted wave Born approximation (rDWBA) formalism was developed, which proves highly successful in describing the experimental data. Comparison of the rDWBA with experiment shows spin asymmetries to be influenced not only by two-electron exchange effects, but also by relativistic contributions, including spin–flip processes and spin–orbit interaction of the projectile electron with the target nucleus. In contrast to the work described in the preceding sections involving beams of transversely polarized electrons, Section 10.6 describes a theoretical study concerning (e,2e) ionization by longitudinally polarized electrons. The interesting result is obtained that non-zero values of spin asymmetry are obtained, as long as the three electrons are not confined to a common plane. Concluding remarks are presented in Section 10.7.

10.2 Experimental considerations

10.2.1 Definition of measured and derived parameters

A beam of electrons is said to be polarized along an arbitrary quantization axis z if, measured along that axis, there exists an imbalance in the relative proportions of spin-up ($m_s = +\frac{1}{2}$) and spin-down ($m_s = -\frac{1}{2}$) electrons. For an ensemble of n electrons, the degree of spin polarization P along the z-axis is defined through the relation:

$$P = \frac{n^\uparrow - n^\downarrow}{n^\uparrow + n^\downarrow}, \tag{10.1}$$

where $n^\uparrow$ is the number of electrons with spin projection $m_s = +\frac{1}{2}$ and $n^\downarrow$ is the number of electrons with spin projection $m_s = -\frac{1}{2}$. For a given electron ensemble, the polarization $\boldsymbol{P}$ (of magnitude P) is oriented in space along the direction in which P assumes the largest positive value.

If the polarization of the electron beam can be inverted, values for the spin-asymmetry function A_J can be derived from measurement through the expression:

$$A_J = \frac{\sigma_J^\uparrow - \sigma_J^\downarrow}{\sigma_J^\uparrow + \sigma_J^\downarrow}. \tag{10.2}$$

Here $\sigma_J^{\uparrow(\downarrow)}$ is the triple differential cross section for ionization by electrons of positive (negative) spin projection measured along the normal to the scattering plane for transitions leading to the ion state J. A_J is a function of the scattering angles and energies of the two continuum electrons resulting from the (e,2e) collision. Experimentally, this quantity can be determined by measuring, respectively, the (e, 2e) count rates $N_J^\uparrow$ and $N_J^\downarrow$ for the beam polarization directed out of and into the scattering plane [1]:

$$A_J = \frac{1}{P} \frac{N_J^\uparrow - N_J^\downarrow}{N_J^\uparrow + N_J^\downarrow}. \tag{10.3}$$

Generalization of this expression for the asymmetry function to cases where *both* the incident beam and the target ensemble are spin polarized can be found in the literature [2, 3] (see Section 10.3).

For the scattering of spin-polarized electrons from an unpolarized ensemble of atoms, the spin-resolved triple differential cross section (TDCS) $\sigma_J{}^{\uparrow(\downarrow)}$ is given by the expression:

$$\sigma_J^{\uparrow(\downarrow)} = \frac{K}{P} \left[(1 + P) \, N_J^{\uparrow(\downarrow)} - (1 - P) \, N_J^{\downarrow(\uparrow)} \right]. \tag{10.4}$$

K is a constant which depends on target gas density, electron beam current, detector efficiencies and other instrumental factors. Spin-averaged TDCS, σ_J are given by:

$$\sigma_J = K \, N_J. \tag{10.5}$$

Here, the count rate N_J is the average count rate for unpolarized electrons and is calculated through the expression $N_J = (N_J^{\uparrow} + N_J^{\downarrow})/2$. For experiments on thin-film targets, K can be accurately determined and experimental (e,2e) cross sections can be determined on an absolute scale. In contrast, for gas targets the determination of K is generally much more difficult. For this reason, the experimental cross sections are generally normalized to theory. Also discussed in this chapter are results for the spin-averaged branching ratio R, describing transitions to the $^2P_{\frac{1}{2}}$ and $^2P_{\frac{3}{2}}$ ion states and defined by the expression:

$$R = \sigma_{\frac{3}{2}}/\sigma_{\frac{1}{2}}$$
$$= N_{\frac{3}{2}}/N_{\frac{1}{2}}. \tag{10.6}$$

In the non-relativistic limit, based solely on the relative number of projection states m_J, a value of 2.0 is expected for this quantity. The fact that significant deviations from this value are observed for heavier target atoms primarily reflects the significant influence of relativity on the radial behaviour of the bound-state wave functions in the residual ion (see Sections 10.3 and 10.4).

10.2.2 Generation of spin-polarized electron beams

For all of the experiments described in this chapter, polarized electrons were produced by the illumination of a GaAs-based photocathode by circularly polarized laser light; subsequent extraction and focusing of these electrons by electron-optical elements formed these electrons into a beam. While the band structure of the photocathodes varied from measurement to measurement, the principle of generation remains the same: angular momentum is transferred from the photon beam to electrons in the crystal along a specific direction in space that defines a quantization axis. Due to the helicity-specific nature of the selection rules governing transitions for photoionization and the specific transitions selected by an appropriate choice of photon wavelength, the emitted photoelectrons are polarized, respectively, either parallel- or anti-parallel to the quantization axis for excitation by right- and left-hand circularly polarized laser light. Transverse beam polarization is achieved by deflecting the electron beam through 90 degrees in an electrostatic field; the field acts on the electrons' momenta, while leaving the direction of the polarization vector P unchanged.

The obtained degree of spin polarization P depends on details of the photocathode structure and on the specific transitions excited by the laser radiation [4, 5] but is unaffected by the helicity of the laser field. Experimentally prepared ensembles of polarized electrons always exhibit a degree of polarization less than 100%. However, an accurate determination of P and the capacity to reverse the orientation of the polarization vector P through the reversal of laser helicity enables contributions from (e,2e) events from the minority spin-projection state of electrons comprising the beam to be subtracted in a statistical manner, as illustrated in Eq. (10.4).

10.3 Low-Z targets and low electron impact energies

The first (e,2e) measurements using spin-polarized electrons were reported by Baum *et al.* [3, 6]. They measured the (e,2e) ionization of the Li(2s) orbital at the impact energies 54.4 eV and $E_0 = 100$ eV. Coplanar scattering geometry was employed, implying that the incident and two scattered electrons share a common plane. The purpose of the measurements was to explore in detail the influence of the exchange interaction on the (e,2e) ionization cross section. This was facilitated through the resolution of spin channels for *both* the projectile electrons *and* the target atoms.

In their experiment, a beam of polarized electrons intersected a beam of spin-polarized lithium atoms, and cross sections were determined as a function of the *relative* orientation of electronic- and atomic-beam polarizations (parallel or antiparallel) and also as a function of the energies and emission angles of the outgoing electrons. The polarized beam of lithium atoms was formed by the action of a hexapole magnet in whose field the spins of the nuclei and electrons are decoupled. In this field, atoms with the projection state $m_J = +\frac{1}{2}$ were focused, while those with $m_J = -\frac{1}{2}$ were defocused and removed from the beam. In conjunction with a radio-frequency (RF) transition region, an atomic beam polarization of 0.5 was achieved. Polarized electrons were generated by photoemission from a strained gallium arsenide crystal.

Zhang *et al.* [7] performed non-relativistic DWBA calculations and found good agreement with the experimental results. In terms of the direct and exchange scattering amplitudes f and g of [8] we have

$$\sigma^{\uparrow\uparrow} = 2\pi^4 \frac{k_f k_s}{k_0} |f - g|^2 = \sigma_T$$

$$\sigma^{\uparrow\downarrow} = 2\pi^4 \frac{k_f k_s}{k_0} (|f|^2 + |g|^2) = \frac{1}{2}(\sigma_S + \sigma_T), \qquad (10.7)$$

where σ_S and σ_T are, respectively, the cross sections for singlet and triplet scattering,

$$A = \frac{\sigma^{\uparrow\downarrow} - \sigma^{\uparrow\uparrow}}{\sigma^{\uparrow\downarrow} + \sigma^{\uparrow\uparrow}} = \frac{r - 1}{r + 3}, \qquad (10.8)$$

where $\sigma^{\uparrow\downarrow}$ and $\sigma^{\uparrow\uparrow}$ represent cross sections for antiparallel and parallel alignment of the atomic and electronic polarization vectors and

$$r \equiv \frac{\sigma_S}{\sigma_T}. \qquad (10.9)$$

The calculations of Zhang *et al.* [7] were performed in the non-relativistic DWBA where

$$f^{T,S} = \langle \chi^-(\boldsymbol{k}_f, \boldsymbol{r}_f)\chi^-(\boldsymbol{k}_s, \boldsymbol{r}_s)| \frac{1}{r_{sf}} |\chi^{(+)T,S}(\boldsymbol{k}_f, \boldsymbol{r}_f)\psi_{2s}(\boldsymbol{r}_s)\rangle,$$

$$g^{T,S} = \langle \chi^-(\boldsymbol{k}_f, \boldsymbol{r}_s)\chi^-(\boldsymbol{k}_s, \boldsymbol{r}_f)| \frac{1}{r_{sf}} |\chi^{(+)T,S}(\boldsymbol{k}_f, \boldsymbol{r}_f)\psi_{2s}(\boldsymbol{r}_s)\rangle. \qquad (10.10)$$

In their approximation, different f, g amplitudes are obtained for triplet, T, and singlet, S, scattering. This is because the distorted waves $|\chi^{(+)T,S}(\boldsymbol{k}_f, \boldsymbol{r}_f)\rangle$ in the incident channel $(e^- + \mathrm{Li}(1s^2, 2s))$ are not the same for the two spin states. The difference arises in the *exchange* part of the elastic scattering; see [9, 10]. In their approach, Zhang *et al.* [7] had

$$\sigma_T = 2\pi^4 \frac{k_f k_s}{k_0} |f^T - g^T|$$

$$\sigma_S = 2\pi^4 \frac{k_f k_s}{k_0} |f^S + g^S| \qquad (10.11)$$

and used a local exchange approximation in the calculation of the distorted waves in the elastic channels [9, 11, 12].

In 2001, Lower *et al.* [2, 13] presented an analogous study on the (e,2e) ionization of spin-polarized sodium atoms by spin-polarized electrons to investigate spin and angular-momentum-transfer effects in the (e,2e) ionization process. As for the lithium experiment, the experiment involved removal of valence electrons in an ensemble of spin-polarized hydrogen-like atoms. In contrast, however, measurements were performed not only on the spin-polarized 3s ground state of the atom, but also on a spin-polarized and orbitally orientated 3p excited state. For ionization of the ground-state ensemble, as for the lithium experiment, the use of spin-polarized reaction participants enabled partial cross sections for singlet and triplet scattering to be extracted from measurement. In both experiments, the (e,2e) cross section depended on the *relative*, but not on the absolute, orientation of the atomic and electronic polarization vectors with respect to the scattering plane. For ionization of the oriented excited state, however, the measurement showed that the (e,2e) cross section depended not only on the relative orientation of the electron and atomic polarization vectors, but also on their *absolute* orientation with respect to the scattering plane. This outcome is a result of coupling between the spin and orbital angular momentum in the target.

Both the studies of Baum *et al.* and Lower *et al.* were performed on a light target at low impact energies. Under such conditions, continuum relativistic effects could be neglected and the observed spin asymmetries could be explained solely through the nature of the exchange interaction. By resolving partial (e,2e) cross sections for the two relative spin orientations of projectile and target, they were able to directly determine relative contributions to the (e,2e) cross section from singlet and triplet scattering processes and show how their respective strengths depend sensitively on the ejection angles and energies of the two scattered electrons in the final state. The results underline the fact that while the spin–orbit interaction diminishes with the strength of the nuclear field and hence the atomic number of the target atom, the exchange interaction does not, and plays a profound role in determining the (e,2e) cross section for all atoms, independent of atomic number. Indeed, even at high impact energies, two-electron exchange scattering still plays an important role under kinematics where the scattering angles and energies of the electrons comprising the (e,2e) electron pair are approximately equal.

10.4 High-Z targets and low electron impact energies

In the previous section we described how spin asymmetries arise when polarized electrons are scattered from polarized target atoms due to the effects of electron exchange. However, that both the target ensemble and the electron beam both be polarized is not a prerequisite to observe spin asymmetries under all conditions; spin asymmetries may arise even in the scattering of polarized electrons from *unpolarized* atoms. In this section we describe a mechanism by which this can occur; the so called 'fine structure effect'. However, for the majority of spin-resolved (e,2e) experiments performed to date involving closed-shell targets, as we shall see, significant deviations from the predictions of this model are observed; a more sophisticated description is required which takes into account many-electron exchange effects and target relativistic effects.

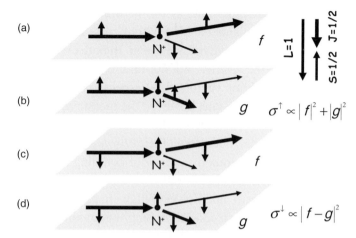

Figure 10.1 Schematic explaining principles of fine-structure effect (see text for details). Ionization of a noble-gas atom N leading the $^2P_{\frac{1}{2}}$ ion state is considered. In the example, the orbital-angular-momentum projection state of the residual ion is assumed to be $m_L = -1$ along the z-axis. The vertical arrows represent, respectively, the spin projections of the incident electron, the residual ion and the two exiting electrons. Scattering of the projectile to the left or to the right is respectively described by the probability amplitudes f and g. The cross sections describing (e,2e) ionization by spin-up electrons (panels (a) and (b)) and spin-down electrons (panels (c) and (d)) are described by incoherent and coherent sums of scattering amplitudes, respectively.

The 'fine structure effect' for electron impact ionization was first proposed by Hanne [14, 15], in analogy to a similar effect already observed in the electron impact *excitation* of atoms [16]. It was predicted to occur in the (e,2e) ionization of closed-shell atoms by polarized electrons, if the fine-structure levels of the residual ion, created through the removal of an electron from an orbital of non-zero orbital angular momentum, are resolved. The predicted spin asymmetry remains finite in the non-relativistic limit, i.e., light targets (negligible fine-structure splitting in the residual ion) and low impact energies.

To assist in describing the mechanism underlying the fine-structure effect we consider, in Fig. 10.1, the particular case of the (e,2e) valence P-shell ionization of a noble-gas atom N leading to excitation of the $^2P_{\frac{1}{2}}$ ion state under coplanar scattering geometry, where the incident and scattered electrons share a common plane (the $x - y$ plane). We will make a number of simplifying assumptions and then consider how modification of this model is required to describe most of the actual experiments performed to date.

For the considered scattering geometry there is generally a net transfer of orbital angular momentum from the projectile electrons to the residual ions along an axis perpendicular to the scattering plane, i.e., $< L_z > \neq 0$, where $< L_z >$ is the expectation value of the operator. The acquired orbital-angular-momentum orientation O results from the Coulomb interaction between the continuum electrons and the residual ions. The degree of orientation $|O|$ and its sign are independent of the spin projection state of the projectile. As a result of the collision-induced orientation and coupling between orbital and spin angular momenta, the ions become spin polarized along the z-axis. The degree of orientation and hence the degree of ion spin polarization depends on the scattering angles and energies of the (e,2e) electron pair.

As a simplifying assumption, we assume that we can describe the collisions in the non-relativistic limit. This implies that LS coupling provides a valid description and that spin–flip processes and other continuum relativistic effects can be neglected. Under such conditions the total spin and its projection are conserved throughout the reaction. Secondly, we assume that exchange only occurs between the projectile and the electron it ejects, in other words, that exchange between the continuum electrons and the bound electrons can be neglected. Finally, we assume that the degree of ion orientation is 100% and that O is directed along the negative z-axis, i.e., $m_L = -1$ and $m_s = +\frac{1}{2}$ along the z-axis.

In panels (a) and (b) in Fig. 10.1, we consider the case of ionization by spin-up electrons and in panels (c) and (d) by spin-down electrons; the scattering geometries in all panels are equivalent. The projectile electron scatters either to the left (panels (a) and (c)) in a process described by the probability amplitude f ('direct' scattering), or to the right (panels (b) and (d)) described by the probability amplitude g ('exchange' scattering). In panels (a) and (b), both 'direct' and 'exchange' processes are distinguishable, due to the different spin projection states of the two scattered electrons (a spin-sensitive detector could in principle identify which of the two processes had occurred). Thus, the (e,2e) cross section $\sigma^\uparrow$ is described by the *incoherent* sum $|f|^2 + |g|^2$. In panels (c) and (d), scattering to the left and right is again described by f and g, however, the two processes are now indistinguishable as the electron-spin projection states are identical. Thus the (e,2e) cross section $\sigma^\downarrow$ is described by the *coherent* sum $|f - g|^2$. As in the general case $|f|^2 + |g|^2 \neq |f - g|^2$, it is apparent that a non-zero up/down spin asymmetry $A_{\frac{1}{2}} = (\sigma^\uparrow - \sigma^\downarrow)/(\sigma^\uparrow + \sigma^\downarrow)$ is observed. These arguments can easily be generalized for an arbitrary degree of ion orientation and to describe excitation of the $^2P_{\frac{3}{2}}$ ion states (Jones *et al.* [15]). They showed that in the non-relativistic limit, $A_{\frac{1}{2}} = -2A_{\frac{3}{2}}$, where $A_{\frac{3}{2}}$ is the asymmetry describing ionization to the $^2P_{\frac{3}{2}}$ ion state and that a zero value of spin asymmetry is obtained if the two fine-structure levels are unresolved. The first spin-polarized (e,2e) measurements on atoms were reported by Guo *et al.* [17] and Hanne *et al.* [18] for xenon target atoms at an impact energy of 147 eV. The aim of those measurements was to establish whether finite spin asymmetries are observed when fine-structure states are resolved and, if so, to what extent does the fine-structure model or other relativistic effects account for the experimental findings. For tests of the fine-structure effect, a choice of a lower Z noble gas would have been more desirable as it had already been established that for atoms of large atomic number LS coupling does not provide an accurate description of angular momentum coupling [19] and target wave functions are strongly influenced by relativistic effects [20]. However, the large 1.3 eV fine-structure splitting of the Xe^+ $^2P_{\frac{1}{2}}$ and $^2P_{\frac{3}{2}}$ ion states was easily resolvable by experiment. Large spin asymmetries were observed and found to approximately follow the relation $A_{\frac{1}{2}} = -2A_{\frac{3}{2}}$, suggesting that the fine-structure-effect model provided at least a partial description of the underlying physics, even for the case of the ionization of a heavy atom. However, the strong deviation of the branching ratios from two is not consistent with the simple 'fine-structure' model. The experimental results were compared with distorted wave Born approximation (DWBA) calculations formulated within the framework of the Schrödinger equation. Relativistic scattering effects were partially accounted for by describing the target wave function with Dirac–Fock wave functions and including a Thomas spin–orbit term in calculation of the distorted waves. Calculated asymmetries were in reasonable agreement with the experimental data. It was shown that replacing the Dirac–Fock target wave functions with Hartree–Fock wave functions and neglecting the spin–orbit term in calculation led to very similar results. While this could be interpreted as suggesting that continuum spin–orbit effects do not play a significant role at these energies, the asymmetry parameter may not provide a sensitive measure of such relativistic effects.

More comprehensive (e,2e) measurements on xenon at 147 eV impact energy were presented by Dorn *et al.* [21]. The experimental results were compared to DWBA calculations in which, in contrast to those reported earlier [17], not only exchange between the incident and ejected electron, but also exchange between the two scattered electrons and the residual ion were included. It was found that this latter exchange effect was particularly important is describing the form of the spin asymmetry functions for larger scattering angles of the fast-scattered electron. Interestingly, Lechner *et al.* [22], using a density functional treatment of exchange, concluded that the apparent relativistic signatures seen in this work were exclusively the result of the exchange potential employed.

Using highly efficient toroidal electrostatic analyzers, higher precision (e,2e) results on the xenon target were presented by Panajotović *et al.* [23] and Bellm *et al.* [24]. In those works, refinements were made to DWBA calculations in an attempt to investigate the origin of disparities between experiment and calculation and bridge the gaps. In [23], DWBA calculations were performed in which post-collision-interaction (PCI) effects, describing the Coulomb interaction between the two scattered electrons, were treated to all orders of perturbation theory. However, this refinement in theory did not substantially reduce discrepancies between theory and measurement.

In [24], new measurements were performed and compared to further refined DWBA calculations in which exchange and polarization effects were accounted for by a full Hartree–Fock calculation for the bound and continuum-electrons. In this way, the non-local aspects of exchange were accounted for, in contrast to previous DWBA calculations in which the effects of the exchange interaction were described through a local exchange-potential. To better assess the sensitivity of calculation to the manner in which exchange effects are treated, calculations employing a local exchange-potential were also performed for comparison and the reaction kinematics were chosen at a lower impact energy (112 eV) and for equal energy of the two scattered electrons, conditions under which two-electron exchange effects are known to be enhanced. The experimental and theoretical results both showed large values of spin asymmetry. While the more accurate treatment of exchange produced slightly better agreement with measurement, the degree of improvement was less than anticipated, indicating that the static local-exchange approximation provides an accurate representation of the physics of exchange scattering and the effects of atomic (ionic) charge-cloud polarization, even at this relatively low impact energy.

For the same energies for the scattered electrons as in [24], analogous fine-structure measurements and calculations [25] were also performed on krypton target atoms. By choosing a lower Z target atom for which bound and continuum relativistic effects should have less influence, it was hoped to obtain improved agreement with the non-relativistic DWBA calculations. Asymmetry measurements were performed for transitions leading to the spin–orbit split $Kr^+ \, 4p^5 \, {}^2P_{\frac{1}{2}}$ and ${}^2P_{\frac{3}{2}}$ ion states and reasonable agreement between theory and experiment was achieved, but not notably better than for the xenon case. Transitions leading to the $Kr^+ \, 4s^1 \, {}^2S_{\frac{1}{2}}$ ion state were also measured for which, due to the absence of fine structure, non-zero values of spin asymmetry can only arise through Mott scattering. The fact that, within statistical error, a null spin-asymmetry result was achieved for this transition suggested that continuum spin–orbit effects would also not be contributing to any significant degree to the asymmetries for the $4p^5 \, {}^2P_{\frac{1}{2}}$ and ${}^2P_{\frac{3}{2}}$ ion states. Perhaps the remaining discrepancies between the DWBA calculations and the low energy (e,2e) measurements discussed thus far can be traced to the fact that all involved evaluation of the first-order Born term only; the extent to which second-order Born effects might contribute was not investigated.

While the experiments discussed above exhibited something of the character one would expect from the fine-structure-effect model, theory allows other physical processes, including many-electron exchange and relativistic wave function effects, to contribute. This motivated Bellm *et al.* [26] to pursue a novel approach to untangle disparate physical mechanisms. It was guided by the theoretical work of Kampp *et al.* [27] who identified (e,2e) kinematics under which non-relativistic DWBA and relativistic DWBA calculations (rDWBA) [28, 29] yielded almost identical cross sections. (A detailed discussion of the rDWBA and its application to high-energy experiments is presented in the following section.) As the DWBA calculations of Kampp *et al.* included exchange in the elastic channels and their rDWBA calculations (which take account of Mott scattering effects to all orders [30]) did not, this suggested that they had identified a kinematics where neither exchange processes involving electrons in the residual ion nor relativistic effects in the collision process play a significant role. While Bellm *et al.* deemed it infeasible to perform measurements under this kinematics, due to the low cross sections involved, they were able to identify a new kinematics where the main features of the Kampp *et al.* choice still held and for which the count rates were sufficiently high to yield reliable cross sections (see Bellm *et al.* [26] for details). The experiment comprised the inner-shell ionization of argon atoms at an impact energy of 909 eV leading to the $2p^5\,^2P_{\frac{1}{2}}$ and $^2P_{\frac{3}{2}}$ ion states, which are separated by an energy of 2.1 eV. The energies of the fast and slow scattered electrons were 600 and 60 eV, respectively. As anticipated, under these chosen conditions where relativistic and few-body exchange effects were suppressed, the experimental results and theory yielded spin asymmetries consistent with the relation $A_{\frac{1}{2}} = -2A_{\frac{3}{2}}$ and branching ratio values close to 2.0 (see Eq. (10.6)) supporting the model proposed by Jones *et al.* [15].

10.5 High-Z targets and high electron impact energies

A series of high-energy (e,2e) experiments were performed by the group of W. Nakel in Tübingen in the 1990s. The aim of this work was to better understand the ionization process under conditions where relativistic effects strongly contribute. Under such high-energy conditions, the angular- and energy distribution of the scattered electrons and the magnitude of the ionization cross section are profoundly affected by relativistic effects. Their choice of heavy target atoms and a focus on inner-shell processes further enhanced continuum relativistic effects due to the strong interaction between the target nuclei and the bound and continuum electrons. The use of a solid target film, as opposed to a molecular target beam, enabled the measured (e,2e) cross sections to be accurately determined on an absolute scale (determination of factor K in Eq. (10.4)), providing a stringent test for theory; in contrast, the vast majority of (e,2e) cross sections derived from gaseous targets are presented on a relative scale. Furthermore, focusing on the ionization of inner shells, whose orbital structure is largely unaffected by those of surrounding atoms, allows the measurement to be modeled as scattering from an aggregate of isolated atoms. The thickness of the foil was chosen sufficiently small that multiple scattering of the continuum electrons before and after the (e,2e) collision could be neglected. The first of these measurements were on the K-shell ionization of silver and gold [31, 32].

In 1995, (e,2e) experiments employing spin-polarized electrons were subsequently presented by this group. The experiment involved the the K-shell ionization of silver by a beam of transversally spin-polarized electrons [33, 34]. The outgoing electrons were detected in Ehrhardt geometry with $\theta_1 = -10°$, $E_0 = 300$ keV, $E_1 = 200$ keV. (e,2e) cross sections were measured as a function of the emission angle of the slow scattered electron for a fixed scattering angle of the fast and for the beam polarization directed, respectively, into or out of the scattering plane. Within statistical error no measurable degree of spin asymmetry was observed in the binary collision region, however, spin asymmetries of up to 15% were observed in the recoil region. This observation, supported by calculation, was explained along the following lines: the binary peak has a large contribution from a direct collision between the incident electron and the ejected target electron, where the nucleus remains essentially a 'spectator'. The recoil peak, which involves a large momentum transfer from projectile to target (more than can be accounted for by the momentum of the ejected target electron prior to the collision), can only be explained if the electron–nucleus interaction is taken into account.

To describe the high-energy experimental results of the Nakel group, a fully relativistic scattering theory must be employed. To this end, Keller *et al.* [33] set the relativistic distorted wave Born approximation formalism (rDWBA) of [8] in terms of the density matrix. The reader is refered to [33] for the details. From this point on, when we talk about the rDWBA, then we shall mean that approximation as defined in [33]. In Fig. 10.2, we compare their experimental asymmetry results with an rDWBA calculation and two semi-relativistic Coulomb–Born calculations [35, 36]. Agreement with the rDWBA is encouraging. Now, as is nearly always the case in such measurements, we see a maximum in the asymmetry in the direction of minimum TDCS, however, for this measurement we see significant asymmetries in the experimentally accessible region where the cross section is large. We note that as we are taking ratios, the parameter A is largely independent of the absolute size of the cross sections, $\sigma_j^\uparrow$ and $\sigma_j^\downarrow$. That, for a K-shell ionization (no fine-structure effect at play), A exhibits a non-zero value at all is an indication of relativistic effects; in the non-relativistic limit, spin and orbital angular-momentum projections are separately conserved and it is straightforward to see that $A = 0$ [33].

Recall that the rDWBA approximation makes no allowance for exchange in the elastic scattering channels, while the DWBA used by Zhang *et al.* is entirely non-relativistic, hence we see that the spin asymmetries observed by [34] are relativistic

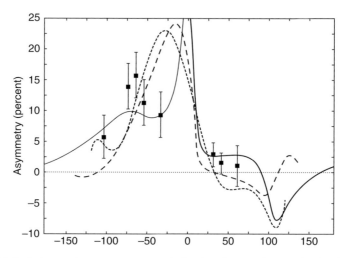

Figure 10.2 Spin up–down asymmetry A for the electron impact ionization of the K shell of silver as a function of the scattering angle, θ_s, of the slow outgoing electron of energy $E_s = 74.5$, keV, where $E_0 = 300$ keV, $\theta_f = -10°$. Experiment: [34]; dotted curve, semi-relativistic Coulomb–Born approximation of [35]; dashed curve, semi-relativistic Coulomb–Born approximation of [36]; solid line, rDWBA calculation [39].

in origin while those observed by [3] are due to exchange. In Fig. 10.3, we compare the relative spin-averaged TDCS observed in the Prinz–Besch–Nakel experiment with the results of the two semi-relativistic Coulomb–Born calculations [35, 36] and the rDWBA [33]. Due to the normalization of the data at the maximum, only small discrepancies between the various theories are visible in the binary region. In the region of the secondary maximum, at about $-50°$, only the rDWBA is capable of reproducing the magnitude and the trend in the TDCS. There are puzzling discrepancies between the Coulomb–Born calculations that not only fail to yield the experimental structure but inexplicably fail to agree; the only difference between the two calculations is the use of a semi-relativistic Coulomb bound-state wave function in [35] and an exact relativistic Coulomb bound-state wave function in [36]. It would be surprising if this were the source of the discrepancy, since the effect of using a semi-relativistic bound state within such an approximation has been studied for silver [37], and found to be negligible.

It is instructive to consider the Z dependence of the asymmetry. Following [33], we exhibit in Fig. 10.4 a series of rDWBA calculations for copper ($Z = 29$), silver ($Z = 47$), gold ($Z = 79$) and uranium ($Z = 92$). The impact energy was fixed by taking the following parameters for gold ($E_0 = 300$ keV, $E_f = 200$ keV, $\theta_f = -10°$). The ejection angle was held fixed. The ratio of impact energy to

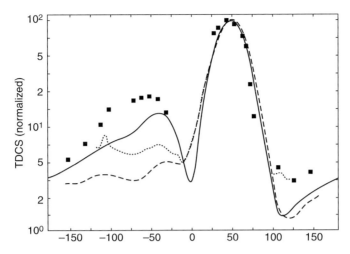

Figure 10.3 Relative TDCSs for the asymmetry experiment (data normalized at binary maximum). $E_s = 74.5$ keV, where $E_0 = 300$ keV, $\theta_f = -10°$. Experiment: [34]; dotted curve, semi-relativistic Coulomb–Born approximation of [35]; dashed curve, semi-relativistic Coulomb–Born approximation of [36]; solid line, rDWBA calculation [33].

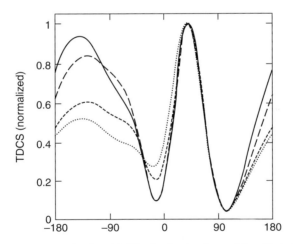

Figure 10.4 rDWBA results for TDCSs for ionization of the K shell of different target species, normalized to unity at the binary maximum. Impact energy for gold 300 keV, constant ratios of $E_0/E_b = 3.72$ and $E_f/E_s = 3$ in all systems, and $\theta_f = -10°$. Full curve, uranium ($Z = 92$); long-dashed curve, gold ($Z = 79$); short-dashed curve, silver ($Z = 47$); dotted curve, copper ($Z = 29$).

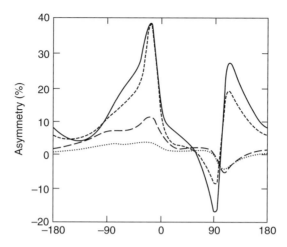

Figure 10.5 Spin asymmetry corresponding to the same kinematics as in Fig. 10.4.

binding energy, E_0/E_b, and the ratio of fast to slow electron energies, E_f/E_s, were held constant for all the other atoms. The ratio of recoil to binary peak is significantly enhanced as Z is increased. This is a clear indication of increasing nuclear influence. Significantly, the corresponding spin asymmetry, Fig. 10.5, is also noticeably enhanced. Recall that the spin–orbit interaction is, to lowest order in $1/c^2$, the spin–orbit interaction energy and is proportional to $r^{-1}dV(r)/dr$, where $V(r)$ is the radial symmetric scattering potential. We would, therefore, intuitively expect an enhancement in spin asymmetry due to spin–orbit interactions of the continuum electrons. It needs to be emphasized once again, however, that in the rDWBA calculations one treats everything fully relativistically and while one automatically includes 'spin–orbit' interactions, one also includes all higher order effects in the elastic scattering. In other words, the rDWBA gives a correct description of the elastic (Mott) scattering of all the unbounded electrons in the strong field of the atom. These processes can change the polarization of the scattered electrons and give rise to the asymmetries that Prinz *et al.* have measured.

   In a later publication, Prinz and Keller [38] presented experimental results for triple differential cross sections and spin asymmetries for the *L*-shell ionization of gold atoms. Individual magnetic sublevels were not able to be resolved due to limitations in energy resolution. Their relativistic rDWBA calculations [33, 39] were able to accurately reproduce both the experimentally determined shape of the (e,2e) cross section and the measured spin asymmetry for *L*-shell ionization, suggesting it could reliably describe collisions under these kinematic conditions. To better understand the origins of the observed spin asymmetry for the *unresolved* states of the *L*-shell hole, which exhibited values up to 8%, the rDWBA calculation was then used to calculate asymmetries for the individual magnetic sublevels. These calculations predicted asymmetries for the $^2P_{\frac{1}{2}}$ and $^2P_{\frac{3}{2}}$ states which closely followed the relation $A_{\frac{1}{2}} = -2A_{\frac{3}{2}}$ for scattering angles in the neighbourhood of the binary peak. In the recoil region, however, in which large angle scattering results from stronger interactions with the nucleus, substantial deviations from this relation were predicted, which were attributed to continuum spin–orbit effect. Over most of the angular range, the calculations showed that the sub-shell averaged asymmetry is dominated by the contribution of the $2p^1\,^2S_{\frac{1}{2}}$ state, as the contributions from the $^2P_{\frac{1}{2}}$ and $^2P_{\frac{3}{2}}$ asymmetries largely cancel out.

In a subsequent publication, with improved energy resolution [40] and an unpolarized beam of electrons, Kull *et al.* presented experimentally determined absolute (e,2e) cross sections for the $2p_{\frac{3}{2}}$ sub-shell of gold. The binding energies of the three sub-shells $2s_{\frac{1}{2}}$, $2p_{\frac{1}{2}}$, $2p_{\frac{3}{2}}$ are 14.4, 13.7 and 11.9 keV, respectively, and using a primary electron energy of 300 keV they succeeded in separating the $2p_{\frac{3}{2}}$ shell from the $2s_{\frac{1}{2}}$ and $2p_{\frac{1}{2}}$ shells, which were not resolved. This experiment was performed in asymmetric geometry and attention was focused on the Bethe ridge,

$$k_{\text{recoil}} = k_0 - k_f - k_s = 0$$

and its neighbourhood. The great advantage of the Bethe ridge is that the TDCS exhibits a strong dependence on the target state. This may be seen most clearly by considering the plane wave impulse approximation (see, e.g., [8]) in which the p orbital structure is reflected in the split binary peak, although this double peak structure is also seen in more sophisticated calculations. In Fig. 10.6, we show a comparison between the absolute experimental TDCS for the $2p_{\frac{3}{2}}$ state for the K-shell ionization of gold, compared with the rDWBA calculations of Keller *et al.* [29], and the semi-relativistic rDWBA calculations of Jakubaßa-Amundsen [36]. The characteristic p-state shape is exhibited by both approximations and the experimental data. Agreement with both calculations and measurements is reasonably good.

By using a heavier uranium target in which the fine-structure splitting for L-shell ionization was larger (3.8 keV compared to 0.6 keV for gold), Besch *et al.* [41] were able to energetically isolate transitions to the U$^+$ $2p^5\,^2P_{\frac{3}{2}}$ ion state. Spin asymmetries measurements were limited to this fine-structure transition as the energy resolution was insufficient to resolve transitions to the $2p^5\,^2P_{\frac{1}{2}}$ from those to the $2s^1$ $^2S_{\frac{1}{2}}$ ion state (0.8 keV separation). As a consequence, a full test of fine-structure-effect phenomena could not be accomplished. rDWBA calculations well described the experimental asymmetry data, which was limited to the neighbourhood of the binary peak and whose origin was attributed by the authors to a fine-structure-effect mechanism. However, when the rDWBA-calculated asymmetries are observed over the full angular range encompassing also the recoil collision region [33], a more

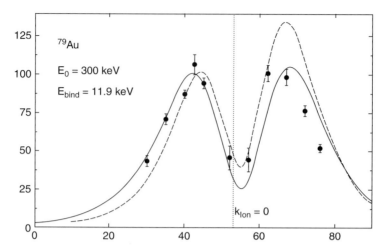

Figure 10.6 Absolute TDCS (mb/sr^2 keV) for the $2p_{\frac{3}{2}}$ ionization of gold in coplanar asymmetric geometry as a function of θ_s ($E_0 = 300$ keV, $E_b = 11.9$ keV, $E_f =$
210 keV, $\theta_f = -27.3°$). Experiment: [40]; solid line, rDWBA [29]; dashed line,
semi-relativistic Coulomb–Born [36]; dotted line indicates position of Bethe ridge,
$\boldsymbol{k}_{ion} = 0$.

complicated relation between $A_{\frac{1}{2}}$ and $A_{\frac{3}{2}}$ is predicted than $A_{\frac{1}{2}} = -2A_{\frac{3}{2}}$, applying to a 'pure' fine-structure effect. The observed deviations reflect the influence
of continuum spin–orbit interaction and other relativistic effects in the collision
process.

10.6 Longitudinally polarized electrons

Naively, one might expect that the TDCS would be invariant under the reversal of the
initial polarization if longitudinal polarized electrons were used as the projectile.
However, in [42] it was argued that this might not always be the case. In their
calculations, the momentum of the incoming particle was aligned along the z
direction and the momentum $\boldsymbol{k}_f$ of the fast outgoing electron defined the $x - z$
plane, as shown in Fig. 10.7. They considered the ionization of a $1s$ electron from
uranium for $E_0 = 300$ keV, with $E_f = 4E_s$, where the slow outgoing electron was
detected in the $x - y$ plane perpendicular to the incoming beam direction, i.e., with
$\theta_s = 90°$. The spin asymmetry was taken to be

$$A = \frac{\sigma^+ - \sigma^-}{\sigma^+ + \sigma^-}, \tag{10.12}$$

where σ^+ is the TDCS for ionization with the spin of the incoming electron
quantized along the positive z direction; σ^- had the spin quantized in the opposite

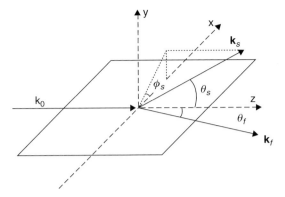

Figure 10.7 The scattering geometry used for the calculations in [42].

direction. They calculated the spin asymmetry for a uranium $1s$ electron, where θ_f was fixed to be $-5°$ and ϕ_s varied. Results are presented in Fig. 10.8(a) showing an rDWBA calculation and a model calculation when the distorted wave for the slow outgoing electron was replaced by a Dirac plane wave, thereby neglecting the effects of the target atom potential on the wave function for this electron. The first thing to note from their results is that a significant spin asymmetry is predicted in the rDWBA; the noticeable reduction in the model calculation suggests that the effect of the target potential on the slow electron is significant in generating the spin asymmetry. The spin asymmetries predicted in Fig. 10.8(b) for another kinematical arrangement are also seen to be non-zero. Here, the slow electron travels along the $+y$-axis ($\theta_s = \phi_s = 90°$) and the azimuthal angle of the fast scattered electron (θ_f) is varied. The underlying idea was to keep the post-collisional kinematics of the slow outgoing electron constant and to vary the recoil momentum of the ion. Both an rDWBA calculation and a model calculation with a plane wave for the slow outgoing electron are shown. These results again show that the effect of the potential of the target atom on the slow outgoing electron is the largest factor generating spin asymmetries. It has been shown in a full mathematical treatment [30] that the spin asymmetry arises from two sources. The first of these is the phase shift, induced by the potential originating from the target atom. The second and less important contribution to the spin asymmetry arises from the presence in the photon propagator of a small contribution of higher order than the Breit interaction.

The leading spin-dependent contributions to relativistic modifications of wave functions arise from the usual spin–orbit coupling, which is of order $1/c^2$ in atomic units, where c is the velocity of light. It is not necessary here to consider any nuclear charge dependencies of this interaction. An electron propagating in the scattering

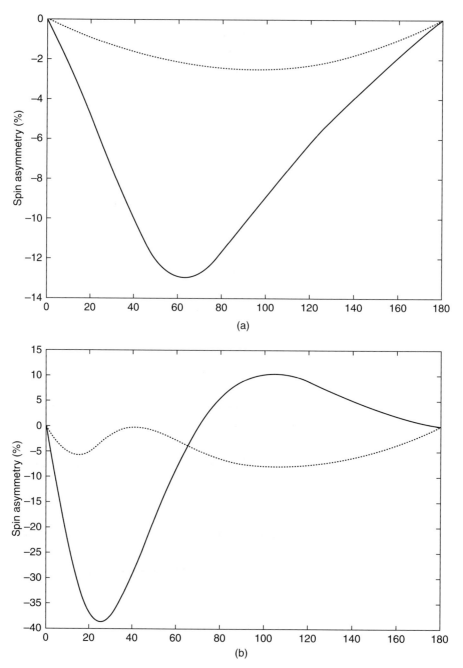

Figure 10.8 Spin asymmetry parameter A for the electron impact ionization of uranium, $1s$, $E_0 = 300$ keV, $E_f = 4$, $E_s = 148$ keV. (a) The TDCS as a function of the elevation angle ϕ_s for fixed $\theta_f = -5°$. (b) TDCS given as a function of θ_f, where the slow electron travels along the y-axis, $\theta_s = \phi_2 = 90°$. In both (a) and (b), the solid line represents an rDWBA calculation and the dotted curve a model calculation, where the distorted wave of the slow ejected electron has been replaced by a Dirac plane wave.

plane is unaffected by this interaction if the spin is also aligned in this plane. Since, furthermore, there is no spin–orbit coupling for a plane wave, one predicts that to order $1/c^2$ there will be no spin asymmetry if the slow outgoing electron is described as a plane wave, even if both the incoming and fast scattered electrons are described using distorted waves. This shows that, to order $1/c^2$, all the spin asymmetry arises from the effects of spin–orbit coupling on the slow outgoing electron. Hence, these effects will only be reproduced by calculations in which this electron is described using distorted waves. This explains why the largest spin asymmetries are predicted from the rDWBA calculations. The calculation in which the slow outgoing electron is described using a plane wave does not predict zero spin asymmetry because there will be higher order spin-dependent relativistic modifications of these wave functions. A slow electron having a momentum component k_s^y and spin aligned along the positive z direction will experience a force due to the usual spin–orbit coupling differing from that arising if the y component of its momentum is reversed to $-k_s^y$. The TDCS $\sigma_+(k_s^y)$ will therefore differ from TDCS $\sigma_+(-k_s^y)$. Reversal of the spin to an alignment along $-z$ therefore reverses the force originating from the spin–orbit coupling so that the TDCS $\sigma_-(k_s^y)$ will be the same as that of $\sigma_+(-k_s^y)$, with $\sigma_-(-k_s^y)$ being equal to $\sigma_+(k_s^y)$. This explains both why (a) there is a spin asymmetry of the order of $1/c^2$, the usual spin–orbit interaction when considering the TDCS for fixed angles, and (b) why the TDCS is certainly invariant to this order under spin reversal if the y momentum component of the slow electron is also reversed. A mathematical proof showing this identity to hold to all orders in 1/c is given in [30]. This was also confirmed by numerical calculations.

It has been shown in a full mathematical treatment [30] that the spin asymmetry arises from two sources. The first of these is the relativistic modification of the distortions of the continuum wave functions induced by the potential originating from the target atom. The second and less important contribution to the spin asymmetry arises from the presence in the photon propagator of a small contribution of higher order than the Breit interaction. In [42] a simple model was suggested. In it, a beam of entirely longitudinal, polarized electrons is scattered from a potential originating from the target atom. The scattered electron thereby acquires a momentum $\mathbf{k}_{\mathrm{tar}}$, assuming it travels freely inside the atom, with an intermediate momentum $\mathbf{k}_I$. Momentum conservation requires that

$$\mathbf{k}_0 = \mathbf{k}_I + \mathbf{k}_{\mathrm{tar}}.$$

This process models the distortion of the wave function of the incoming electron entering the rDWBA. The electron with momentum $\mathbf{k}_I$ then interacts with only the electron to be ionized so that the target momentum $\mathbf{k}_{\mathrm{tar}}$ remains essentially unchanged. Momentum conservation then requires that

$$\boldsymbol{k}_I + \boldsymbol{k}_{\text{tar}} = \boldsymbol{k}_f + \boldsymbol{k}_s + \boldsymbol{k}_{\text{tar}}$$
$$\Rightarrow \boldsymbol{k}_I = \boldsymbol{k}_f + \boldsymbol{k}_s.$$

After this second collision, the scattered electron again travels freely with momentum $\boldsymbol{k}_f$. It is assumed that the atom is unpolarized and that the final spin states of the continuum electrons are not resolved. Considering this ionization step as an interaction between the electron with momentum $\boldsymbol{k}_I$ and the spherically symmetric charge distribution of the orbital containing the s electron to be ionized enables the cross section $\sigma(\theta, \phi)$ to be expressed in terms of the Sherman function $S(\theta)$, as described in Sec. 3.2 of [16]. The angles θ and ϕ are defined in the co-ordinate system in which the z-axis is that line parallel to $\boldsymbol{k}_I$ which passes through the nucleus of the target atom. This approach yields [16],

$$\sigma(\theta, \phi) = I_{\text{f}}(\theta)[1 + S_{\text{f}}(\theta)\boldsymbol{P}_I \cdot \hat{\boldsymbol{n}}_I], \qquad (10.13)$$

where

$$\hat{\boldsymbol{n}}_I = \frac{\boldsymbol{k}_I \times \boldsymbol{k}_f}{\|\boldsymbol{k}_I \times \boldsymbol{k}_f\|}.$$

Here, both $I_{\text{f}}(\theta)$, the total cross section produced by an unpolarized beam, and $S_{\text{f}}(\theta)$ are defined by the wave functions entering the scattering process, which in turn define the asymptotic direct and spin–flip scattering amplitudes as defined in [16]. Both $I_{\text{f}}(\theta)$ and $S_{\text{f}}(\theta)$ are thus independent of both $\hat{\boldsymbol{n}}_I$ and the vector $\boldsymbol{P}_I$ describing the polarization of the electrons of momenta $\boldsymbol{k}_I$ before the ionization process. The result (10.13) shows that any dependence of $\sigma(\theta, \phi)$ on the polarization $\boldsymbol{P}_0$ of the electrons with momenta $\boldsymbol{k}_0$ before any interaction with the target can only arise through the intermediate polarization $\boldsymbol{P}_I$. This polarization generated in the first step is given in the approach presented in Sec. 3.2 of [16], Eq. (3.78):

$$\boldsymbol{P}_I = S_I(\theta)\hat{\boldsymbol{n}} + T_I(\theta)\boldsymbol{P}_0 + U_I(\theta)(\hat{\boldsymbol{n}} \times \boldsymbol{P}_0), \qquad (10.14)$$

where

$$\hat{\boldsymbol{n}} = \frac{\boldsymbol{k}_0 \times \boldsymbol{k}_I}{\|\boldsymbol{k}_0 \times \boldsymbol{k}_I\|}.$$

For the case where the spin of the incident electron is aligned either parallel or antiparallel to the z-axis, $\boldsymbol{P}_0 = a\hat{\boldsymbol{e}}_z$, with $a = \pm 1$. The quantities $T_I(\theta)$ and $U_I(\theta)$, like the Sherman function, are independent of $\boldsymbol{P}_0$ and $\hat{\boldsymbol{n}}$ since they are defined by the direct and spin–flip scattering amplitudes for this first step. The result (10.14) is the simplification of the general result (3.8) of [16] produced by noting that

$P_0 \cdot \hat{n} = 0$. For the coplanar case, momentum conservation shows that

$$k_I = (k_f^x + k_s^x, 0, k_f^z + k_z^x),$$

so that

$$\hat{n} = \frac{(0, k_s^z k_f^x - k_s^x k_f^z, 0)}{\|(0, k_s^z k_f^x - k_s^x k_f^z, 0)\|},$$

and

$$P_I = (aU_I(\theta), S_I(\theta), aT_I(\theta)),$$

but since $P_0 \cdot \hat{n}_I$ is independent of a, it follows that the result (10.13) is independent of a, so the TDCS is independent of the direction of polarization of the incident electron. The more interesting case, however, is that in which the slow outgoing electron has a non-zero y momentum component and travels out of the scattering plane. Clearly, the intermediate momentum vector k_I has a non-zero y momentum component due to the finite y momentum component of the slow outgoing electron. A simple analysis shows that the scalar product of the polarization vector P_I and the unit vector $\hat{n}$ entering the cross section depends on the direction of spin alignment a. Thus, the cross section is not invariant under reversal of initial spin polarization. In this description, the remaining interaction of the outgoing electrons in the field of the ion have been neglected.

It has to be noted that the description of a relativistic (e,2e) process in this model suffers from two simplifications. The first of these is to assume that the inelastic electron–electron interaction can be treated as pure elastic scattering. The second is to assume that the electron, after the first elastic scattering process, travels asymptotically as a free wave, where in fact the electron travels inside the atom. Although this model describes the occurrence of the spin asymmetry in relativistic (e,2e) processes with incoming longitudinally polarized electrons, it does not provide a convenient method for elucidating the orders in $1/c$.

10.7 Conclusion

The (e,2e) process describes a kinematically complete description of electron-impact-induced ionization. In this chapter, we have considered atomic (e,2e) processes where the spin-state of the projectile electron and/or the target atom are resolved. Related experiments performed to date have all involved the application of beams of spin-polarized electrons. For light low-Z targets and low impact energies (the non-relativistic limit), we have described how the resolution of spin channels provides sensitive information on exchange scattering. Under such conditions, non-relativistic scattering theories can adequately describe the ionization process. For high-Z atoms and low impact energies, a semi-relativistic scattering calculation may be sufficient to describe the data if target relativistic and many-electron exchange effects are at least partially accounted for. For heavy targets and high impact energies, spin-resolved studies reveal strong signatures of relativistic interactions in the momenta of the continuum electrons and a fully relativistic scattering formalism is essential to adequately describe the data.

In this chapter we have reviewed experimental and theoretical progress in the area and highlighted important results. The ultimate spin-resolved (e,2e) measurement, which presently lies beyond experimental capabilities, is yet to be performed, namely, one in which the spin projection states of the scattered electrons are additionally determined. If such an experiment can be realized in the future, exciting new insight into the nature and occurrence of spin–flip processes will be achieved.

Acknowledgements   One of the authors (JL) gratefully acknowledges the support of the Mercator programme of the Deutsche Forschungsgemeinschaft.

References

[1] R. Panajotović, J. Lower, E. Weigold, A. Prideaux and D. H. Madison, *Phys. Rev. A*, **73**, 052701 (2006).

[2] J. Lower, E. Weigold, J. Berakdar and S. Mazevet, *Phys. Rev. Lett.*, **86**, 624 (2001).

[3] G. Baum *et al.*, *Phys. Rev. Lett.*, **69**, 3037 (1992).

[4] T. Nakanishi *et al.*, *Phys. Lett. A*, **158**, 345 (1991).

[5] D. T. Pierce *et al.*, *Rev. Sci. Instrum.*, **51**, 478 (1980).

[6] M. Streun *et al.*, *J. Phys. B*, **31**, 4401 (1998).

[7] X. Zhang, Colm T. Whelan and H. R. J. Walters, *J. Phys. B*, **25**, L457 (1992).

[8] Colm T. Whelan, this volume.

[9] Jens Rasch, unpublished PhD thesis, University of Cambridge (1996).

[10] Colm T. Whelan, R. J. Allan, H. R. J. Walters and X. Zhang, in *(e,2e) and Related Processes*, ed. Colm T. Whelan, H. R. J. Walters, A. Lahmam-Bennani and H. Ehrhardt (The Netherlands: Kluwer, 1993), p. 33.

[11] J. B. Furness and I. E. McCarthy, *J. Phys. B*, **6**, 2280 (1973).

[12] M. E. Riley and D. G. Truhlar, *J. Chem. Phys.*, **63**, 2182 (1975).

[13] J. Lower, E. Weigold, J. Berakdar and S. Mazevet, *Phys. Rev. A*, **64**, 042701 (2001).

[14] G. F. Hanne, in *Proceedings of the 6th International Symposium on Correlation and Polarization of Electronic and Atomic Collisions and (e,2e) Reactions*, Adelaide 1991, ed. P. J. O. Teubner and E. Weigold, *IOP Conf. Proc.*, No. 122 (Bristol: Institute of Physics and Physical Society, 1991), p. 15.

[15] S. Jones, D. H. Madison and G. F. Hanne, *Phys. Rev. Lett.*, **72**, 2554 (1994).

[16] J. Kessler, *Polarized Electrons*, 2nd edn. (Berlin: Springer-Verlag, 1985).

[17] X. Guo *et al.*, *Phys. Rev. Lett.*, **76**, 1228 (1996).

[18] G. F. Hanne, *Can. J. Phys.*, **74**, 811 (1996).

[19] M. Dümmler, G. F. Hanne and J. Kessler, *J. Phys. B*, **28**, 2985 (1995).

[20] J. P. D. Cook, I. E. McCarthy, J. Mitroy and E. Weigold, *Phys. Rev. A*, **33**, 211 (1986).

[21] A. Dorn *et al.*, *J. Phys. B*, **30**, 4097 (1997).

[22] U. Lechner, S. Keller, E. Engel, H. J. Lüdde and R. M. Dreizler, in *Electron Scattering from Atoms, Molecules, Nuclei and Bulk Matter*, ed. Colm T. Whelan and N. J. Mason (New York: Kluwer/Plenum, 2005), p. 131.

[23] R. Panajotović, J. Lower, E. Weigold, A. Prideaux and D. Madison, *Phys. Rev. A*, **73**, 052701 (2006).

[24] S. Bellm, J. Lower, Z. Stegen, D. H. Madison and H. P. Saha, *Phys. Rev. A*, **77**, 032722 (2008).

[25] S. Bellm *et al.*, *Phys. Rev. A*, **78**, 062707 (2008).

[26] S. Bellm, J. Lower, M. Kampp and Colm T. Whelan, *J. Phys. B*, **39**, 4759 (2006) see also S. Bellm, J. Lower, Marco Kampp and Colm T. Whelan, *J. Elec. Spect. Rad. Phenom.*, **161**, 6 (2007).

[27] M. Kampp *et al.*, *J. Eur. Phys. D*, **29**, 17 (2004).

[28] W. Nakel and C. T. Whelan, *Phys. Rep.*, **315**, 499 (1999).

[29] S. Keller, R. M. Dreizler, H. Ast, C. T. Whelan and H. R. J. Walters, *Phys. Rev. A*, **53**, 2295 (1996).

[30] N. C. Pyper, Marco Kampp and C. T. Whelan, *Phys. Rev. A*, **71**, 052701 (2005).

[31] H. Ast, S. Keller, Colm T. Whelan, H. R. J. Walters and R. M. Dreizler, *Phys. Rev. A*, **50**, R1 (1994).

[32] Colm T. Whelan, H. Ast, H. R. J. Walters, S. Keller and R. M. Dreizler, *Phys. Rev. A*, **53**, 3262 (1996).

[33] S. Keller, R. M. Dreizler, H. Ast, Colm T. Whelan and H. R. J. Walters, *Phys. Rev. A*, **53**, 2295 (1996).

[34] H. T. Prinz, K. H. Besch and W. Nakel, *Phys. Rev. Lett.*, **74**, 243 (1995).

[35] R Tenzer and N Grün, *Phys. Lett. A*, **194**, 300 (1994).

[36] D. H. Jakubaßa-Amundsen, *J. Phys. B*, **28**, 259 (1995).

[37] D. H. Jakubaßa-Amundsen, *J. Phys. B*, **25**, 1297 (1992).

[38] H-Th. Prinz and S. Keller, *J. Phys. B*, **29**, L651 (1996).

[39] S. Keller, Colm T. Whelan, H. Ast, H. R. J. Walters and R. M. Dreizler, *Phys. Rev. A*, **50**, 3865 (1994).

[40] T. Kull, W. Nakel and C. D. Schröter, *J. Phys. B*, **30**, L815 (1997).

[41] K.-H.Besch, M. Sauter and W. Nakel, *Phys. Rev. A*, **58**, R2638 (1998).

[42] M. Kampp, N. C. Pyper, Colm T. Whelan and H. R. J. Walters, *Phys. Rev. A*, **67**, 044702 (2003).

Index